Springer Proceedings in Energy

More information about this series at http://www.springer.com/series/13370

Ali Nezihi Bilge · Ayhan Özgür Toy
Mehmet Erdem Günay
Editors

Energy Systems and Management

Springer

Editors
Ali Nezihi Bilge
Department of Energy Systems Engineering
Istanbul Bilgi University
Istanbul
Turkey

Mehmet Erdem Günay
Department of Energy Systems Engineering
Istanbul Bilgi University
Istanbul
Turkey

Ayhan Özgür Toy
Department of Industrial Engineering
Istanbul Bilgi University
Istanbul
Turkey

ISSN 2352-2534 ISSN 2352-2542 (electronic)
Springer Proceedings in Energy
ISBN 978-3-319-16023-8 ISBN 978-3-319-16024-5 (eBook)
DOI 10.1007/978-3-319-16024-5

Library of Congress Control Number: 2015932967

Springer Cham Heidelberg New York Dordrecht London

Printed on acid-free paper

Springer International Publishing AG Switzerland is part of Springer Science+Business Media (www.springer.com)

Preface

This book on *Energy Systems and Management* reports selected papers of the International Conference on Energy and Management held during 5–7 June 2014 at Istanbul Bilgi University, Turkey. It was organized by Istanbul Bilgi University Department of Energy Systems Engineering and PALMET Energy to share knowledge on the recent trends, scientific developments, innovations and management methods in energy. Academicians, scientists, researchers and industry specialists studying in the energy field from nine countries contributed through oral and poster presentations.

The book starts with the chapter "An Overview of Energy Technologies for a Sustainable Future", which examines the correlation between population, economy and energy consumption in the past, and reviews the conventional and renewable energy sources as well as the management of them to sustain the ever-growing energy demand in the future. The rest of the chapters are divided into three parts; the first part of the book, "Energy Sources, Technologies and Environment", consists of 12 chapters, which include research on new energy technologies and evaluation of their environmental effects. The second part "Advanced Energy Materials" includes seven chapters devoted to research on material science for new energy technologies. The third part "Energy Management, Economics and Policy" contains ten chapters on planning, controlling and monitoring energy-related processes together with the policies to satisfy the needs of increasing population and growing economy.

This book is designed to provide the reader with an understanding of the current status and the future of energy sources and technologies, as well as their interaction with the environment and the global energy policies. I hope you will find this book useful in energy studies.

I would like to mention that this conference was made possible to celebrate the 100th anniversary of the first Istanbul Electric Power station, the 30th anniversary of PALMET Energy and the first graduates of Energy System Engineering students. I would also like to thank the organizing committee and the scientific committee members for their valuable contributions to the conference. Finally, I express my sincere thanks to Mr. Doğanay Samuray, CEO of PALMET Energy, and his team for their cooperation and full support.

Istanbul, Turkey, November 2014 Prof. Ali Nezihi Bilge

Contents

Chapter 1
An Overview of Energy Technologies for a Sustainable Future

Ayse Nur Esen, Zehra Duzgit, A. Özgür Toy and M. Erdem Günay

Abstract Population and the economic growth are highly correlated with the energy demand. The world population was multiplied by a factor of 1.59 (reaching above 7 billion) from 1980 to 2013, while the total energy consumption of the world was multiplied by 1.84 (getting beyond 155,000 TWh) in the same time interval. Furthermore, the demand for energy is expected to increase even more with an average annual rate of 1.2 % in the near future. However, for the last 30 years, about 85–90 % of the energy demand is supplied by petroleum, natural gas, and coal, even though they are harmful for the environment and estimated to be depleted soon. Hence, building energy policies to satisfy the needs of increasing population and growing economy in a sustainable, reliable, and secure fashion has become quite important. This may involve optimizing the energy supplies, minimizing the environmental costs, promoting the utilization of clean and renewable energy resources and diversifying the type of energy sources. Thus, not only the conventional energy generation technologies must be developed more, but also environmentally friendly alternative energy sources (such as wind, solar, geothermal, hydro, and bio) must become more widespread to sustain the energy needs for the future. However, this requires a significant amount of research on energy technologies and an effective management of the energy sources.

A.N. Esen (✉) · Z. Duzgit · A.Ö. Toy · M.E. Günay (✉)
Istanbul Bilgi University, Istanbul, Turkey
e-mail: ayse.esen@bilgi.edu.tr

M.E. Günay
e-mail: erdem.gunay@bilgi.edu.tr

Z. Duzgit
e-mail: zehra.duzgit@bilgi.edu.tr

A.Ö. Toy
e-mail: ozgur.toy@bilgi.edu.tr

A.N. Bilge et al. (eds.), *Energy Systems and Management*,
Springer Proceedings in Energy, DOI 10.1007/978-3-319-16024-5_1

1.1 Introduction

Energy has become one of the main elements of economic and social development in the modern world, and access to reliable and affordable energy is essential for sustainable development. Energy sources including fossil fuels, renewables, and nuclear, technologies related to the production, conversion, and distribution of energy, and the use of energy such as lighting, heating/cooling, and transportation compose an overall energy system (International Energy Agency 2011a). However, the economic, social, environmental, and policy-related issues raised by unsustainable energy systems lead the search for cleaner and more efficient ways to supply, transform, deliver, and use energy. A well-designed energy system can make a significant contribution to sustainability.

The growth in population has always been one of the key drivers of energy demand: as the world population increases, the demand for energy rises. The world population was about 7.1 billion in 2013, while it was 4.5 billion in 1980, which means that the population was multiplied by 1.59 in such a small time interval (Fig. 1.1). Likewise, the total energy consumption of the world has increased continuously from about 83,000 TWh to almost 155,000 TWh (multiplied by 1.84) in the same years (Fig. 1.2), and it is projected to grow even more at an average annual rate of 1.2 % from 2012 to 2035 (International Energy Agency 2014a). The increase in the world energy consumption is also largely driven by rapid economic growth in the developing countries, which are expected to account for around 90 % of the net increase in the energy demand until the year 2035 (International Energy Agency 2013a).

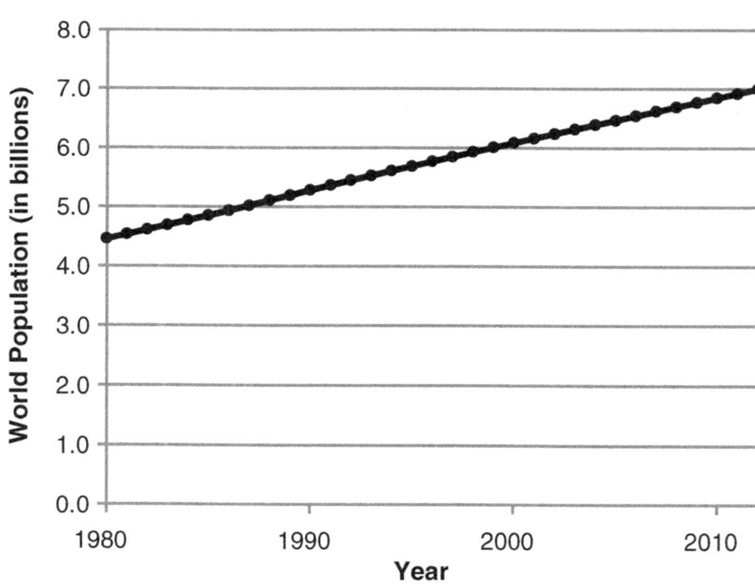

Fig. 1.1 Total world population through years

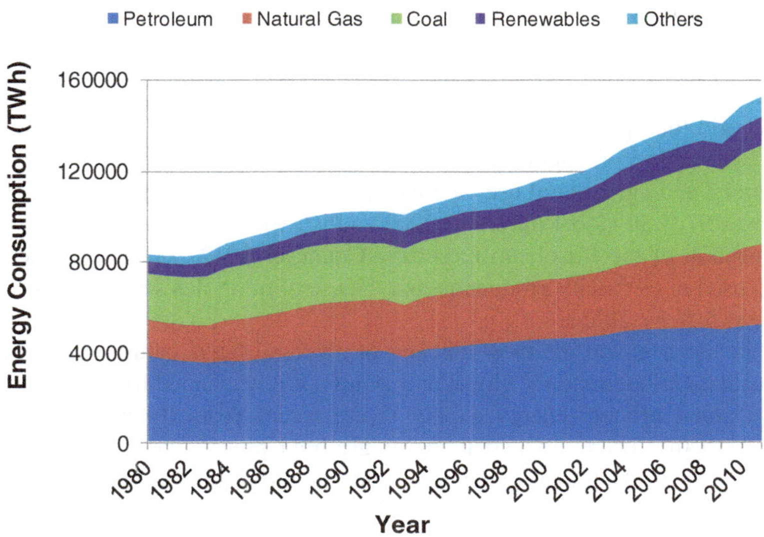

Fig. 1.2 World total energy consumption by different sources through years (US Energy Information Administration 2014)

The energy demand in the world increases year by year, but as indicated by Fig. 1.2, about 85–90 % of this demand is supplied by fossil fuels (petroleum, natural gas, and coal) for the last 30 years, even though they are harmful for the environment and estimated to be depleted soon. Contrary to fossil fuel sources, renewable energy sources, (such as: wind, solar, geothermal, hydro, and bio) employ environmentally friendly technologies, and they can be the alternatives to fossil fuel systems. However, there is only a slight increase in the total share of renewable energy sources in the last 30 years. For example, in the year 2011, the total share of renewable sources was 8.2 %, while it was 6.4 % in the year 1980 (Fig. 1.2), yet this is still not sufficient to substitute for the fossil fuel sources. Thus, not only the conventional energy generation technologies must be developed more, but also renewable energy technologies must become more widespread to sustain the ever-growing energy needs for the future. This requires a significant amount of research on energy technologies and an effective management of the energy sources.

1.2 Energy Sources and Technologies

Energy sources can be separated into three main categories: fossil fuels, nuclear, and renewables. In this section, these energy sources and the related technologies are reviewed briefly.

1.2.1 Fossil Fuels

Petroleum (oil) is the world's leading energy source with the highest share (Fig. 1.2) in the world total primary energy supply (International Energy Agency 2014b). It has a wide range of applications including transportation, industry, residential/ commercial and agricultural use, and electricity generation. The share of transportation was 57 % in global oil use in 2009 and is expected to rise to 60 % by 2035 (Organization of the Petroleum Exporting Countries 2012). Globally, little petroleum is used in electricity production and the use of oil in this sector is projected to decline to 5 % by 2035.

As the cleanest and the most efficient of all fossil fuels, natural gas is making significant contribution to the global energy mix. It is accounted for more than 20 % of world total primary energy supply in the recent years (International Energy Agency 2014b). There are two types of gas-fired power plants, open-cycle gas turbine (OCGT) plants and combined-cycle gas turbine (CCGT) plants. In comparison with coal-fired power plants, CCGT plants offer lower construction costs and emission. Hence, the share of CCGT plants in electricity has been increasing over the past decades.

Shale gas, which is classified as an unconventional source of natural gas, has effectively reshaped the gas industry, especially in the United States (US) (Armor 2013). The USA, China, Argentina, Algeria, Canada, and Mexico account for nearly two-thirds of the assessed, technically recoverable shale gas resources. In 2013, the US Energy Information Administration (EIA) estimated that shale gas resources in 42 countries represent 32 % of the global technically recoverable natural gas resources.

Despite its environmental challenges primarily associated with emissions, coal is still at the center of the global energy system. In 2012, its share in world total primary energy supply reached 29 % (International Energy Agency 2014b). The dominant position of coal in the global energy mix is largely due to its availability in almost every country and relatively low cost. The three types of coal power plants, pulverized coal combustion (PCC), fluidized bed combustion (FBC), and integrated gasification combined cycle (IGCC), are the most widely used ones today. Although PCC power plant causes significant emissions, it dominates the power industry. The efficiency of IGCC power plants is comparable with PCC power plants. They have lower emissions, but the investment costs are high.

The future of fossil fuels depends primarily on the new technologies for better efficiency and environmental performance particularly to eliminate CO_2 and other polluting emissions. In this respect, carbon capturing and sequestration (CCS), which is the process of injection of captured CO_2 to deep underground for permanent storage and sequestration, is a promising technology that could make a significant impact on the emissions from fossil fuels. Although CCS is technically feasible today, it has not been commercially proven on an integrated basis. The high cost of CCS is a major issue like all new low-emission technologies (Davies et al. 2013).

1.2.2 Nuclear

Nuclear energy is one of the alternative energy sources, which uses the heat produced by nuclear fission to generate power. After the first nuclear reactor was commissioned in 1954, today, 437 nuclear power reactors are in operation. The nuclear share in the global power generation in 2011 was estimated at 2,517 TWh, but it slightly decreased to 2,344 TWh at 2012 (U.S. Energy Information Administration 2014). The decline is mostly related to Fukushima Daichii nuclear accident. As explained by Rogner (2013), several nuclear power plants in Japan and in Germany were shut down due to safety concerns following the accident. Nevertheless, some other countries such as Canada, France, and the USA responded the accident with different nuclear policies by introducing safety improvements to their nuclear installations. France still produces 73.3 % of its electricity from nuclear energy. Russia is aiming to supply almost 50 % and India 25 % of their electricity from nuclear energy by 2050. Also, developing countries are continuing their plans to expand nuclear energy. Today, 72 nuclear power reactors are under construction (International Atomic Energy Agency 2014), which demonstrates the renewed global interest in nuclear energy mainly due to its important advantages. Nuclear energy, as a low-carbon and a price-stable energy source, is not subjected to unpredictable fuel costs and has a critical role to fight against climate change (Mari 2014). Most nuclear electricity is generated using the pressurized water reactor (PWR) and the boiling water reactor (BWR) designs, which were developed in the 1950s and improved since. Today, more sophisticated and more efficient types of nuclear power reactors are designed to fulfill the new demands.

1.2.3 Renewables

Renewable energy sources (such as wind, solar, geothermal, hydro, and bio) employ environmentally friendly technologies, and they can be the alternatives to fossil fuel systems. Although the use of renewable sources has become more widespread in electricity generation, heating/cooling, and transportation, there is only a slight increase in the total share of renewable energy sources in the last 30 years (Fig. 1.2). Nevertheless, the number of research studies on renewable energy technologies is increasing exponentially year by year. Some recent review publications together with their brief objectives on renewable energy sources are given in Table 1.1.

1.2.3.1 Bioenergy

Bioenergy refers to renewable energy produced from a variety of biomass to generate electricity, heat, and fuel. Today, it is the largest renewable energy source,

Table 1.1 Recent review publications on alternative energy sources and technologies

Author/Year	Energy source/ technology	Objective
Kralova and Sjöblöm (2010)	Biofuel	Reviews biodiesel sources, oil refining methods, current technologies in biodiesel production
Cheng and Timilsina (2011)	Biofuel	Summarizes the current status of advances in biofuel production technologies and key barriers to their commercial applications
Pereira et al. (2012)	Biomass	Surveys the current applications of biomass gasification technologies
Long et al. (2013)	Biomass	Investigates the biomass sources and estimates their bioenergy potential
Popp et al. (2014)	Biomass	Presents the risks related to availability of land for food crops, energy security, and environment
Raj et al. (2011)	Cogeneration technologies	Points out the present-day cogeneration technologies based on renewable energy sources and their various designs, numerical and simulation models, key development areas, economic and environmental considerations
Viral and Khatod (2012)	Distributed generation (DG) systems	Reviews basics of DG, current status of DG technologies, and their advantages and disadvantages, presents optimizations techniques/methodologies used in optimal planning of DG in distribution systems
Ruiz-Romero et al. (2014)	Distributed generation (DG) systems	Understands the difficulties of integration of DG in electricity distribution network, analyzes the effects of DG on power quality
Akinyele and Rayudu (2014)	Energy storage technologies	Presents current and new energy storage technologies for electrical power applications, discusses technological progress, performance and capital costs assessment of the systems
Mahlia et al. (2014)	Energy storage technologies	Discusses various types of energy storage, compares different types of energy storage, addresses barriers and issues in deploying energy storage system
Zarrouk and Moon (2014)	Geothermal energy	Provides a high-level assessment of the conversion efficiency of geothermal power plants
Bayer et al. (2013)	Geothermal energy	Reviews the potential environmental effects of geothermal power plants during their life cycle
Sternberg (2010)	Hydropower	Addresses hydropower's future, takes attention to environmental impacts of hydropower facilities, presents the technical, political and economic variables to identify the status of hydroprojects
Zimny et al. (2013)	Hydropower	Presents the development of hydropower in the world over the period 1995–2011 on the basis of available international statistical data
Basak et al. (2012)	Microgrid	Presents latest research and development in microgrids

(continued)

Table 1.1 (continued)

Author/Year	Energy source/ technology	Objective
Ellaban et al. (2014)	Renewable energy sources	Presents current status and future projection of major renewable energy sources, as well as their benefits, growth, investment, and deployment, presents the role of power electronics converters as enabling technology for using different renewable energy sources
Panwar et al. (2011)	Renewable energy sources	Overviews applications of major renewable energy gadgets in the scope of CO_2 mitigation for environmental protection
Armor (2013)	Shale gas	Highlights the growing importance and emergence of shale gas as an energy source
Aman et al. (2015)	Solar energy	Presents an overview of solar energy technologies, addresses safety, health, and environmental issues of solar energy systems
Devabhaktuni et al. (2013)	Solar energy	Reviews solar energy technologies, addresses costs of deployment, maintenance, and operation as well as economic policies that promote installation of solar energy systems
Yang and Sun (2013)	Wind energy	Surveys the testing, inspecting, and monitoring technologies for wind turbine blades, discusses development trends, and makes suggestions
Cheng and Zu (2014)	Wind energy	Reviews the wind energy conversion systems (WECS) and technologies, introduces the latest developments and future trends of WECS technologies
Rasuo et al. (2014)	Wind energy	Reviews development of new turbine rotor blades, presents test methods for blade of composite materials

providing 10 % of world total primary energy supply (International Energy Agency 2014b). According to Directive 2009/28/EC (2009), biomass resources are classified as the products, waste, and residues from agriculture, forestry, and related industries, as well as the biodegradable fraction of industrial and municipal waste. Heat production by the direct combustion of biomass is the leading application throughout the world. Numerous conversion technologies to convert biomass into heat and electricity already exist. Biomass can also be converted to liquid or gas biofuels. Bioethanol and biodiesel are the two alternative biofuels to replace petroleum and diesel fuels used in transport (Kralova and Sjöblom, 2010). Conventional biofuel technologies such as sugar-based ethanol and oil-crop-based biodiesel are in use on commercial scale. Advanced biofuel technologies such as bioethanol from lignocellulosic materials and biodiesel from microalgae are still in progress (Cheng and Timilsina 2011). Besides liquid biofuels, biomass gasification for heat and electricity generation as well as hydrogen and ethanol production has been applied widely (Perreira et al. 2012). The main challenge in commercialization of advanced biofuel is their high production costs. The further development of bioenergy can be estimated through studies done in global and in local scale. Long et al. (2013) and Popp et al. (2014) reviewed the existing researches about potential of bioenergy (Table 1.1). Both studies pointed out that there are differences among estimations due to methodologies, assumptions, and datasets. Estimates of global primary bioenergy potential are in the range of 30–1500 EJ/year by 2050.

1.2.3.2 Geothermal Energy

Geothermal energy, defined as the thermal energy generated and stored in the earth, can be used for electricity production, for direct heating purposes and for efficient home heating and cooling through geothermal heat pumps. The main geothermal power plant types are dry steam, flash steam, and binary cycle. They differ in the temperatures and pressures of reservoir. In addition to conventional technologies, projects to commercialize enhanced geothermal system (EGS) and man-made reservoir created where there is hot rock but insufficient or little natural permeability or fluid saturation are in development. Geothermal power plants release large quantities of waste heat due to the lower conversion efficiency than other conventional thermal power plants (Bayer et al. 2013). Zarrouk and Moon (2014) analyzed the conversion efficiency of geothermal power plants based on the data from 94 geothermal power plants worldwide and calculated the average conversion efficiency around 12 % (Table 1.1).

1.2.3.3 Hydropower

Hydropower harnesses the energy of moving water for electricity. There are three main hydropower technologies: run of river, reservoir, and pumped storage. Depending on the hydrology and topography of watershed, hydropower plants vary from small to large in terms of scale. Pumped-storage system, where pump turbines transfer water from bottom to upper reservoir during off-peak hours to be used later, is a practical approach to enhance the hydropower and currently accounts for 99 % of on-grid electricity storage. The vast majority of pumped-storage systems are currently found in Europe, Asia (Japan in particular), and the USA. Hydropower plants play a strategic role in regional geopolitics since water management is one of the greatest global challenges (Sternberg 2010). A detailed review on development of hydropower in the world was presented by Zimny et al. (2013) (Table 1.1).

1.2.3.4 Solar Energy

Solar energy is the most abundant energy source available; however, it represents a small share in the world's current energy mix. Solar photovoltaic (PV), solar thermal electricity (STE), and solar heating/cooling are well-established solar technologies (International Energy Agency 2011b). PV systems directly convert sunlight into electricity. Crystalline silicon and thin-film solar panels are current commercial PV technologies, the first of these dominate solar industry. Thin-film modules can be made flexible and produced in various sizes, shapes, and colors. The main thin films are made of amorphous silicon (a-Si), cadmium telluride (CdTe), copper indium diselenide (CIS), and copper indium gallium diselenide (CIGS), which receives considerable commercial attention among them. Today, STE exists only as concentrating solar power (CSP) plants in arid and semi-arid regions. CSP plants use mirrors to focus sunlight onto a receiver, which heats a transfer fluid to generate electricity through conventional steam turbines. The four types of CSP plants are parabolic trough, fresnel reflector, solar tower, and solar dish. The parabolic trough system is the most commercially available technology. Non-concentrating solar collectors, flat plate or evacuated tube, capture solar energy as heat for heating and cooling purposes. Devabhaktuni et al. (2013) evaluated the global situation and challenges of solar energy systems. It is emphasized that the solar electricity cost is higher than the other renewable technologies; however, it is expected to decline due to advances in technology.

1.2.3.5 Wind Energy

Wind energy is widely available throughout the world. Wind turbines can be located onshore or offshore. Today, the majority of wind power is still generated in onshore wind farms. Due to higher and more consistent wind speeds at sea, offshore wind turbines can harness more frequent and powerful winds than onshore wind turbines; however, the capital costs as well as the technical challenges are higher than onshore. The main parts of a wind turbine are base, tower, nacelle, and blades. Blades are the most critical component among them. The energy conversion efficiency increases with larger blades; therefore, today, blade diameter ranges from 20 m to 100 m. In order to extend life cycle of blades and minimize the operation risks, proper testing, inspecting, and monitoring must be applied (Yang and Sun 2013). Several wind energy conversion technologies have been developed to reduce cost, and to enhance efficiency and reliability. Cheng and Zhu (2014) presented a review on the common types and future trends of wind energy conversion systems (WECS) (Table 1.1). The variable-speed wind turbines with three blades are currently the most popular of WECS. As pointed out by Rašuo et al. (2014), wind turbine construction requires an extensive collaboration of materials and manufacturing techniques. Much development of existing composite material components is ongoing in respect of innovative wind turbines.

A wide range of renewable energy sources and technologies have been used for heat and electricity generation. The advantages and disadvantages as well as current status and future prospects of renewable energy sources and technologies were summarized by Ellaban et al. (2014). One of the most promising commercially available technologies is cogeneration, which produces electrical and thermal energy from the same primary energy source with a higher efficiency. Several renewable technologies can be operated in cogeneration mode, which can accelerate the integration of renewable energy technologies (Raj et al. 2011).

In the near future, a further increase in the renewable share of the global electricity mix is expected. Renewable sources generate power only intermittently and with variable output. Thus, electrical power systems face with difficulties when integrating renewable sources into the power grids. Grid systems require smart, efficient power transmission and distribution networks. In that respect energy storage, distributed generation and microgrid technologies have become important in the evolution of electricity markets to increase the smart grid development. Energy storage systems offer possible solutions to meet peak demands, to improve power reliability, and to reduce costs. Akinyele and Rayidu (2014) and Mahlia et al. (2014) presented a comprehensive review on available energy storage systems technologies such as capacitors, flywheel energy storage, superconducting magnetic energy storage, lead–acid batteries, lithium-ion batteries, nickel cadmium batteries, metal–air batteries, compressed-air energy storage, pumped-hydrostorage, thermal energy storage, and high-energy batteries (Table 1.1). Distributed generation (DG), also called on-site generation, is mostly demanded by solar and wind industry to reduce infrastructure costs. A review on DG systems was carried out by Viral and Khatod (2012) and Ruiz-Romero et al. (2014). Microgrids are small-scale power grids to meet local energy demand by ensuring supply control. Basak et al. (2012) presented a literature survey on operation and control techniques, power quality issues, and protection and stability features of microgrids (Table 1.1).

1.3 Energy and Environment

There is an intimate relation between energy and environment. The harvesting, generating, distribution, and use of energy sources have serious impacts on environment in many different ways. No form of energy source is completely "clean"; only some energy sources have smaller impact than others. Before any power plant construction begins, an environmental review may be required to categorize potential environmental effects. Power plants should be designed to minimize the potential effect upon ecological system. Environmental impacts associated with energy involve climate change, destruction of natural ecosystems, and pollution of air, soil, and water.

Over the years, the extensive use of fossil fuels cause the CO_2 emission into the atmosphere, which leads the beginning of global warming. According to "CO_2 Emissions from Fuel Combustion" (International Energy Agency 2013b), global CO_2 emissions were 31.3 $GtCO_2$ in 2011 and coal is accounted for 44 % of it. In recent years, researchers have focused on developing new methods, technologies, and tools to reduce CO_2 emissions. It is projected to reduce CO_2 emissions from coal to 5.7 $GtCO_2$ by 2035 through the use of more efficient power plants, CCS technologies, and other energy sources such as renewables and nuclear.

Renewables are considered as clean energy sources, and optimal use of these resources minimizes environmental impacts. A comprehensive review on the scope of CO_2 emission mitigation through use of solar energy, wind energy, and bioenergy was presented by Panwar et al. (2011). It was pointed out that renewable energy has great potential to reduce CO_2 emissions depending on the use and availability of sources. It is still important, however, to understand the environmental impacts associated with generating power from renewables.

Biomass power plants emit CO_2, NO_x, and small amount of SO_2, but cause less pollution than fossil fuel power plants. The solid waste produced, called ash, must be disposed properly as it contains varying levels of toxic metals depending on the source and area.

The environmental effects of geothermal power plants are related to land use, geological hazards, emissions, wastes, and water use. These effects depend on the type and size of the geothermal power plant. A comprehensive overview of environmental impacts of geothermal power plants was presented by Bayer et al. (2013). Geothermal power plants may cause geological hazards such as induced seismicity and ground deformations. They release larger quantities of waste heat because of lower conversion efficiencies in comparison with other power plant types.

While hydropower does not cause air pollutant emissions, environmental impacts of hydropower can be significant. The extent of the impact depends on the project size. The dam and reservoir may harm the aquatic habitat and the species present.

Environmental impacts of solar panels can be considered in three stages, manufacturing, operation, and decommissioning. The negative impacts of solar energy on environment were reviewed by Aman et al. (2015). During the operation of solar panels, no emissions are released; however, manufacturing process produces some toxic materials and chemicals such as cadmium, lead, and arsenic depending on the composition of panel. Consequently, if used solar panels are decommissioned improperly, they can be environmental threats due to the toxic materials in their compositions.

Wind energy produces no air or water pollution because no fuel is burned to generate electricity. However, improperly installed wind turbines may create soil erosion problems. Wind farms can also have noise impacts, depending on the number of wind turbines on the farm. The most serious environmental impact from wind energy may be bird mortality; several researches on bird mortality were conducted (Zimmerling et al. 2013; Everaert and Stienen 2007). Improvements in wind turbine design and siting helps to reduce these impacts.

Nuclear power plants do not emit greenhouse gases; therefore, nuclear energy can play an important role to reduce the impacts of global warming (van der Zwaan 2013). However, the production of radioactive wastes, spent fuels, and the risk of accidents are still the drawbacks of nuclear energy from environmental perspective.

1.4 Energy Management, Economics, and Policy

Energy is one of the trending topics being discussed by everyday citizens, journalists, academicians, business world, and politicians. Hence, in recent years, due to energy awareness, there has been a greater interest in the issue of energy management, economics, and policies.

Energy management is concerned with the planning, controlling, and monitoring energy-related processes so as to conserve energy resources, protect climate, and save energy-related costs. Energy management is an interdisciplinary approach that includes technical, economic, geopolitical, and political issues regarding the production and consumption of energy and investments, research, and development of energy systems.

Since fossil fuel sources cannot be replenished once depleted and they are the most widely used sources, the shift from non-renewable energy to renewable energy is critical from the viewpoint of energy management. New targets must be set for renewable energy, and these clean energy resources should be promoted both legally and commercially through supports such as discounts, privileges, and subventions.

Each non-renewable or renewable energy resource has its own advantages and disadvantages. In order to generalize, the downside of non-renewable energy resources is that their supply is limited and they cause the release of carbon dioxide and greenhouse gas emissions into atmosphere while burning, which in turn contributes to global warming. The advantage of renewable resources is their unlimited replenishment and cleanness. However, the major obstacle of renewable energy resources is the requirement of high initial investment.

According to Renewables (2014), Global Status Report, EU-wide target for renewable energy is 20 % and Chinese target is 20 % by 2020, where it was 8.5 and 9 % in 2011, respectively. On the other hand, Atiyas et al. (2012) stated that Turkey has introduced the 30 % target for renewables by 2023, the hundredth anniversary of the Republic of Turkey which is said to be an achievable target.

With respect to economics, due to increasing demand in spite of limited supply, market is faced with escalating energy prices. Based on BP Outlook 2035 (2014), industry will continue to remain the dominant source of growth for primary energy consumption and will account for more than half of the growth of energy consumption 2012–2035. Therefore, especially from the viewpoint of industry, cheap energy is necessary, which leads to lower production costs and higher profits.

Increasing population and economic development triggers new energy policies to regulate consumption and demand for energy. The consequences of energy policies affect individuals, companies in private and public sector, and society as a whole. Energy policies address legislation, regulations, incentives, and guidelines regarding energy production, distribution, consumption, conservation, and diversification.

The objective of energy policies is to satisfy needs of increasing population and growing economy in an economic, sustainable, reliable, and secure fashion. Recent energy policies may involve optimizing sustainability of energy supply and environmental costs, promoting the utilization of clean and renewable energy resources, diversifying energy sources, avoiding dependence on energy imports from a single source or country and encouraging investments in power industry.

According to European Environment Agency (2011), increasing energy utilization efficiency and increasing utilization of domestic renewable sources is seen as a key policy for Turkey due to current dependence on scarce sources. According to WWF (2013), for promoting renewable energy development, China sets its national objectives so as to improve energy structure, energy supply diversification, energy security, environmental protection, and sustainable development of the economy and society, whereas India tends to energy security, low-carbon planning, energy availability and access, energy affordability, and energy equity.

In terms of projections for energy demand in near future, based on (Energy Information Administration 2013), more than 85 % of the increase in global energy requirement from 2010 to 2040 is expected to occur among the developing countries outside the Organization for Economic Cooperation and Development (non-OECD), due to strong economic growth and increasing populations, whereas most of the OECD member countries are assumed to be more mature energy consumers, with slower anticipated economic growth and little or no anticipated population growth (Fig. 1.3).

In specific, according to Ernst & Young (2013), as a member of OECD countries, Turkey has seen the fastest growth in energy demand in the OECD in the past two years and demand is expected to double by 2023. This is interpreted as both a challenge and an opportunity such that the challenge is to ensure that supply keeps pace with such rapid future growth and the opportunity is to invest. Hence, anticipated growth in energy sector promises new opportunities locally and globally.

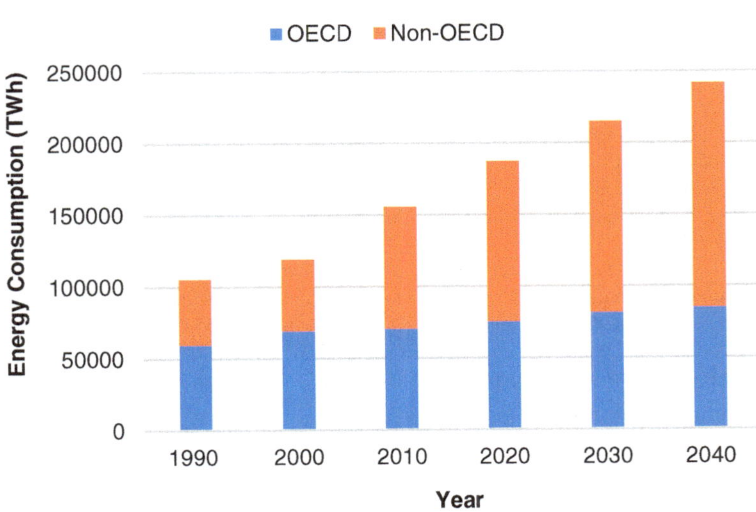

Fig. 1.3 Projection for world total energy consumption (Energy Information Administration 2013)

1.5 Conclusions

In this work, the correlation of the population and the economic growth with the energy demand was analyzed; almost all the energy generation technologies together with their environmental effects were reviewed; and finally, effective methods for the management of energy sources for a sustainable future were examined. It was reported that the population is now more than 7 billion, while the energy demand has reached over 155,000 TWh in the world, both of which are expected to increase even more in the near future. It was found that most of the energy demand is supplied by fossil fuel-related technologies depending on petroleum, natural gas, and coal, even though they are harmful for the environment and estimated to be depleted soon. Renewable energy sources employing environmentally friendly technologies were recommended as the alternatives to fossil fuel systems; however, the use of them was found to be still not sufficient to substitute the fossil fuel sources. To conclude, not only the conventional energy generation technologies must be developed more, but also renewable energy technologies must become more widespread to sustain the ever-growing energy needs for the future.

Acknowledgments We would like to thank Cemre Karaman for her valuable contributions for this work.

References

Akinyele, D. O., & Rayudu, R. K. (2014). Review of energy storage technologies for sustainable power networks. *Sustainable Energy Technologies, 8*, 74–91.

Aman, M. M., Solangi, K. H., Hossain, M. S., Badarudin, A., Jasmon, G. B., Mokhlis, H., et al. (2015). A review of safety, health and environmental (SHE) issues of solar energy system. *Renewable and Sustainable Energy Reviews, 41*, 1190–1204.

Armor, J. N. (2013). Emerging importance of shale gas to both the energy & chemicals landscape. *Journal of Energy Chemistry, 22*, 21–26.

Atiyas, I., Cetin, T., & Gülen, G. (2012). *Reforming Turkish energy markets political economy, regulation and competition in the search for energy policy*. Berlin: Springer. ISBN 978-1-4614-0289-3.

Basak, P., Chowdhury, S., Halder nee Dey, S., & Chowdhury, S. P. (2012). A literature review on integration of distributed energy resources in the perspective of control, protection and stability of microgrid. *Renewable and Sustainable Energy Reviews, 16*, 5545–5556.

Bayer, P., Rybach, L., Blum, P., & Brauchler, R. (2013). Review on life cycle environmental effects of geothermal power generation. *Renewable and Sustainable Energy Reviews, 26*, 446463.

BP. (2014). *BP energy outlook 2035*. Retrieved from http://www.bp.com/content/dam/bp/pdf/Energy-economics/Energy-Outlook/Energy_Outlook_2035_booklet.pdf.

Cheng, J. J., & Timilsina, R. G. (2011). Status and barriers of advanced biofuel technologies: A review. *Renewable Energy, 36*, 3541–3549.

Cheng, M., & Zhu, Y. (2014). The state of the art of wind energy conversion systems and technologies: A review. *Energy Conversion and Management, 88*, 332–347.

Davies, L. L., Uchitel, K., & Ruple, J. (2013). Understanding barriers to commercial-scale carbon capture and sequestration in the United States: An empirical assessment. *Energy Policy, 59*, 745–761.

Devabhaktuni, V., Alam, M., Depuru, S. S. S. R., Green, R. C, I. I., Nims, D., & Near, C. (2013). Solar energy: Trends and enabling technologies. *Renewable and Sustainable Energy Reviews, 19*, 555–564.

Ellaban, O., Abu-Rub, H., & Blaabjerg, F. (2014). Renewable energy resources: Current status, future prospects and their enabling technology. *Renewable and Sustainable Energy Reviews, 39*, 748–764.

Energy Information Administration (EIA). (2013). *International energy outlook 2013*. Retrieved from http://www.eia.gov/forecasts/archive/ieo13/pdf/0484(2013).pdf.

Ernst & Young. (2013). *Ernst & Young's attractiveness survey: Turkey 2013: The shift, the growth and the promise*. Retrieved from http://www.ey.com/Publication/vwLUAssets/Turkey_attractiveness_survey_2013/$FILE/turkey_attractiveness_2013.pdf.

European Environment Agency (EEA). (2011). *Survey of resource efficiency policies in EEA member and cooperating countries Country Profile: Turkey*. Retrieved from http://www.eea.europa.eu/themes/economy/resource-efficiency/turkey-2014-resource-efficiency-policies.

Everaert, J., & Stienen, E. W. M. (2007). Impact of wind turbines on birds in Zeebrugge (Belgium). *Biodiversity and Conservation in Europe, 16*, 3345–3359.

International Atomic Energy Agency. (2014) http://www.iaea.org/. Accessed on November 12, 2014.

International Energy Agency. (2011a). *Advantage energy: emerging economies, developing countries and the private-public sector interface*. Retrieved from http://www.iea.org/publications/freepublications/publication/advantage_energy.pdf.

International Energy Agency. (2011b). *Solar energy perspectives*. Retrieved from http://www.iea.org/publications/freepublications/publication/solar_energy_perspectives2011.pdf.

International Energy Agency. (2013a). *World energy outlook 2013 factsheet*. Retrieved from http://www.iea.org/media/files/WEO2013_factsheets.pdf.

International Energy Agency. (2013b). *CO_2 emissions from fuel combustion highlights* (2013th ed.). Retrieved from http://www.iea.org/publications/freepublications/publication/co2emissionsfromfuelcombustionhighlights2013.pdf.

International Energy Agency. (2014a). *World energy investment outlook*. Retrieved from http://www.iea.org/publications/freepublications/publication/WEIO2014.pdf.

International Energy Agency. (2014b). *Key world energy statistics*. Retrieved from http://www.iea.org/publications/freepublications/publication/KeyWorld2014.pdf.

Kralova, I., & Sjöblöm, J. (2010). Biofuels-renewable energy sources: A review. *Journal of Dispersion Science and Technology, 31*(409), 409–425.

Long, H., Li, X., Wang, H., & Jia, J. (2013). Biomass resources and their bioenergy potential estimation: A review. *Renewable and Sustainable Energy Reviews, 26*, 344–352.

Mari, C. (2014). The costs of generating electricity and the competitiveness of nuclear power. *Progress in Nuclear Energy, 73*, 153–161.

Mahlia, T. M. I., Saktisahdan, T. J., Jannifar, A., Hasan, M. H., & Matseelar, H. S. C. (2014). A review of available methods and development on energy storage; technology update. *Renewable and Sustainable Energy Reviews, 33*, 532–545.

Official Journal of the European Union. (2009). Directive 2009/28/EC of the European Parliament and of the Council. Retrieved from http://eur-lex.europa.eu/legal-content/EN/TXT/PDF/?uri=CELEX:32009L0028&from=EN.

Organization of the Petroleum Exporting Countries. (2012). *World oil outlook 2012*. Retrieved from http://www.opec.org/opec_web/static_files_project/media/downloads/publications/WOO2012.pdf.

Panwar, N. L., Kaushil, S. C., & Kothari, S. (2011). Role of renewable energy sources in environmental protection: A review. *Renewable and Sustainable Energy Reviews, 15*, 15131524.

Perreira, E. G., da Silva, J. G., de Oliveira, J. L., & Machado, C. S. (2012). Sustainable energy: A review of gasification technologies. *Renewable and Sustainable Energy Reviews, 16*, 47534762.

Popp, J., Lakner, Z., Harangi-Rakos, M., & Fari, M. (2014). The effect of bioenergy expansion: Food, energy, and environment. *Renewable and Sustainable Energy Reviews, 32*, 559–578.

Raj, T. N., Iniyan, S., & Goic, R. (2011). A review of renewable energy based cogeneration technologies. *Renewable and Sustainable Energy Reviews, 15*, 3640–3648.

Rašuo, B., Dinulović, M., Veg, A., Grbović, A., & Bengin, A. (2014). Harmonization of new wind turbine rotor blades development process: A review. *Renewable and Sustainable Energy Reviews, 39*, 874–882.

REN21 (the Renewable Energy Policy Network for the 21st Century). (2014). *Renewables 2014 global status report*. Retrieved from http://www.ren21.net/portals/0/documents/resources/gsr/2014/gsr2014_full%20report_low%20res.pdf.

Rogner, H. H. (2013). World outlook for nuclear power. *Energy Strategy Reviews, 1*, 291–295.

Ruiz-Romera, S., Colmenar-Santos, A., Mur-Pérez, F., & L ópez-Rey, Á. (2014). Integration of distributed generation in the power distribution network: The need for smart grid control systems, communication and equipment for a smart city—Use cases. *Renewable and Sustainable Energy Reviews, 38*, 223–234.

Sternberg, R. (2010). Hydropower's future, the environment, and global electricity systems. *Renewable and Sustainable Energy Reviews, 14*, 713–723.

U.S. Energy Information Administration. (2014). http://www.eia.gov/. Accessed on November 12, 2014.

Van der Zwaan, B. (2013). The role of nuclear power in mitigating emissions from electricity generation. *Energy Strategy Reviews, 1*, 296–301.

Viral, R., & Khatod, D. K. (2012). Optimal planning of distributed generation systems in distribution system: A review. *Renewable and Sustainable Energy Reviews, 16*, 5146–5165.

WWF International in collaboration with the World Resources Institute (WRI). (2013). *Meeting renewable energy targets: Global lessons from the road to implementation*. Retrieved from http://awsassets.panda.org/downloads/meeting_renewable_energy_targets__low_res_.pdf.

Yang, B., & Sun, D. (2013). Testing, inspecting and monitoring technologies for wind turbine blades: A survey. *Renewable and Sustainable Energy Reviews, 22*, 512–526.

Zarrouk, S. J., & Moon, H. (2014). Efficiency of geothermal power plants: A worldwide review. *Geothermics, 51*, 142–153.

Zimmerling, J. R., Pomeroy, A. C., d'Entremont, M. V., & Francis, C. M. (2013). Canadian estimate of bird mortality due to collisions and direct habitat loss associated with wind turbine developments. *Avian Conservation and Ecology, 8*(2), 10.

Zimny, J., Michalak, P., Bielik, S., & Szczotka, K. (2013). Directions in development of hydropower in the world, in Europe and Poland in the period 1995–2011. *Renewable and Sustainable Energy Reviews, 21*, 117–130.

Part I
Energy Sources, Technologies and Environment

Chapter 2
Thermal Pollution Caused by Hydropower Plants

Alaeddin Bobat

Abstract Thermal pollution is the change in the water temperatures of lakes, rivers, and oceans caused by man-made structures. These temperature changes may adversely affect aquatic ecosystems especially by contributing to the decline of wildlife populations and habitat destruction. Any practice that affects the equilibrium of an aquatic environment may alter the temperature of that environment and subsequently cause thermal pollution. There may be some positive effects, though, to thermal pollution, including the extension of fishing seasons and rebounding of some wildlife populations. Thermal pollution may come in the form of warm or cold water being dumped into a lake, river, or ocean. Increased sediment build-up in a body of water affects its turbidity or cloudiness and may decrease its depth, both of which may cause a rise in water temperature. Increased sun exposure may also raise water temperature. Dams may change a river habitat into a lake habitat by creating a reservoir (man-made lake) behind the dam. The reservoir water temperature is often colder than the original stream or river. The sources and causes of thermal pollution are varied, which makes it difficult to calculate the extent of the problem. Because the thermal pollution caused by Hydropower Plants (HPPs) may not directly affect human health, it is neglected in general. Therefore, sources and results of thermal pollution in HPPs are ignored in general. This paper aimed to reveal the causes and results of thermal pollution and measures to be taken in HPPs.

2.1 Introduction

Manufacturing is the activity of "**utility creation**" according to economists. Economic output is divided into physical (tangible) goods and intangible services. Consumption of goods and services is assumed to produce utility. Goods are items that can be seen

A. Bobat (✉)
University of Kocaeli, Kocaeli, Turkey
e-mail: bobatus@gmail.com

A.N. Bilge et al. (eds.), *Energy Systems and Management*,
Springer Proceedings in Energy, DOI 10.1007/978-3-319-16024-5_2

and touched, such as a book, a pen, and a folder. Services are provided for consumers by other people, such as doctor, dentist, haircut, and eating out at a restaurant (Zwikael and Smyrk 2011). Services are something completely different from goods. Services are intangible commodities that cannot be touch, felt, tasted, etc. They are the opposite of goods, where goods are something that can be traded for money; services are when you hire a person or someone to do something for you in exchange of money. Services are usually hired or rented; they cannot be owned like goods can. Since it requires people and one cannot legally own a person in today's world, services can only be for hire (Humphreys et al. 2001). In this regard, electric production can be acknowledged as a service output.

All the manufacturing systems that produce utility (goods and services) perform in accordance with inputs-transformation-outputs (ITO) model (Fig. 2.1). The factors of production such as capital, manpower (labour), raw material (land or natural resources), and management (or entrepreneur) constitute the main inputs of this model. These factors are processed in a certain time, and goods and services are obtained as outputs at the end of process. However, efficiency of this production process is less than 100 % because of waste or deficiency (Ely and Wicker 2007). On the other hand, the waste or deficiency may involve significant social, economic, and environmental costs. For instance, the existing wastes can cause the environmental problems in ecosystem.

According to the first law of thermodynamics, energy and matter available in a system or an environment can be transformed (changed from one form to another), and they can disperse around but they can neither be created nor destroyed. The clean-up costs of hazardous waste, for example, may outweigh the benefits of a product that creates it. The same law is acceptable for energy production. Moreover, the hazardous waste, it can be waste heat or thermal pollution/alteration in the energy production, can cause the environmental problems in aquatic ecosystem. Therefore, natural resources or natural systems have been deteriorated or consumed/used ex parte by human in a way.

According to the second law of thermodynamics, every process emits some heat or waste to the environment at the end of process (Toossie 2008). One of the biggest sources of the thermal pollution in water comes from electric power plants where water passes through the condenser and returned to the environment as an altered

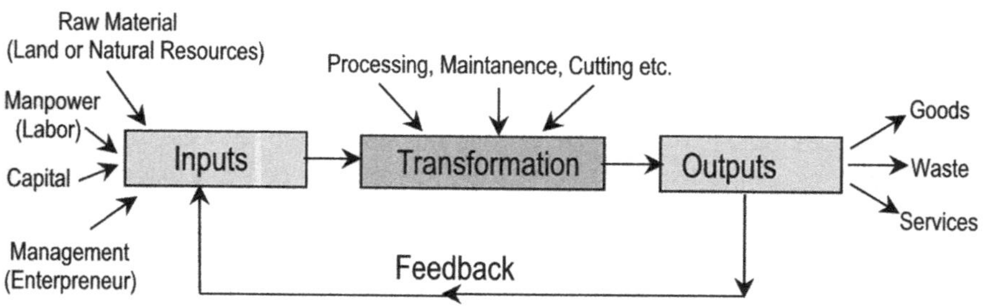

Fig. 2.1 General manufacturing model

temperature. The production of energy in hydropower plants (HPPs) also resembles general manufacturing model—ITO—and complies with the first and second laws of thermodynamics. The water changing physical, chemical, biological properties, and flowing regime during energy production process causes the problems in aquatic environment as well. As to ITO model, main inputs are running water as a natural resource, equipment, capital, manpower, etc. in HPPs. The water changing physical, chemical, and biological features in reservoir, penstock, turbine, pipelines, and cooling system reveals thermal pollution/alteration in aquatic environment more or less while electrical energy is obtained as an output in ITO model (Fig. 2.2).

In fact, healthy water bodies exhibit ecological integrity, representing a natural or undisturbed state. Ecological integrity is a combination of three components: chemical integrity, physical integrity, and biological integrity. Chemical integrity includes the chemical components such as dissolved oxygen, organic matter inputs, nutrients, groundwater and sediment quality, hardness, alkalinity, turbidity, metals, and pH; physical integrity includes the physical features such as sunlight, flow, habitat, temperature, gradient, soils, precipitation runoff, channel morphology, local geology, groundwater input, in-stream cover, and bank stability; and biological integrity includes the function and structures of biological communities (EPA 2002). When one or more of these components is degraded by any man-made structures, the health of the water body will be negatively affected and, in most cases, the aquatic life living there will reflect the degradation (Fig. 2.3).

Thermal pollution is the change in the water temperatures of lakes, rivers, and oceans caused by man-made structures or industries. It is also called as "waste heat" in a sense. Waste heat is an inevitable by-product of the power plants. In a fossil-fuel power plant, the amount of heat energy rejected is approximately 60 %. The amount rejected by a nuclear power plant is even larger—close to 70 % (Haider 2013).

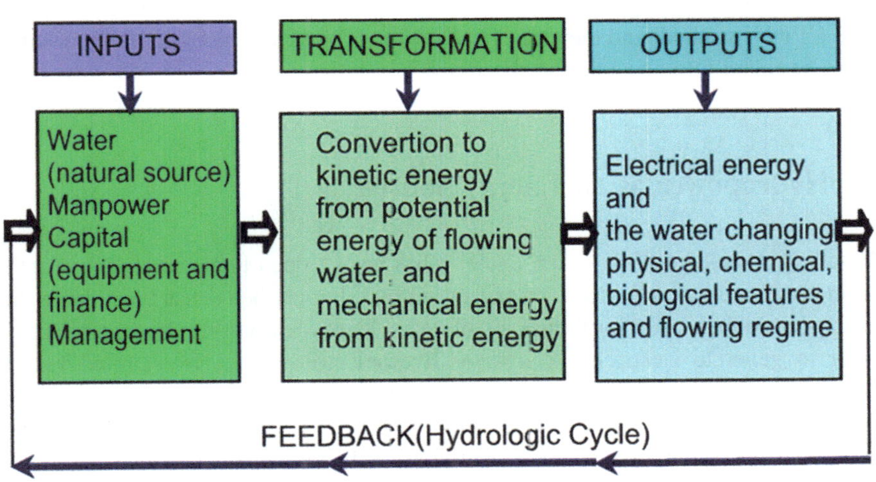

Fig. 2.2 ITO system in HPPs

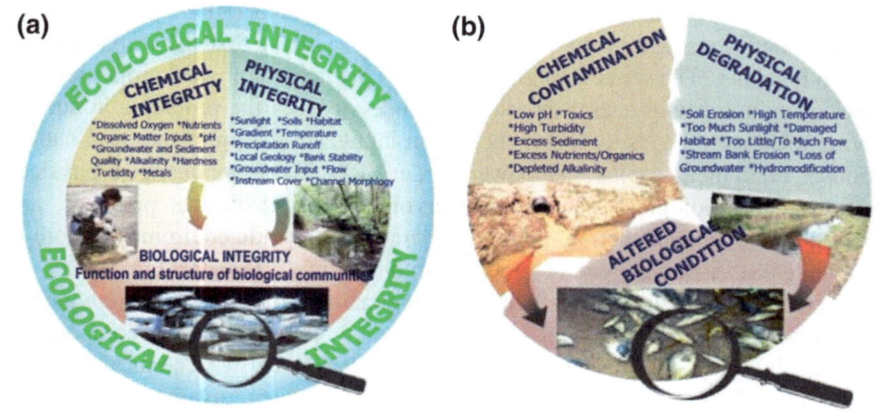

Fig. 2.3 **a** Healthy water bodies and **b** unhealthy water bodies (from EPA 2002)

The medium that receives this no longer-needed heat is the coolant from where water was drawn to keep the equipment cool. That is why power plants are almost always built near rivers, lakes, or seashores for a ready supply of cooling water. This practice of dumping the waste heat in the form of modified water into its natural source is called thermal pollution, and the sources and causes of thermal pollution are varied, which makes it difficult to calculate the extent of the problem. Also, because the negative effects of thermal pollution may not directly affect human health, it is not as well known as other types of pollution. The nuclear power industry is tightly regulated; therefore, the impact of nuclear power plants on the environment, including its production of thermal pollution, usually in the form of warm water, is better documented. But, the thermal pollution caused by HPPs has not been well known and documented, and therefore, sources and results of thermal pollution in HPPs are ignored in general. In this chapter, thermal pollution coming from HPPs is studied, and the causes and results of thermal pollution, and measures are to be taken into account and discussed.

2.2 Hydropower as an Energy Source

By taking advantage of the water cycle, it has been tapped into one of the nature's engines to create a useful form of energy. In fact, human being has been using the energy in moving water for thousands of years. Today, exploiting the movement of water to generate electricity, known as hydroelectric power, is regarded as the largest source of renewable power in worldwide.

The most common method of generating hydroelectric power is by damming rivers to store water in reservoirs. Building-up behind a high dam, water accumulates potential energy. The water in the reservoir is considered stored energy. When the gates open, the water flowing through the penstock becomes kinetic energy because it is in motion. Hydroelectric energy is produced by the force of falling water. Upon its release, the flow turns turbines, which then generate electricity. The capacity to produce this energy is dependent on both the available flow and the height from which it falls. In order to generate electricity from the kinetic energy in moving water, the water has to be moving with sufficient speed and volume to turn a generator. To increase the force of moving water, **impoundments or dams** are used to raise the water level, creating a "**hydraulic head**", or height differential. When water behind a dam is released, it runs through a pipe called a penstock and is delivered to the turbine. When the water reaches the end of the penstock, it turns a water wheel or "**turbine**" at enormous speeds. The turbine rotates, via a connected shaft to an electrical generator, and this generator creates electricity. It is the turbine and generator working in combination that converts "**mechanical energy**" into "**electric energy**". The current is then passed onto the transformer, converting it to a small current at a high voltage, and through the transmission lines to substations where the voltage will be reduced and the electricity distributed to customers. High voltage is needed because a large amount of energy is needed to transport electricity over long distances (Fig. 2.4) (HowStuffWorks 2001). Moreover, to avoid the excess warming of the turbine, cooling water coming from adjacent reservoir or ground is used (USDI 2005; Kumar et al. 2011).

Hydroelectric generation can also work without dams, in a process known as **diversion**, or **run-of-river** (RoR). Portions of water from fast-flowing rivers, often at or near waterfalls, can be diverted through a penstock to a turbine set in the river or off to the side. Another RoR design uses a traditional water wheel on a floating platform to capture the kinetic force of the moving river. While this approach is inexpensive and easy to implement, it does not produce much power.

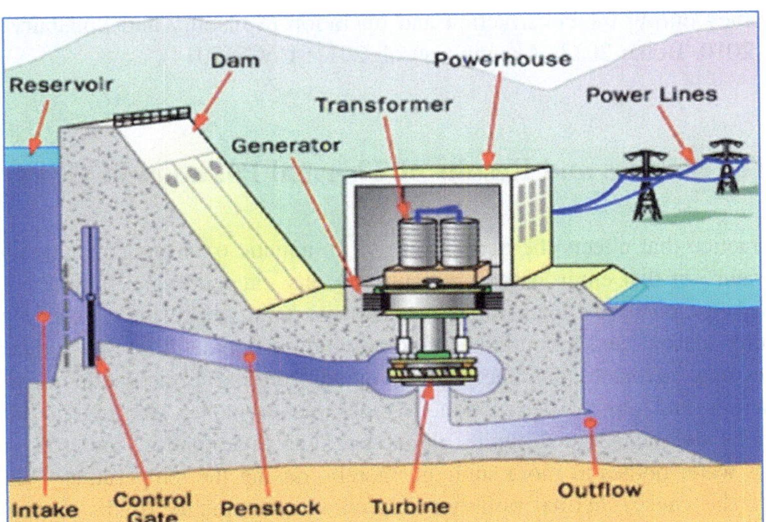

Fig. 2.4 A view from *inside* part of reservoir hydropower plant (from HowStuffWorks 2001)

Another type of hydropower, though not a true energy source, is pumped storage. In a pumped storage plant, water is pumped from a lower reservoir to a higher reservoir during off-peak times, using electricity generated from other types of energy sources. When the power is needed, it is released back into the lower reservoir through turbines. Inevitably, some power is lost, but pumped storage systems can be up to 80 % efficient. Future increases in pumped storage capacity could result from the integration of hydropower and wind power technologies. Researchers believe that hydropower may be able to act as a battery for wind power by storing water during high-wind periods (DoE 2004; EERE 2006).

Compared to other power plants working on fossil fuels, HPPs have the lowest operational cost (only the initial cost is high and construction takes a long time), the longest operational life, the highest efficiency rates, and the cheapest electricity generated (Davis 2006). The ability to meet power demand fluctuations is an advantage of HPPs with reservoirs. In this regard, HPPs are one of the most responsive factors (easy to start and stop) of any electric power generating source. Furthermore, HPPs are preferred because of their environment-friendly (they do not emit any direct pollutant), clean, (partly) renewable, and reliable technologies. HPPs stop the flooding and harmful effects of the rivers, store irrigation, and drinking water and give chance to fish farming and produce revenue (Yüksek and Kaygusuz 2006). Reservoir lakes can be used for recreation, water-based activities such as boating, skiing, fishing, and hunting. HPPs can set up in many sizes and have relatively low maintenance costs. Moreover, they are the domestic source of energy and can become tourist attractions in their own right. However, they have some disadvantages, such as disappearing habitats and species, melting deltas, decreasing ground water, changing water quality, drying natural lakes, influencing physical and biological environment, economical unproductiveness, landscape destruction, deforestation, microclimate changes as well as socio-economic degeneracy during the construction and operation (Spilsbury and Spilsbury 2008; Bobat 2010; Bobat 2013; Aksungur et al. 2011; UoCS 2014).

2.3 The Sources and Results of Thermal Pollution in HPPs

Any practice that affects the equilibrium of an aquatic environment may alter the temperature of that environment and subsequently cause thermal pollution. There may be some positive effects, though, to thermal pollution, including the extension of fishing seasons and rebounding of some wildlife populations. But, temperature changes may adversely affect ecosystems by contributing to the decline of wildlife populations and habitat destruction. Besides power plants, thermal pollution is caused by deforestation, drought, global warming, evaporation, and soil erosion exposes water bodies to more sunlight, thereby raising the temperature. Whatever may be the cause, thermal pollution degrades water quality of the source by a process that changes its ambient temperature.

Water temperature has a direct or indirect influence on aquatic water ecosystems, and it regulates fish distribution in freshwaters. Thus, there is limited potential for transfer of fish between various water environments. Water temperature, quantity, and quality play a critical role and determine the distribution of fish species, stock catch, and diversification of aquaculture, i.e. presence of frogs, crabs, shrimps, and molluscs in the water body. This species cannot survive beyond 19 °C temperature of water (Miller and Stillman 2012).

In fact that the decrease in water amount and the change in water temperature reduce the amount of dissolved oxygen and nutrients negatively denature fish and other macroscopic/microscopic organisms, and their quantitative and qualitative nature. As a result of this, these alterations in stream negatively affect all the food chain from freshwater to marine environment (Poff et al. 1997).

Freshwater streams are complex ecosystems that sustain a range of thermally sensitive organisms, but are at risk due to encroachment and degradation by increasing power generation, industrialization, and urbanization. A freshwater stream is defined as a stream that supports trout and other cold water fish species (Eaton et al. 1995). Trout are native to cold water streams and very sensitive to temperature changes. Trout species prefer to avoid water temperatures exceeding 21 °C (Coutant 1977). Once temperatures rise above that level, mortality rates begin to increase (Lee and Rinne 1980). Feeding, spawning, overall health, and growth of cold water fish are also adversely affected (Edwards et al. 1979). For example, the survival of the endemic trout species (*Salmo trutta labrax*) of Eastern Black Sea Region (Fig. 2.5a) is closely related to water amount in the downstream since it has liking for gravity, cold, and oxygen-rich aquatic environment. And this species cannot spawn in the water of which temperature exceeds 12 °C (Aksungur et al. 2007). Water temperature also influences the early development of aquatic organisms. Furthermore, it affects the larvae and eggs of fish in freshwaters. For instance, trout eggs may not hatch if the water is too warm. Even if they hatch, they would not survive for a long time because aquatic juveniles are the least tolerant to abrupt

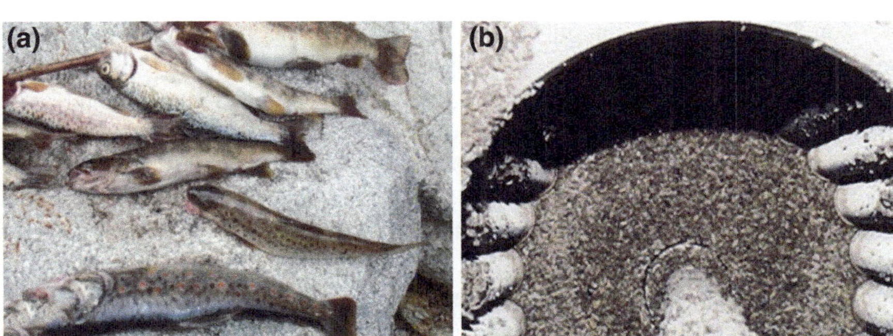

Fig. 2.5 **a** Endemic *brown* trout and **b** bioaccumulation of Zebra Mussel in screens

temperature changes. It was observed that trout will actually select cool water even if it is low in oxygen (Matthews et al. 1994; Matthews and Berg 1997). These findings suggest that mitigating thermal pollution in developing freshwater stream watersheds is essential for protecting sustainable trout fisheries. Water temperature affects the overall biological and chemical composition of a stream (Pluhowski 1970; Paul and Meyer 2001; Poole and Berman 2001). It influences nutrient cycles and productivity within fluvial systems (Allan and Castillo, 2007). The anthropogenic thermal degradation associated with land development has been found to permanently alter water temperature regimes (Nelson and Palmer 2007). Water temperature was cited as one of the most important environmental factors that affect assemblages of cold water macro-invertebrates (Wang and Kanehl 2003). These unique systems are often drastically altered, or even destroyed, when traditional development occurs within a cold water stream watershed.

HPPs with dams may change a running water habitat into a lake system by creating a reservoir (man-made lake) behind the dam. The water temperature in reservoir is occasionally colder than the original stream or river. On the contrary, downstream of dams and HPPs has generally warm water than upstream because of passing of water from pipelines, penstock, turbine, and cooling system. Despite the change in water temperature emerging from construction and operation of HPPs is not occurred as high as that in fossil-fuel and nuclear power plants, it is too important to affect lifecycle and survival of aquatic organisms. For instance, on the Euphrates River of Turkey, due to cascade of HPPs with reservoirs, Zebra Mussel (*Dreissena polymorpha* Pallas) has extremely reproduced under changing ecological conditions. The alteration in water temperature of reservoir especially has supported the lifecycle and spawning of Zebra Mussel. So, it has caused the significant technical, economic, and ecological damages reducing and/or blocking water flow (Fig. 2.5b) in HPPs (Bobat et al. 2002; Bobat 2003; Bobat et al. 2004). Also, the mitigations taken against populations of Zebra Mussel, such as hypoplants at the entrance of the drinking water systems, cause TOC risk during the water treatment plant process.

The construction of dams and reservoirs for water storage, power generation, and diversion for other usage can affect flow and depth of the water. Furthermore, the speed and depth of flow can play the important role in transfer and dispersion of nutrients and dissolved gases such as CO_2 and O_2 in aquatic environment and it can influence the respiration and reproduction activities of some aquatic species (Allan and Flecker 1993). The changes in the speed, depth, outflow, and timing of stream generally damage aquatic organisms (Power et al. 1995). Together with other factors cited above, the friction of water through penstock and pipelines, the crashing of water to turbine blades, and use of water for cooling increase water temperature in both reservoir and tailrace to some extent. Since water can absorb thermal energy while experiencing only small changes in temperature, most aquatic organisms have developed enzyme systems that operate in a narrow temperature range. In some cases, this change can devastate even heat-tolerant species that are inured to warmer waters. The temperature change of even one to two degree Celsius can cause significant change in organism metabolism and other adverse cellular biology effect. Principal adverse changes can include rendering cell walls less permeable to necessary osmosis, coagulation of cell protein, and alteration of enzyme metabolism. These cellular level effects can adversely affect reproduction and also cause mortality.

Menon et al. (2000) opined that the species that are intolerant to warm water conditions may disappear, while others preferring this condition may thrive so that the structure of the community changes. Respiration and growth rates may be changed. An increase of temperature may result in the loss of sensitive species. They have also stressed on the influence of certain parameters, which exert profound influence on the distribution and availability, in which temperature is stated to be first. HPPs have been adversely affected the natural stocks of cold water fish. The activities pertaining to the projects under construction are responsible for increase in the silt load and destruction of fish food organisms in streams. The increased silt load along with changed temperature regime of channel adversely affected the feeding and spawning of fish (Qureshi et al. 2010).

The presence of dissolved oxygen is probably the single most important factor in the biology of aquatic systems, and a great variety of physical and biological interactions stem from it. But, as the temperature of water increases, dissolved oxygen content in water decreases. Since metabolism requires oxygen, some species may be eliminated entirely if the water temperature rises by 1–2 °C. Additionally dissolved oxygen is the key to assimilation of organic wastes by microorganisms. Warming a water body will impair this assimilation (Resh et al. 1988).

Thermal pollution not only kills heat-intolerant fish, but also plants, thereby disrupting the web of life dependent on the aquatic food chain. Also, the elimination of heat-intolerant species may allow less desirable heat-tolerant species to take over.

Fish are often disturbed, migrate, and spawn in response to temperature cues. When water temperature is artificially changed, the disruption of aquatic organisms' normal activities and patterns can be catastrophic. There may be large-scale migration to an environment more favourable to their survival. The addition of new species of fish will change the eco-balance of the migrated area.

Thermal pollution can also increase the susceptibility of aquatic organisms to parasites, toxins, and pathogens, making them vulnerable to various diseases. If thermal pollution continues for a long time, it can cause huge bacteria and plant growth leading to algae bloom that will subsequently result in even less oxygen in the water. Algae have unfavourable effects on aquatic life.

2.4 Discussion, Conclusion, and Recommendations

Thermal pollution occurs as a natural process required by first and second laws of thermodynamics for any thermodynamic cycle to operate. The first-law efficiency is defined based on the first-law principle of conversion of one form of energy to another, without any consideration to the quality of the energy resource. First-law efficiency is input/output efficiency, i.e. the ratio of energy delivered in a desired form and the energy that must be expended to achieve the desired effect. It does not differentiate between the qualities of the energy sources. The second-law efficiency compares the efficiency of an actual device (first-law efficiency) to that of the same or a similar device operated under ideal conditions. The second-law efficiency of all ideal devices is equal to one, and for all real devices is smaller than one. The Kelvin-Plank statement of the second law says no power plant can be 100 % efficient. They will always reject some heat to the environment. The Clausius statement of the second law implies that heat transfer by itself will always be from a high temperature to a lower temperature. This means that the waste heat rejected from the power plant will always be at a higher temperature than the environment (Toossie 2008).

The same process also performs in HPPs. Water is often used as a cooling medium and a main source of kinetic energy in HPPs and returned to the environment warmer than when it started in accordance with first and second laws of thermodynamics. Therefore, the changes in water temperature of reservoir and tailrace cause thermal alterations.

The release of water changing temperature into the environment is a controversial issue. In some cases, it is said that heat addition causes no harm and even improves conditions, whereas in others, the whole ecosystem is changed. Some say the best fishing is near a thermal outfall while others claim the hot water is killing the fry and larva. Whatever the stand, it is generally agreed that some species prefer warmer water while others prefer colder water. It is also known that some species, especially juveniles, can only stand elevated temperatures for a given period of time, the higher the temperature, the shorter of the lethal exposure time. In general, discharging warm water into a cooler body of water will result in the change of biolife in the neighbourhood of the discharged from cold water species to warm water species. The size of the effected region can be from a few centimetres to several thousand metres from the discharge. As a result, the problem becomes how to reduce the amount of heat rejected to the environment and the impact it has on it.

All hydroelectric structures affect a freshwater's ecology mainly by inducing a change in its hydrologic characteristics (flow regime, temperature, pH, DO, and so on) and by disrupting the ecological continuity of sediment transport and fish migration through the building of dams, dikes, and weirs. However, the extent to which a stream's physical, chemical, and biological characteristics are modified depends largely on the type of HPPs, whereas RoR HPPs do not alter a stream's flow regime extremely the creation of a reservoir for storage hydropower entails a major environmental change by transforming a fast-running fluvial ecosystem into a still-standing lacustrine one. The extent to which a HPP has adverse impacts on the riverbed morphology, on water quality, and on fauna and flora is highly site-specific and to a certain degree dependent on what resources can be invested into mitigation measures.

Water quality issues related to reservoirs depend on several factors: climate, reservoir morphology and depth, water retention time in the reservoir, water quality of tributaries, quantity and composition of the inundated soil and vegetation, and rapidity of impounding, which affects the quantity of biomass available over time. Also, the operation of the HPP and thus the reservoir can significantly affect water quality, both negatively and positively. Water quality issues can often be managed by site selection and appropriate design, taking the future reservoir morphology and hydraulic characteristics into consideration. The primary goals are to reduce the submerged area and to minimize water retention in the reservoir. The release of poor-quality water (due to thermal stratification, turbidity, and temperature changes both within and downstream of the reservoir) may be reduced by the use of selective or multi-level water intakes. This may also help to reduce oxygen depletion and the volume of anoxic waters. Since the absence of oxygen may contribute to the formation of methane during the first few years after impoundment, especially in warm climates, measures to prevent the formation of anoxic reservoir zones will also help mitigate potential methane emissions.

The choice of location is playing an important role in HPPs, especially to the relation of water course and the habitat in surrounding. The careful environmental feasibility studies are needed to prevent the negative impacts to the environment. The deep studies need to take to decide the best location to minimize the negative effects of changing the water course and also the effect to the immigration of the fish and other water inhabitants. The place will not require flooding large areas also need to be chosen, and the environmentally sensitive places, such as the natural conservation, also need to be avoided.

The developing countries and Turkey especially have the richest biodiversity, and the richness of freshwater in many regions has not been discovered yet. For these reasons, HPPs under construction and in operation have the potential to cause major environmental problems. Whereas many natural habitats are successfully transformed for human purposes, the natural value of certain other areas is such that they must be used with great care or left untouched. The choice can be made to preserve natural environments that are deemed sensitive or exceptional. To maintain biological diversity, the following measures have proven to be effective: establishing protected areas; choosing a reservoir site that minimizes loss of ecosystems; managing invasive species through proper identification, education, and eradication; and conducting specific inventories to learn more about the fauna, flora, and specific habitats within the studied area. Moreover, the efficiency of HPPs can be improved by replacing obsolete or worn turbine or generator parts with new equipment, fine-tuning performance, reducing friction losses of energy, and automating operations. These efficiency improvements are expected to have only very minor and short-term environmental impacts. Replacement of turbine gates and runners (the surfaces against which water is impinged) can be environmentally beneficial because more efficient turbines generally kill fewer fish, and because new turbine parts can be designed to facilitate aeration at dams that release water with low dissolved oxygen (DO) concentrations.

Since investors are usually interested in construction costs and incomes to be obtained, the physical, chemical, and biological features are ignored in general. In fact, the alterations in freshwater are vital for both sustainability of aquatic ecosystem and return of investment. For this reason, it is also important to compare the environmental effects of HPPs with alternatives.

Water availability is crucial for many energy technologies, including hydropower, and energy is needed to secure water supply for agriculture, industries, and households, particularly in water-scarce areas in developing countries. This mutual dependence has leaded to the understanding that the water-energy nexus must be addressed in a holistic way, especially regarding climate change and sustainable development. Providing energy and water for sustainable development will require improved regional and global water governance, and since hydroelectric facilities are often associated with the creation of water storage facilities, hydropower is at the crossroads of these issues and can play an important role in enhancing both energy and water security. Therefore, hydropower development is part of water management systems as much as energy management systems, both of which are increasingly becoming climate driven.

References

Aksungur, M., Alkan, A., Zengin, B., Tabak, İ., & Yılmaz, C. (2007). Karadeniz Alabalığının Tatlısu Ortamındaki Göçü Üzerine Bazı Çevresel Parametrelerin Etkisi. *Ekoloji, 17*(65), 28–35.

Aksungur, M., Ak, O., & Özdemir, A. (2011). Nehir tipi hidroelektrik santrallerinin sucul ekosisteme etkisi: Trabzon Örneği. *Journal of Fisheries Sciences, 5*(1), 79–92.

Allan, J. D., & Castillo, M. M. (2007). *Stream ecology: Structure and function of running waters* (2nd ed.). Dordrecht: Springer.

Allan, J. D., & Flecker, A. S. (1993). Biodiversity conservation in running waters. *BioScience, 43*, 32–43.

Bobat, A., Hengirmen, M., & Zapletal, W. (2002). Problems of zebra mussel at dams and hydro projects on the Euphrates River, *Hydro 2002: Development, Management and Performance, Antalya, Proceedings Book*, 475–484.

Bobat, A. (2003). Hidroelektrik Santrallarda Ekolojik Bir Sorun: Zebra Midye, *Türkiye 9. Enerji Kongresi*, Bildiriler Kitabı, Cilt I, 327–349.

Bobat, A., Hengirmen, M., & Zapletal, W. (2004). Zebra mussel and fouling problems in the Euphrates basin. *Turkish Journal of Zoology, 28*, 161–177.

Bobat, A. (2010). Yavaş ve Sessiz Olur Akarsuların Ölümü-III: Baraj ve HES'lerin Etkileri (The streams die slowly and quietly—III: The Positive and Negative Effects of Dams and HPPs), December 15, 2010. www.enerjienergy.com

Bobat, A. (2013). The triple conflicts in hydro projects: Energy, *economy and environment, Fresenius Environmental Bulletin, 22*(7a), 2093–2097.

Coutant, C. C. (1977). Compilation of temperature preference data. *Journal of the Fisheries Research Board of Canada, 34*, 740–745.

Davis, J. (2006). *Advances in hydropower technology can protect the environment, alternative energy sources* (pp. 82–87). Detroit: Greenhaven.

DOE (U.S. Department of Energy). (2004). Hydropower: Setting a course for our energy future, DOE/GO-102004-1981. http://www.nrel.gov/docs/fy04osti/34916.pdf

Eaton, J. G., McCormick, J. H., Goodno, B. E., O'Brien, D. G., Stefany, H. G., Hondzo, M., & Sheller, R. M. (1995). A field information-based system for estimating fish temperature tolerances. *Fisheries, 20*(4), 10–18.

Edwards, R. W., Densem, J. W., & Russell, P. A. (1979). An assessment of the importance of temperature as a factor controlling the growth rate of brown trout in streams. *Journal of Animal Ecology, 48*(2), 501–507.

EERE (Office of Energy Efficiency and Renewable Energy). (2006). Wind energy systems integration. Washington, DC: U.S. Department of Energy. http://www1.eere.energy.gov/windandhydro/wind_sys_integration.html

Ely, R. T., & Wicker, G. R. (2007). *Elementary principles of economics: Together with a short sketch of economic history, chapter II: The factors of production*. Whitefish: Kessinger Publishing, LLC.

EPA. (2002). Biological assessments and criteria: Crucial components of water quality programs. USA-EPA, Office of Water, 822-F-02-006.

Haider, Q. (2013). Thermal pollution of water by power plants. The Daily Star, September 14, 2013.

HowStuffWorks. (2001). How hydropower plants work. http://science.howstuffworks.com/environmental/energy/hydropower-plant1.htm. Access: September 11, 2014.

Humphreys, B. R., Maccini, L. J., & Schuh, S. (2001). Input and output inventories. *Journal of Monetary Economics, 47*, 347–375.

Kumar, A., Schei, T., Ahenkorah, A., Caceres Rodriguez, R., Devernay, J. M. Freitas, M., Hall, D., Killingtveit, A., & Liu, Z. (2011). Hydropower. In O. Edenhofer, R. Pichs-Madruga, Y. Sokona, K. Seyboth, P. Matschoss, S. Kadner, T. Zwickel, P. Eickemeier, G. Hansen, S. Schlömer & C. von Stechow (Eds.), *IPCC special report on renewable energy sources and climate change mitigation*. Cambridge: Cambridge University Press.

Lee, R. M., & Rinne, J. N. (1980). Critical thermal maxima of five trout species in the Southwestern United States. *Transactions of the American Fisheries Society, 109*, 632–635.

Matthews, K. R., Berg, N. H., Azuma, D. L., & Lambert, T. R. (1994). Cool water formation and trout habitat use in a deep pool in the sierra Nevada, California. *Transactions of the American Fisheries Society, 123*(4), 549–564.

Matthews, K. R., & Berg, N. H. (1997). Rainbow trout responses to water temperature and dissolved oxygen stress in two Southern California stream pools. *Journal of Fish Biology, 50* (1), 50–67.

Menon, A. G. K., Singh, H. R., Kumar, N. (2000). Present eco-status of cold water fish and fisheries. In H. R. Singh & W. S. Lakra (Eds.), *Coldwater fish and fisheries* (pp. 1–36). New Delhi: Narendra Publishing House.

Miller, N. A., & Stillman, J. H. (2012). Physiological optima and critical limits. *Nature Education Knowledge, 3*(10), 1.

Nelson, K. C., & Palmer, M. A. (2007). Stream temperature surges under urbanization and climate change: Data and responses. *Journal of the American Water Resources Association, 43*(2), 440–452.

Paul, M. J., & Meyer, J. L. (2001). Streams in the urban landscape. *Annual Review of Ecology Systematics, 32*, 333–365.

Pluhowski, E. J. (1970). Urbanization and its effect on the temperature of streams on Long Island, New York. U.S. Geological Survey, Professional Paper 627-D, New York City, NY.

Poff, L. N., Allan, D., Bain, M. B., Karr, J. R., Prestaggard, K. L., Richter, B. D., et al. (1997). The natural flow regime: a paradigm for river conservation and restoration. *BioScience 47*, 769–784.

Poole, G. C., & Berman, C. H. (2001). An ecological perspective on in-stream temperature: Natural heat dynamics and mechanisms of human-caused thermal degradation. *Environmental Management, 27*(6), 787–802.

Power, M. E., Sun, A., Parker, M., Dietrich., W. E., Wotton J. T. (1995). Hydraulic food-chain models: An approach to the study of web dynamics in large rivers. *BioScience 45*, 159–167.

Qureshi, T. A., Qureshi, T. A., Chalkoo, S. R., Borana, K., & Manohar, S. (2010). Effect of thermal pollution on the hydrological parameters of river Jhelum (J & K). *Current World Environment, 5*(2), 292–2827.

Resh, V. H., Brown, A. V., Covich, A. P., Gurtz, M. E., Li, H. W., Minshall, G. W., et al. (1988). The role of disturbance in stream ecology. *Journal of the North American Benthological Society, 7*, 433–455.

Spilsbury, R., & Spilsbury, L. (2008). *The pros and cons of water power*. New York: Rosen Central Publication.

Toossie, R. (2008). *Energy and the environment: Sources, technologies, and impacts*. Irvine: VerVe Publishers. ISBN 978-1-4276-1867-2.

UoCS (Union of Concerned Scientists). (2014). How hydroelectric energy works, environmental concerns. Access: May 06, 2014. http://www.ucsusa.org/clean_energy/our-energy-choices/renewable-energy/how-hydroelectricenergy.html

US Department of Interior, Bureau of Reclamation, Power Resources Office. (2005). Hydroelectric power. http://www.usbr.gov/power/edu/pamphlet.pdf

Wang, L., & Kanehl, P. (2003). Influences of watershed urbanization and instream habitat on macroinvertebrates in cold water stream. *Journal of the American Water Resources Association, 39*(5), 1181–1196.

Yüksek, O., & Kaygusuz, K. (2006). Small hydropower plants as a new and renewable energy source. *Energy Sources, Part B: Economics, Planning, and Policy, 1*(3), 279–290.

Zwikael, O., & Smyrk, J. (2011). The input-transform-outcome (ITO) model of a project. In O. Zwikael & J. Smyrk (Eds.), *Project management for the creation of organisational value* (pp. 11–35). London: Springer.

Chapter 3
Comparing Spatial Interpolation Methods for Mapping Meteorological Data in Turkey

Merve Keskin, Ahmet Ozgur Dogru, Filiz Bektas Balcik, Cigdem Goksel, Necla Ulugtekin and Seval Sozen

Abstract Determining the potentials of the renewable energy sources provides realistic assumptions on useful utilization of the energy. Wind speed and solar radiation are the main meteorological data used in order to estimate renewable energy potential. Stated data is considered as point source data since it is collected at meteorological stations. However, meteorological data can only be significant when it is represented by surfaces. Spatial interpolation methods help to convert point source data into raster surfaces by estimating the missing values for the areas where data is not collected. Besides the purpose, the total number of data points, their location, and their distribution within the study area affect the accuracy of interpolation. This study aims to determine optimum spatial interpolation method for mapping meteorological data in northern part of Turkey. In this context, inverse distance weighted (IDW), kriging, radial basis, and natural neighbor interpolation methods were chosen to interpolate wind speed and solar radiation measurements in selected study area. The cross-validation technique was used to determine most efficient interpolation method. Additionally, accuracy of each interpolation method were compared by calculating the root-mean-square errors (RMSE). The results prove that the number of control points affects the accuracy of the interpolation. The second degree IDW (IDW2) interpolation method performs the best

M. Keskin (✉) · A.O. Dogru · F.B. Balcik · C. Goksel · N. Ulugtekin · S. Sozen
Istanbul Technical University, Istanbul, Turkey
e-mail: keskinmer@itu.edu.tr

A.O. Dogru
e-mail: dogruahm@itu.edu.tr

F.B. Balcik
e-mail: bektasfi@itu.edu.tr

C. Goksel
e-mail: goksel@itu.edu.tr

N. Ulugtekin
e-mail: ulugtek@itu.edu.tr

S. Sozen
e-mail: sozens@itu.edu.tr

A.N. Bilge et al. (eds.), *Energy Systems and Management*,
Springer Proceedings in Energy, DOI 10.1007/978-3-319-16024-5_3

among the others. Thus, IDW2 was used for mapping meteorological data in northern Turkey.

3.1 Introduction

As the world population increases, the demand of the energy and the energy consumption by industry and households also increase. This situation leads to increases in energy imports and rising carbon dioxide emissions. Under these circumstances, it is pretty clear that people need to give up consuming energy sources which will run out in the near future and facilitate the renewable energy sources such as wind, solar, and hydropower. Meteorological data such as precipitation, temperature, and wind speed can be considered as important renewable energy data, and they are measured at a specific location that is mostly a meteorological monitoring station. Such data are widely used for monitoring meteorological conditions in a region. However, point source data should be represented by surfaces as grid, contours, triangulated irregular networks, or points for covering the regions that they belong to (Childs 2004; Luo et al. 2008).

A variety of spatial interpolation methods are used to estimate unsampled locations of a region by using sampled points in order to represent point source data by raster surfaces. Raster data derived by using suitable spatial interpolation method are the main component of the meteorological maps. Therefore, spatial interpolation can be considered as one of the important step of the mapping meteorological data. There are several spatial interpolation methods such as kriging, inverse distance weighted, and natural neighbor. However, the main point is to determine the interpolation method, which calculates best estimated values of unsampled areas for each specific case. The accuracy of the spatial interpolation methods varies depending on the total number of point sources, their locations, distribution and measured values.

This study is developed and being conducted within the context of EnviroGRIDS Project supported by European Union 7th Framework Programme. The main aim of this study was to determine optimum spatial interpolation method for mapping meteorological data in Sakarya Catchment located along the Black Sea cost of Turkey. In this context, three different interpolation methods, which are extensively available in commercial or open source GIS software and suggested by the previous studies (Borga and Vizzaccaro 1997; Goovaerts 2000; Attorre et al. 2007) were selected for performance assessments in this study.

3.2 Methodology and Data

In this study, inverse distance weighted (IDW), natural neighbor (NN), and kriging were considered as the main spatial interpolation methods to be compared in terms of their practical accuracy. Spatial interpolation is the procedure for estimating the value of properties at unsampled sites within the area covered by existing observations (Shepard 1968). Interpolation techniques are considered as deterministic and geostatistical techniques. Deterministic techniques create surfaces based on measured points or mathematical formulas, while geostatistical interpolation techniques are based on statistics and are used for more advanced prediction surface modeling (Childs 2004).

IDW is an interpolation technique that estimates cell values from a set of weighted sample points with measurement values. As it is seen in the Eq. 3.1, the interpolated values of unsampled points are estimated as a function of sampled point values $u_i = u(x_i)$ and weights, $w_i(x)$ (Shepard 1968). N denotes the total number of sampled points.

$$u(x) = \sum_{i=0}^{N} \frac{w_i(x)u_i}{\sum_{j=0}^{N} w_j(x)} \tag{3.1}$$

Weights are determined for each sampled point as a function of distance between known (x) and unknown (x_i) points, d, and power parameter, p that is a positive real number (see Eq. 3.2). The choice of value for p is therefore a function of the degree of smoothing desired in the interpolation, the density and distribution of samples being interpolated, and the maximum distance over which an individual sample is allowed to influence the surrounding ones. In this study, power number is considered as 2 in practice so applied methodology is abbreviated as IDW2. It is possible to imply that as the distance between sampled and unsampled point's increases, less weight is calculated for that point, so that this method assumes that each measured point has a local influence that diminishes with distance (Luo et al. 2008; Waters 1988).

$$w_i(x) = \frac{1}{d(x, x_i)^p} \tag{3.2}$$

NN interpolation finds the closest subset of input samples to a query point and applies weights to them based on proportionate areas in order to interpolate a value (Sibson 1981). The basic equation of NN method is given in Eq. 3.3, where $G(x, y)$ is the estimate at (x, y), w_i are the weights, $f(x_i, y_i)$ are the known data at (x_i, y_i), and N is the total number of sampled points. The natural neighbor method proposes a measure for the computation of the weights and the selection of the interpolating neighbors. The basic properties of NN are that it is local, using only a subset of samples that surround a query point, and that interpolated heights are guaranteed to be within the range of the samples used. It does not infer trends and will not produce peaks, pits, ridges, or valleys that are not already represented by the input samples. NN adapts locally to the structure of the input data, requiring no input from the user pertaining to search radius, sample count, or shape. It works equally well with regularly and irregularly distributed data (Childs 2004; Watson 1992).

$$G(x,y) = \sum_{i=1}^{N} w_i f(x_i, y_i) \tag{3.3}$$

Natural data are difficult to model using smooth functions because normally random fluctuations and measurement error combine to cause irregularities in sampled data values. Kriging is stochastic technique similar to IDW and was developed to model those concepts. Kriging is an interpolation method that optimally predicts data values by using data taken at known nearby locations. It uses linear combinations of weights at known points to estimate the value at an unknown point (Luo et al. 2008). However, in this method, the spatial correlation is taken into account while estimating the surface. This correlation is determined by using semi-variance function as stated in Eq. 3.4 where $N(h)$ denotes the number of pairs of sampled points with a distance of h. The complete formulation of Kriging methodology is provided by several literatures (Attore et al. 2007; Goovaerts 2000).

$$\gamma(h) = \frac{1}{2N(h)} \sum_{i=1}^{N(h)} [Z(s_i) - Z(s_i + h)]^2 \tag{3.4}$$

There are several types of kriging: Ordinary kriging is the most common method which assumes that there is no constant mean for the data over an area, while universal kriging assumes that an overriding trend exists in the data and that it can be modeled (Borga and Vizzaccaro 1997). Ordinary kriging is used in this study to estimate the surfaces, because there is no trend that can be modeled in the meteorological data in hand.

In this study, above-stated spatial interpolation methods were applied on interpolating the point sourced precipitation, temperature, and wind speed data (recorded at 10 m.) at 36 meteorological monitoring stations in Sakarya Catchment and its neighboring stations. The data were provided by Turkish State Meteorological Service. Study area covering the stations is indicated in Fig. 3.1, which also represents the distribution of meteorological monitoring stations in study area. Each data represents the monthly average calculated by the unique daily values collected in January, 2005. January was selected by considering precipitation amount in a year; however, this study can be extended to a seasonal projection.

Geometric data used for mapping meteorological data were vector data covering study area in Turkey. Vector data of countries neighboring to study area were also used as the complementary data for communicating the reference information about the location of study area in thematic maps. ESRI World Vector Dataset was used for this purpose. Geographic Coordinate System European 1950 was used as geographic reference, and European Datum 1950 (ED50) was selected as the datum of the vector data (Fig. 3.1).

Fig. 3.1 Distribution of meteorological monitoring stations in the study area

As the methodology of this study, above-stated spatial interpolation techniques were implemented by using precipitation, temperature, and wind speed data in a Geographic Information System (GIS) environment. GIS provides logical solutions to visualize existing situation, produce maps, and manage geospatial data. In the study, mainly default settings were accepted as the parameters of each interpolation methods. Performances of the applied methods were assessed in terms of three different extents mainly considering the assessment of the root-mean-square errors (RMSE) of estimated values. In this context firstly, inertial overall accuracy assessment results were considered to compare the accuracies of the applied interpolation methods. Secondly, cross-validation technique was used, and four monitoring stations were selected as control station in Turkey. Interpolation methods were applied for the study area separately by using the data of remaining monitoring stations. Control stations belonging to Sakarya Catchment are Eskisehir Bolge, Nallihan and Sivrihisar. Locations of monitoring stations and measured meteorological data at these stations were considered while selecting control stations. Estimated values for control stations were compared with measured data, and RMSEs were calculated depending on differences at each control station. Finally, performances of the spatial interpolation methods in terms of accuracy were determined by calculating the RMSE regarding the errors obtained at each control station (Keskin and Ozdogu 2011).

Table 3.1 Accuracy assessment statistics of each method

Station ID	Station Name	Precipitation (mm)				Temperature (°C)				Wind speed (m/s)			
		Measured	Interpolated			Measured	Interpolated			Measured	Interpolated		
			IDW	Kriging	NN		IDW	Kriging	NN		IDW	Kriging	NN
17128	Eskişehir Bölge	1.64	3.92	3.64	3.86	2.54	3.83	3.41	3.54	2.40	1.63	1.60	1.63
17679	Nallihan	5.31	3.79	2.90	3.18	3.23	4.19	4.19	3.89	1.33	1.47	1.55	1.45
17726	Sivrihisar	3.40	2.99	3.30	3.42	3.19	3.41	3.32	3.41	2.08	2.00	2.02	2.10
Station ID	Station name	Precipitation (mm)				Temperature (°C)				Wind speed (m/s)			
			IDW	Kriging	NN		IDW	Kringing	NN		IDW	Kriging	NN
17128	Eskişehir Bölge	ERRORS	−2.29	−2.00	−2.22	ERRORS	−1.29	−0.87	−1.00	ERRORS	0.77	0.80	0.77
17679	Nallihan		1.52	2.41	2.13		−0.96	−0.96	−0.66		−0.14	−0.22	−0.12
17726	Sivrihisar		0.41	0.10	-0.02		−0.23	−0.13	−0.23		0.08	0.05	−0.02
		RMSE	1.60	1.81	1.78		0.94	0.75	0.70		0.46	0.48	0.45

3.3 Results and Discussion

For assessing the accuracies of the applied interpolation methods, each interpolated value at selected control stations was compared with the measured ones. While interpreting the results, individual errors may be misleading about the performance of interpolation methods. That is why one has to consider the overall RMSE values to make a better evaluation. As it is represented in the Table 3.1, accuracy assessment statistics points out that IDW has the best performance with the lowest RMSE which is 1.60 for precipitation. For temperature data, natural neighbor and kriging resulted with a lowest and very close RMSE values. However, it is hard to pick the best performance for wind speed, because each method almost gave the same RMSEs.

The accuracy assessment process applied by selecting limited number of control stations can be affected by several parameters. Two significant parameters affecting these results are distribution of the control stations on the study area and the measurement value at each control station. For increasing the results of the accuracy assessment control points located at centralized positions in study area should be selected. Additionally, measurements recorded at selected control stations should be close to mean value of all records. Thus, cross-validation technique may not give reliable results when it is applied with limited number of control station (Doğru et al. 2011).

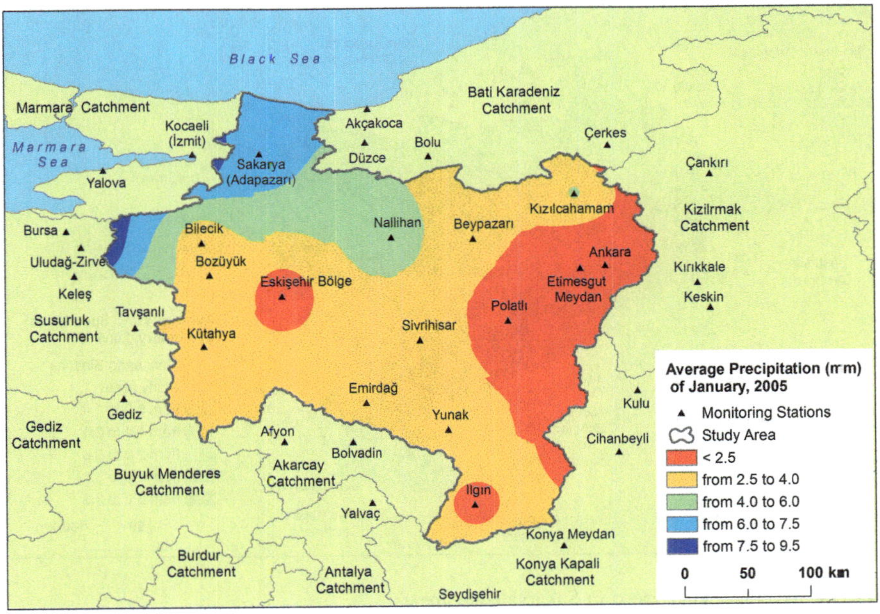

Fig. 3.2 Average precipitation in Sakarya Catchment

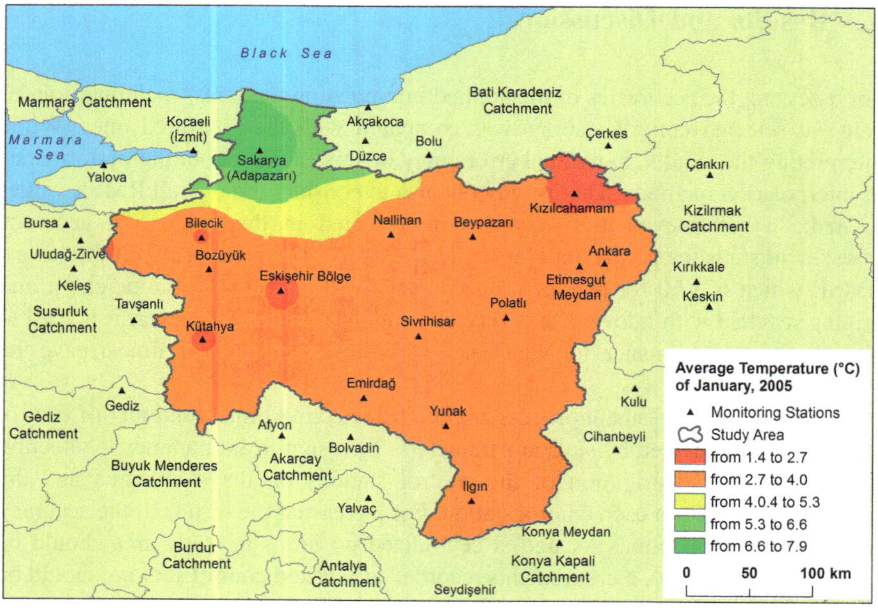

Fig. 3.3 Average temperature in Sakarya Catchment

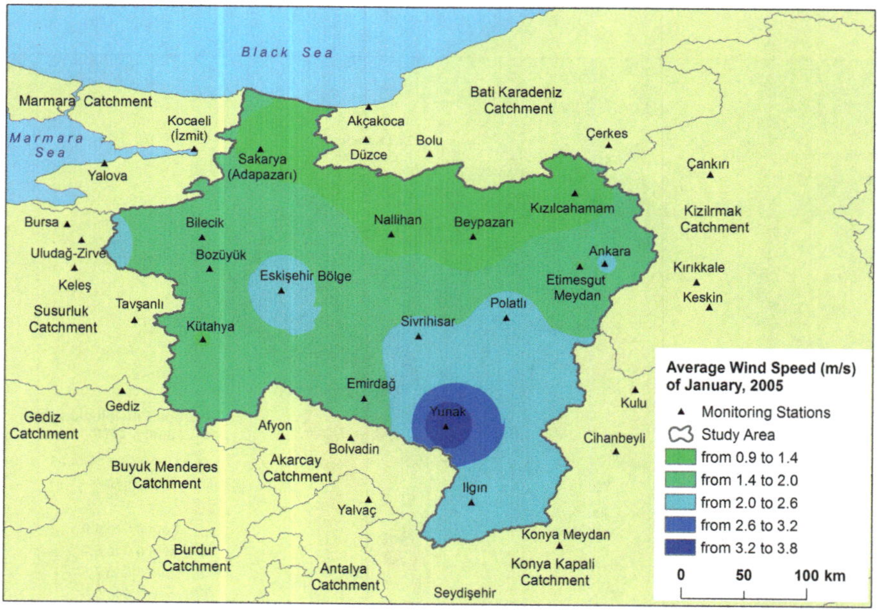

Fig. 3.4 Average wind speed in Sakarya Catchment

As the final step of the study, thematic maps indicating the precipitation, temperature, and wind speed distribution in Sakarya Catchment were produced by considering the cartographic principles. Following figures display the average precipitation, temperature, and wind speed data, respectively, by using IDW method (Fig. 3.2, 3.3 and 3.4).

3.4 Conclusion

This study was conducted to evaluate the performances of IDW, Kriging, and NN spatial interpolation methods, which are widely used in GIS software for presenting the point source data as raster surfaces. The default settings of each interpolation method were considered in the study implemented by using data collected at monitoring stations distributed over Sakarya Catchment. Accuracy assessment studies present different results for the most accurate estimations for different data sets. This gives an idea that the efficiency of the applied interpolation methods is strongly related with the density and the distribution of the monitoring stations over study area. Additionally, determination of the control stations and statistical characteristics of measurement values (range or distribution of the data, etc.) and some other environmental factors such as topography are considered as significant parameters influencing the accuracy of the interpolation. This study was conducted with the parameters of 2-D interpolation methods. However, it can be extended to 3-D by integrating the effect of topography.

On the other hand, accuracy assessment method also affects the accuracy results. To obtain more respectable results, another accuracy assessment methods can be applied such as cross-validation technique with larger extent. In the context, each monitoring station is considered as control station and its value is estimated by interpolating remaining monitoring station with different interpolation methods.

Moreover, the accuracy of the applied interpolation methods can be increased by changing their settings with default ones. Therefore, it is possible to obtain different results by applying the same methods to data sets with different values or density and distribution.

As a result, since the efficiency of interpolation has unsteady characteristic, accuracies of different interpolation methods should be examined before mapping point source data set obtained for each case, although point sources have the similar distribution over study area.

Acknowledgements This study was conducted within the EnviroGRIDS Project, which is an EU Funded 7th Framework Program Project under Grant Agreement 226740.

References

Attorre, F., Alfo, M., De Sanctis, M., & Bruno, F. (2007). Comparison of interpolation methods for mapping climatic and bioclimatic variables at regional scale. *International Journal of Climatology, 27*, 1825–1843. doi:10.1002/joc.1495.

Borga, M., & Vizzaccaro, A. (1997). On the interpolation of hydrologic variables: formal equivalence of multiquadratic surface fitting and kriging. *Journal of Hydrology, 195*(1–4), 160–171. doi:10.1016/S0022-1694(96)03250-7.

Childs, C. (2004). *Interpolating SURFACES IN ArcGIS spatial analyst* (pp. 32–35). ArcUser, July–September. Internet version available at http://www.esri.com.

Doğru A. O., Keskin, M., Ozdogu, K., İliev, N., Ulugtekin, N. N., Balcik, F. B., Goksel, C., & Sozen, S. (2011, October 31– November 04). Meteorolojik Verilerin Değerlendirilmesi ve Sunulması için Enterpolasyon Yöntemlerinin Karşılaştırılması (Comparison of Interpolation Methods for Evaluation and Presentation of Meteorological Data). Paper presented at TMMOB Coğrafi Bilgi Sistemleri Kongresi (Geographic Information Systems Congress), Antalya, Turkey. Internet version available at http://www.scribd.com/doc/82334134/CBS-Kongresi-31-EK%C4%B0M-4-KASIM-2011.

Goovaerts, P. (2000). Performance comparison of geostatistical algorithms for incorporating elevation into the mapping of precipitation. *Journal of Hydrology, 228*(1), 113–129. doi:10.1016/S0022-1694(00)00144-X.

Keskin, M., & Ozdogu, K. (2011). *Comparison of interpolation methods for meteorological data*. B.Sc thesis, ITU Geomatics Engineering, Istanbul.

Legates, D. R., & Willmott, C. J. (1990). Mean seasonal and spatial variability in global surface air temperature. *Theoretical Application in Climatology, 41*(1–2), 11–21. doi:10.1007/bf00866198.

Luo, W., Taylor, M. C., & Parker, S. R. (2008). A comparison of spatial interpolation methods to estimate continuous wind speed surfaces using irregularly distributed data from England and Wales. *International Journal of Climatology, 28*(7), 947–959. doi:10.1002/joc.1583.

Shepard, D. (1968). A two-dimensional interpolation function for irregularly-spaced data. In *Proceedings of the 1968 ACM National Conference*. doi:10.1145/800186.810616.

Sibson, R. (1981). A brief description of natural neighbor interpolation. In: Barnett, V. (Ed.), *Interpreting multivariate data* (pp. 21–36). Chichester, New York: Wiley.

Waters, N. M. (1988). Expert systems and systems of experts. In: Coffey, W.J., (Ed.), *Geographical systems and systems of geography: Essays in honour of William Warntz* (pp. 173–187). London: Department of Geography, University of Western Ontario. doi:10.1177/030913258901300311.

Watson, D. (1992). Contouring: A guide to the analysis and display of spatial data. *Computer Methods in the Geosciences, 10*, 321 (Pergamon Press, Oxford).

Chapter 4
Energy Storage with Pumped Hydrostorage Systems Under Uncertainty

Ahmet Yucekaya

Abstract Energy storage is becoming an important problem as the difference between supply and demand becomes sharper and the availability of energy resources is not possible all the time. A pumped hydrostorage system (PHSS) which is a special type of hydroelectric power plant can be used to store energy and to use the water more efficiently. When the energy demand and the energy price are high (peak hours), the water at upper reservoir is used to generate electricity and the water is stored in the lower reservoir. Revenue is gained from the power sale to the market. When the demand and the energy price are low (off-peak hours), the water at lower reservoir is pumped back to the upper reservoir. Cheap electricity is used to pump the water. The hourly market price and water inflow are uncertain. The main objective of a company is to find an operation schedule that will maximize its revenue. The hourly electricity prices and the water inflow to the reservoir are important parameters that determine the operation of the system. In this research, we present the working mechanism of the PHSS to store energy and to balance the load changes due to demand.

4.1 Introduction

Electrical energy storage is still expensive and not technologically efficient. The electricity is still stored in other forms such as magnetic, mechanical, and chemical energy. If a large amount of energy need to be stored, the most efficient and economic options are pumped hydrostorage system (PHSS) and compressed air storage systems (CAES). PHSS systems are more efficient compared to CAES, and they have lower cost and longer economic life.

There are at least two water reservoirs in this system, the large reservoir is at the upper level and the smaller one is in the lower level. There are electricity generation

A. Yucekaya (✉)
Kadir Has University, Istanbul, Turkey
e-mail: ahmety@khas.edu.tr

A.N. Bilge et al. (eds.), *Energy Systems and Management*,
Springer Proceedings in Energy, DOI 10.1007/978-3-319-16024-5_4

units and pumping units between two reservoirs. The water is used to generate electricity, and it is released from the upper reservoir to the lower level reservoir. This electricity is generated and sold when the electricity is most needed and the electricity price is high. These periods are called on-peak times and are usually daytime until the midnight. The electricity price is low on off-peak time. The water in the lower reservoir is stored or used to generate electricity in generation units on the small reservoir. Or it can be pumped again to the upper reservoir to be used to generate electricity later on-peak times. Hence, the water can be recycled whenever it is needed. In other words, the potential energy is stored in one reservoir to be converted to electrical energy later (Kapsalli and Kaldellis 2010). Figure 4.1 presents a typical PHSS system. A PHSS can also be used to balance the load changes as it has a fast ramp-up rate and quick response.

First, PHSS power plants were built in 1890 in Italy and Sweden. But their commercial usage was started in early 1930s and major developments took place after the World War II. Their importance has increased after the liberalization and deregulation of power markets as they became economic (Yang 2012).

In Yang and Jackson (2011), it is mentioned that pumped hydrostorages are the best low-carbon electricity resource because that the other options cannot flexibly adjust their output to match instable electricity demand. PHSSs are more efficient than other low-carbon electricity resources which are nuclear, wind, and solar system. It is indicated that PHSSs provide many commercial advantages for the producers. PHSSs serve to stabilize the electricity grid through peak shaving, reserve generation, load balancing, and frequency control. It is also noted that while global warming increasing rapidly, developers are working new PHSS projects and in 2014, its expected capacity worldwide is 76 GW.

Zhu et al. (2006), mentioned that hydropower plants are prepared with an optimum plan for the least operation cost, maximum reservoir management, and

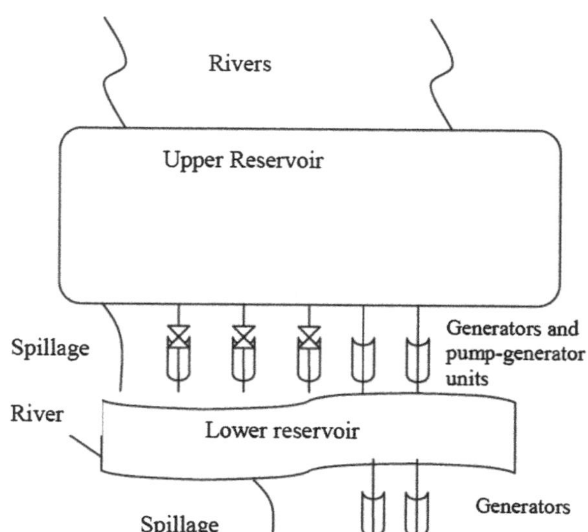

Fig. 4.1 A systemic view of PHSS

maximum proportion of electricity generation. In Zhao and Davison (2009a), a simple model for the operation of pumped hydrostorage facility is developed. It is assumed that there are time-varying but deterministic electricity prices and constant water inflows. The objective is to optimize the energy and profit produced by the plants (Zhao and Davison 2009b).

Ikudo (2009) develops an optimization model for the scheduling of a pumped hydropower plant. The model considers uncertainty of market prices and inflow rates. The dynamic programming model maximizes the expected profit from the operation of the plant given that there are Markov-based inflow rates and many price scenarios.

Through the literature, it is shown that the PHSSs are more efficient and essential low-carbon-level resources. However, it is also obvious that an efficient and effective operation of the system is needed given that there are deregulated power prices and variable water inflow. The remainder of the paper is as follows. Section 4.2 provides the model formulation and the solution methodology. Section 4.3 includes a case study developed for a PHSS system. The conclusions are given in Sect. 4.4.

4.2 Model Development and Solution Methodology

The objective of a PHSS operator is to schedule the system to maximize its profit, given power sale price and purchase price. There are also operational constrains that need to be considered. As a result, a model that includes all constraints and proves a generation and pumping schedule that will maximize the profit. The notation of the model is provided in appendix. The operational purpose is to maximize the profit of operation of PHSS which is revenue of power sale minus cost of power purchase as follows:

$$\text{Max} \quad \sum_{t=1}^{T}\left(\sum_{i=1}^{I} M_t^1 Q_{i,t}^1 + \sum_{j=1}^{J} M_t^2 Q_{j,t}^2 - \sum_{k=1}^{K} M_t^1 P_{k,t}\right) \tag{4.1}$$

Assume a PHSS system given in Fig. 4.1. The first part in the objective function is the sales revenue gained from the upper reservoir which is generated power by each unit multiplied with the power price. The second part is the revenue gained from the generator units at the lower reservoir. The last part is the cost of purchased power for pumping units which is calculated by multiplying the power amount and purchase price.

There are also operational constraints that need to be included as follows:

$$0 \le Q_{i,t}^1 \le Q_{i,\text{cap}}^1 \quad \forall\ i,t \tag{4.2}$$

$$0 \le Q_{i,t}^2 \le Q_{i,\text{cap}}^2 \quad \forall\ i,t \tag{4.3}$$

$$0 \leq P_{k,t} \leq P_{k,\mathrm{cap}} \quad \forall \ k, t \tag{4.4}$$

$$\sum_{i=1}^{I} Q_{i,t}^{1} \sum_{k=1}^{K} P_{k,t} = 0 \quad \forall \ k, t \tag{4.5}$$

$$V_t^1 = V_{t-1}^1 + \sum_{k=1}^{K} E_k^0 P_{k,t} - \sum_{i=1}^{I} E_i^1 Q_{i,t}^1 + I_t^1 - S_t^1 \quad \forall \ t \tag{4.6}$$

$$V_t^2 = V_{t-1}^2 - \sum_{k=1}^{K} E_k^0 P_{k,t} + \sum_{i=1}^{I} E_i^1 Q_{i,t}^1 + \sum_{j=1}^{J} E_j^1 Q_{j,t}^1 + I_t^2 + S_t^1 - S_t^2 \quad \forall \ t \tag{4.7}$$

$$V_{\mathrm{min,t}}^1 \leq V_t^1 \leq V_{\mathrm{max,t}}^1 \quad \forall \ t \tag{4.8}$$

$$V_{\mathrm{min,t}}^2 \leq V_t^2 \leq V_{\mathrm{max,t}}^2 \quad \forall \ t \tag{4.9}$$

$$I_t^1 = I_t^a + I_t^b \quad \forall \ t \tag{4.10}$$

Equation (4.2) and (4.3) limit the amount of power generation for units in upper reservoir and lower reservoir with their capacities, respectively. Equation (4.4) shows the limit for pumping units. These constraints basically ensure that capacity limits are not violated. Some water tunnels are used both for generation and pumping water. It is not possible to generate electricity and pump water at the same time. Hence, Eq. (4.5) ensures that only one of these activities take place at any given time. Equation (4.6) represents the amount of water in upper reservoir at time t. The amount of water in upper reservoir at time t is equal to accumulation of amount of water at time $t-1$, amount of water that is pumped to from the lower reservoir, and amount of water that added with rivers. Some of the water is used to generate electricity, and some water can be spilled when too much water exists in the reservoir. Equation (4.7) represents the amount of water in lower reservoir at time t. This amount is equal to the accumulation of amount of water at time $t-1$, amount of water that is added from the upper reservoir (with generation units and spillage), and amount of water that is added with the river. The amount of water that is pumped up back used to generate electricity, and spilled is deducted from this amount. Equations (4.8) and (4.9) represent the volume limits for both reservoirs. Equation (4.10) shows that inflow to first reservoir consists of inflow of two rivers.

The amount of water that in each river is uncertain throughout a year. The seasonal effects play an important role to determine the water level. On the other hand, the electricity price at time is also uncertain. The market mechanisms which include supply and demand determine the market price. A solution methodology that will help a PHSS operator to schedule the system is needed.

Main inputs of the model are the inflow rates of rivers for both reservoirs, the market prices for electricity, and characteristics of generating units. Figure 4.2

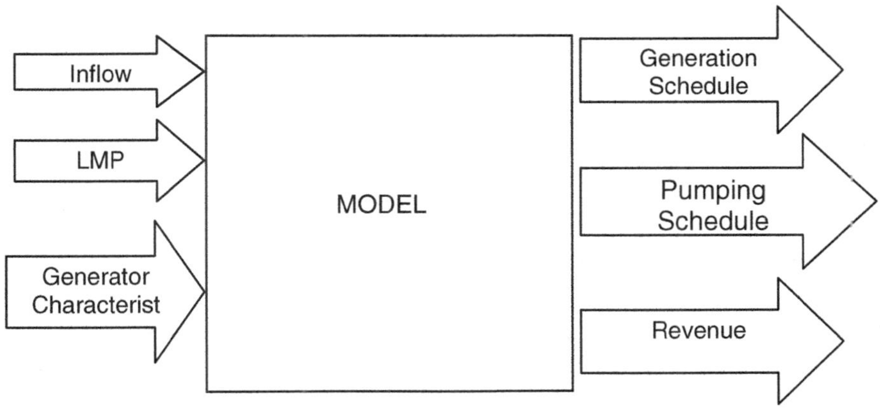

Fig. 4.2 Model inputs and outputs

provides the model inputs and outputs. The model should include the inputs and turn out an optimum generation and pumping schedule that will maximize the profit given that there is sales revenue and electricity cost. It is expected that generation should occur at the times when the electricity price is high to maximize the sales revenue. It is also expected that the pumping should occur when the price is low to minimize the cost.

The market prices are determined hourly in deregulated power markets. Hence, the hourly generation and pumping decisions are made. The flow of the solution methodology is determined as given in Fig. 4.3. The algorithm first estimates hourly electricity prices based on the historical prices. Then, hourly inflow rates are determined based on the historical estimations.

Fig. 4.3 General flow diagram

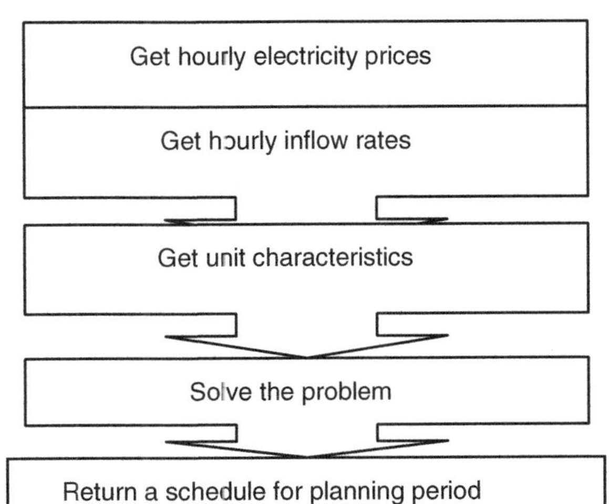

Finally, the model is solved using a General Algebraic Modeling System (GAMS), which is a high-level modeling system for mathematical programming problems. The software is suitable to build large-scale and complex modeling applications. CPLEX 12.5 solver is used which is included in GAMS. A generation and pumping schedule is proposed as an output of the solution algorithm.

4.3 Numerical Analysis

The model is applied for the Smith Mountain Reservoir which is a PHSS in Virginia, USA. The project was build to generate hydroelectric power with two reservoirs in the near by the Roanoke River, Virginia. The reservoir, which has two reservoirs namely Smith Mountain and Leesville lake, was built on the Roanoke River in the 1960s. Table 4.1 provides data for both reservoirs. They occupy about 600 miles of shoreline and get round about 25,000 surface acres of water for different uses.

The Smith Mountain is the larger and upper reservoir. The reservoir is fed by Roanoke, Back, and Blackwater rivers. The capacity of 5 generation units is 636 MW. The water in the Smith Mountain is utilized to drive the turbines and generate electricity. This water is stored in Leesville for generating electricity or to pump back to the Smith Mountain. Smith Mountain has 5 generation units and 3 pumping units. Table 4.2 shows the unit characteristics. The pumping units use the same tunnel with the generation units.

Table 4.1 Data for the reservoirs

Reservoir	Smith Mountain	Leesville
Surface area (km^2)	83	13.2
Max depth (mt)	76	9.4
Shore length (km)	800	32
Length (km)	96	160
Surface elevation (mt)	242	187

Table 4.2 Unit and pump characteristics for Smith Mountain

Unit	Capacity (MW)	Yield (M^3/MWh)
Generators		
1	64.5	0.269
2	177	0.273
3	109	0.279
4	178	0.271
5	68	0.280
Pumps		
1	77	0.189
3	127	0.213
5	77	0.190

Table 4.3 Unit characteristics for Leesville Lake

Unit	Capacity (MW)	Yield (M^3/MWh)
1	22	0.736
2	22	0.736

The Leesville Lake is the smaller and lower reservoir. The lake is fed by Pigg River and the water flowed from the Smith Mountain Reservoir. It has two generation units as given in Table 4.3. The water is either used for electricity generation in these units or to pump back to Smith Mountain to use it later.

An important input to the solution algorithm is the hourly market prices for the electricity. The electricity prices are called locational marginal prices (LMP) in USA. The historical prices for the year 2009 are given in Fig. 4.4. Historical prices are stochastic decision variables, and they show seasonal variations, depend on load, temperature, and work hours (Levine 2007).

The market prices effect the results of the optimization model as the objective is to maximize the profit which is revenue minus cost. However, high market prices can be exercised if the water is available in the reservoirs. The amount of water added to the two reservoirs with the inflows of four rivers needs to be considered in the model. Figure 4.5, 4.6, 4.7 and 4.8 provides hourly inflows in 2009 (8760 h) starting from the beginning of the year. The water inflows also show variations through the year. Winter and fall are the periods when much rainfall and runoff are occurred. Summer is the period with lowest inflow.

An optimum generation and pumping schedule can be found with the information of the market price and water level in the reservoirs. For the analysis, we assume that the system operator would like to determine a schedule for the January given that they have historical power prices and water inflows. The system operator forecasts hourly power prices and water inflows and needs to determine the

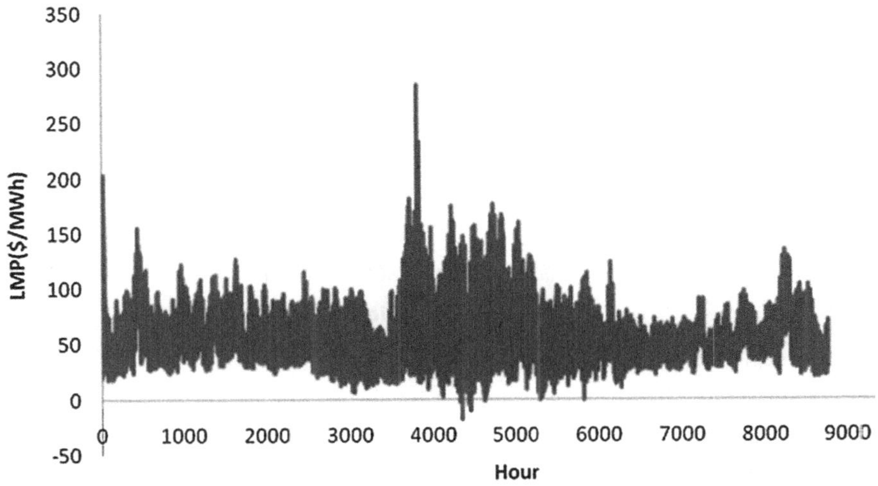

Fig. 4.4 LMP at Smith Mountain and Leesville Lake

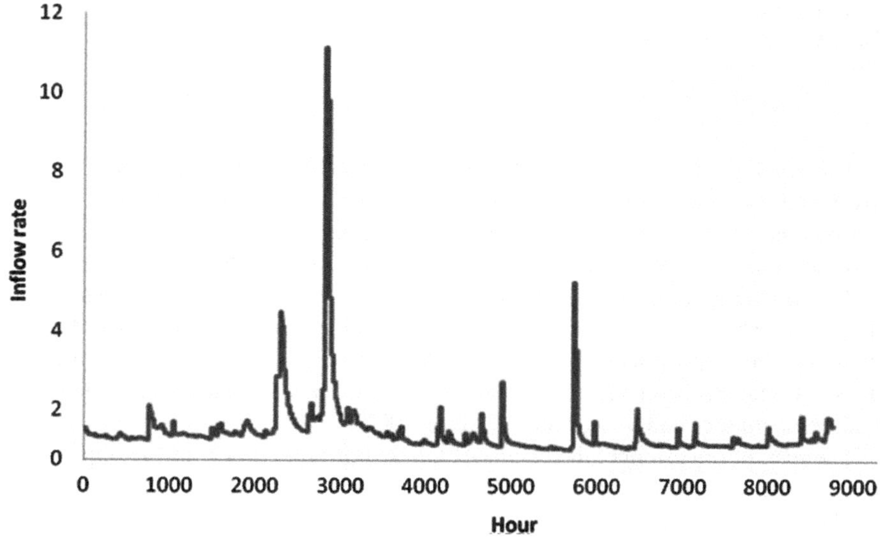

Fig. 4.5 Hourly inflows for Raonake (m^3/h)

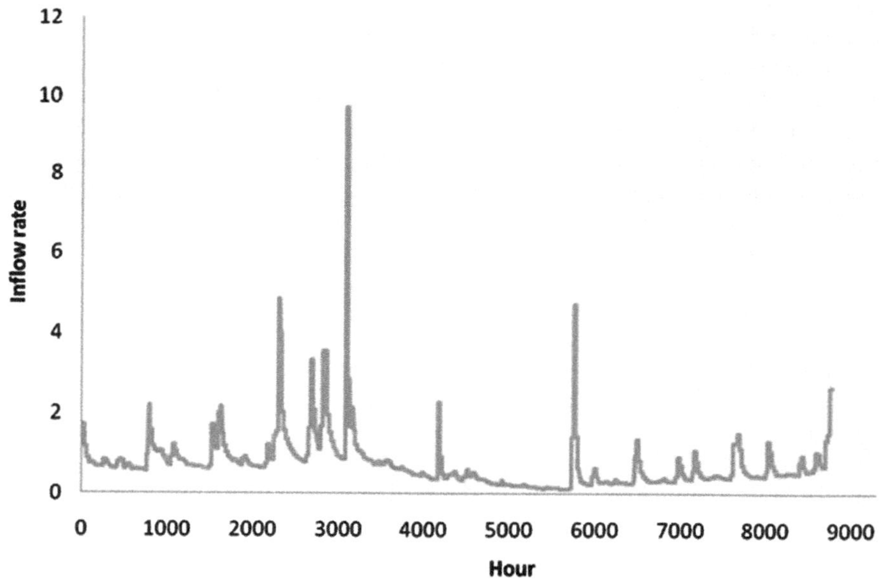

Fig. 4.6 Hourly inflows for Pigg (m^3/h)

optimum generation and pumping schedule. At the beginning of the month, the system operator run the algorithm using the price and flow estimates and determines an operation schedule for one month. Figure 4.9 shows the generation schedule for

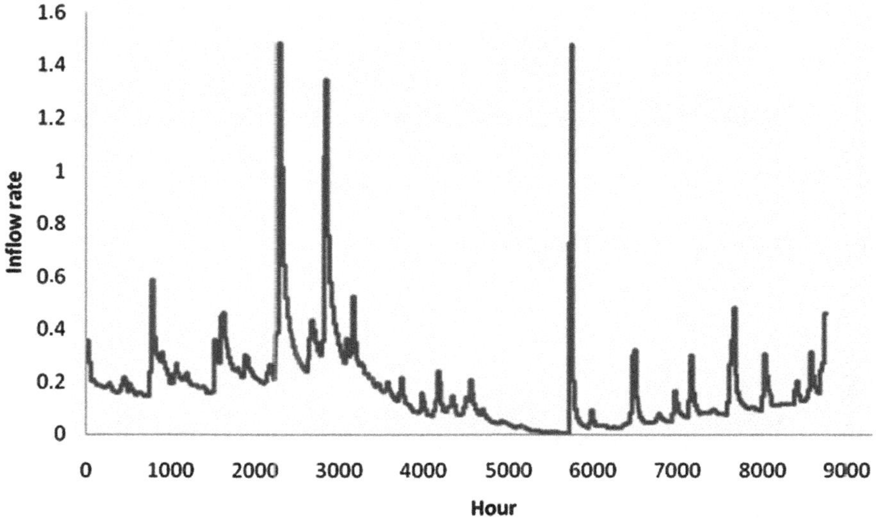

Fig. 4.7 Hourly inflows for Blackwater (m^3/h)

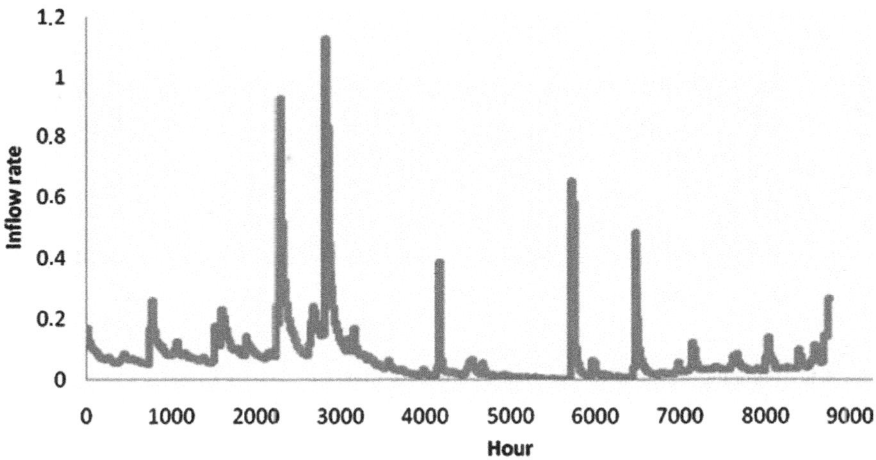

Fig. 4.8 Hourly inflows for Back (m^3/h)

the first unit at the Smith Mountain. When it is economic, the electricity is generated and the water in Smith Mountain Reservoir is discharged to the Leesville Lake.

The discharged water to the Leesville Lake is used for electricity generation, or it can be pumped back to the Smith Mountain Reservoir. Figure 4.10 shows the generation schedule in first unit of the Leesville Lake. When the electricity price was high enough to generate profit, the generation unit was run and the electricity

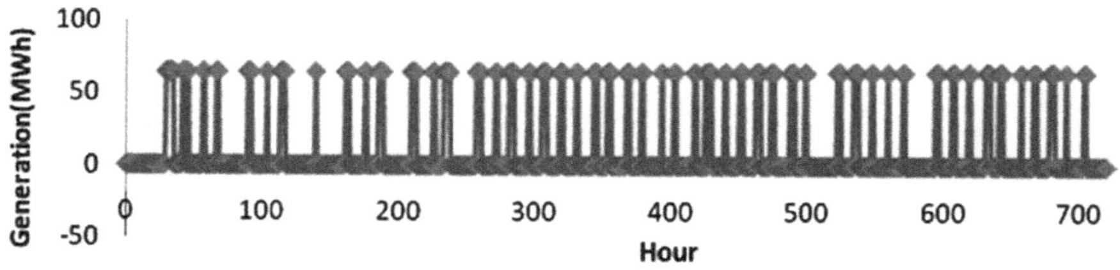

Fig. 4.9 Generation for Smith Mountain unit

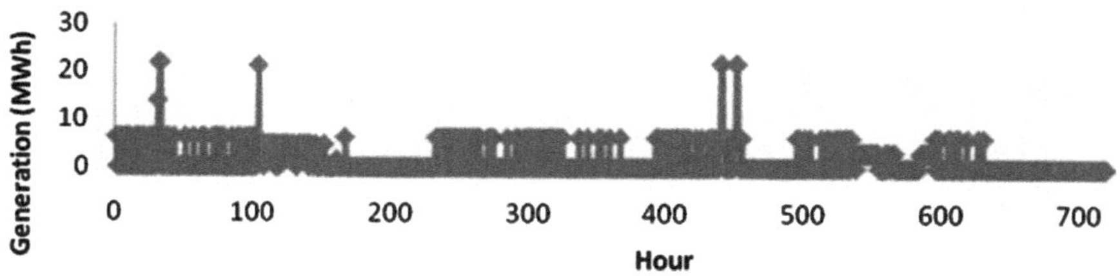

Fig. 4.10 Generation for Leesville unit

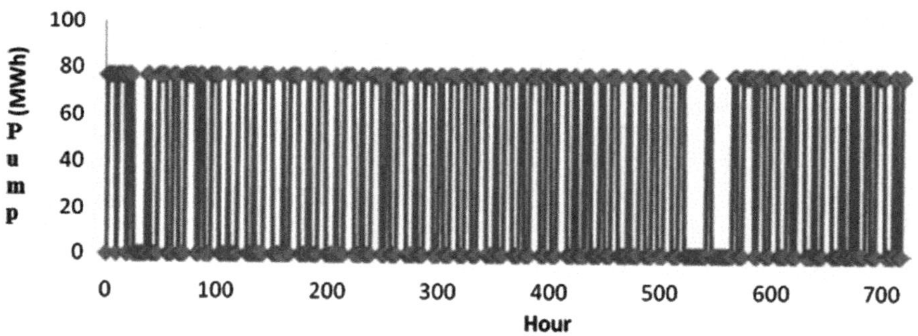

Fig. 4.11 Pumping schedule for the Smith Mountain unit

was sold to the market. The water is then discharged, and it can no longer be used for power generation.

If the power price is low, it is not wise to generate electricity as the electricity can be sold with higher prices in other times. If it is economic, the water in the Leesville Lake can be pumped back to the Smith Mountain Reservoir using cheap electricity prices. Figure 4.11 provides the pumping schedule for the planning month. It is also important to note that pumping can only be occurred when there is no electricity generation. The power prices are low in nights and weekends and high in daytimes because of the demand. It is expected to have generation in daytime and pumping in nighttime.

4.4 Conclusion

In this paper, a model is developed for the scheduling a PHSS under uncertain electricity prices and water inflow rates. The objective is to maximize the profit of the system operator, and the model returns an operation schedule that includes generation for the units in upper and lower reservoir and a scheduling for the pumping units. The model is validated for the Smith Mountain PHSS. It is shown that if the estimated hourly prices and inflow rates which are forecasted based on the historical data are used, an operation schedule can be found for the system operator. The research is open for further development. In order to extend the research, different scenarios for market prices and inflows can be generated and expected profit can be obtained. The model returned promising results and hence shows that it can be used by power companies for scheduling of PHSS.

Appendix

Notation

M_t^1	Power market price during hour t in first reservoir (\$/MWh)
M_t^2	Power market price during hour t in second reservoir (\$/MWh)
$Q_{i,t}^1$	Power generated at first reservoir by unit i in hour t (MW)
$Q_{i,t}^2$	Power generated at second reservoir by unit j in hour t (MW)
$P_{k,t}$	Power used for pumping by unit k during hour t (MW)
$Q_{i,\text{cap}}^1$	Capacity of first reservoir of unit i (MWh)
$Q_{i,\text{cap}}^2$	Capacity of second reservoir of unit j (MWh)
$P_{k,\text{cap}}$	Capacity of pumping unit k (MWh)
V_t^1	Volume of first reservoir at the end of hour t (ft^3)
V_t^2	Volume of second reservoir at the end of hour t (ft^3)
I_t^1	Inflow to first reservoir during hour t (ft^3)
I_t^2	Inflow to second reservoir during hour t (ft^3)
I_t^a, I_t^b	Inflow from rivers a and b at hour t (ft^3)
S_t^1	Spillage from first reservoir in hour t (ft^3)
S_t^2	Spillage from second reservoir in hour t (ft^3)
E_k^0	Yield (efficiency) of pump k (ft^3/MW)
E_i^1	Yield (efficiency) of generator i in first reservoir (ft^3/MW)
E_i^2	Yield (efficiency) of generator j in second reservoir (ft^3/MW)
$V_{\text{min,t}}^1, V_{\text{max,t}}^1$	Minimum and maximum limits of first reservoir in hour t (ft^3)
$V_{\text{min,t}}^2, V_{\text{max,t}}^2$	Minimum and maximum limits of second reservoir in hour t (ft^3)

References

Ikudo, A. (2009), Maximizing gross margin of a pumped storage hydroelectric facility under uncertainty in price and water inflow (M.Sc. Thesis, The Ohio State University), from http://etd.ohiolink.edu/send-pdf.cgi/Ikudo%20Akina.pdf?osu1243970453.

Kapsalli, M., & Kaldellis, J. K. (2010). Combining hydro and variable wind power generation by means of pumped-storage under economically viable terms. *Applied Energy, 87*(11), 3475–3485.

Levine, J. G. (2007). Pumped hydroelectric energy storage and spatial diversity of wind resources as methods of improving utilization of renewable energy sources (Thesis, University of Colorado), from http://www.colorado.edu/engineering/energystorage/files/MSThesis_JGLevine_final.pdf.

Yang, C. (2012). Pumped hydroelectric storage, wiley encyclopedia of energy, from http://www.duke.edu/~cy42/PHS.pdf.

Yang, C. J., & Jackson, R. B. (2011). Opportunities and barriers to pumped-hydro energy storage in the United States. *Renewable and Sustainable Energy Reviews, 15*, 839–844.

Zhao, G., & Davison, M. (2009a). Optimal control of hydroelectric facility incorporating pump storage. *Renewable Energy, 34*(4), 1064–1077.

Zhao, G., & Davison, M. (2009b). Valuing hydrological forecasts for a pumped storage assisted hydro facility. *Journal of Hydrology, 373*(3–4), 453–462.

Zhu, C. J., Zhou, J. Z., Yang, J. J., & Wu, W. (2006). Optimal scheduling of hydropower plant with uncertainty energy price risks. In *International Conference on Power System Technology*. October 22–26, 2006.

Chapter 5
Telelab with Cloud Computing for Smart Grid Education

Pankaj Kolhe and Berthold Bitzer

Abstract As the demand for energy increases, the need to generate and distribute energy to the customers with greater efficiency also increases. Introduction of smart grids provides platform for the utilities to collect and analyze consumption data in real time. This helps them to define the generation profile and offer competitive energy prices to the customers. Customer on the other hand can use the knowledge of his own consumption profile to define and tune his energy usage. Education about smart grid environment which involves software, hardware devices, and network technologies for data collection and analysis is important for both utilities and customers. Research and development in Internet technologies promote remote laboratory as a cost-effective solution for users located across the globe. Cloud computing platform can further reduce costs involved in data storage and software used. This paper presents the idea of developing the remote laboratory located at South Westphalia University in Soest, Germany, further by integrating cloud computing and smart grid simulation environment. This will educate people across the globe by offering them hands on experience on smart grid technology and thus will contribute to the field of power engineering education.

5.1 Introduction

Advancements in Internet technologies have brought some significant and cost-effective changes in the field of technology and education. The relevance of Web-based learning and teaching has increased in many research fields. As per the requirement of the application, user can effectively use virtual laboratories that are based on software simulation of physical processes. However, the laboratories that

P. Kolhe (✉) · B. Bitzer
South Westphalia University of Applied Sciences, Soest, Germany
e-mail: kolhe@fat-soest.de

B. Bitzer
e-mail: bitzer.berthold@fh-swf.de

A.N. Bilge et al. (eds.), *Energy Systems and Management*,
Springer Proceedings in Energy, DOI 10.1007/978-3-319-16024-5_5

have attracted user attention are remote laboratories which deal with real physical processes and not simulations. These real physical processes or systems can be accessed over Internet from any remote part of the world. Examples of these laboratories both virtual and remote could be found in Schmid (1998), Henry (1998), Bhandari and Shor (1998), Hahn and Spong (2000).

In this paper, remote laboratory or Telelab will be used to introduce users to the smart grid technology and provide hands on experience. The hybrid power system model consisting of solar panel, wind energy system, solar charge controller, current sensor, inverter, and battery system is available in the laboratory of South Westphalia University in Soest and could be accessed from any remote location for experiment purposes. The idea is to use the cloud computing facilities as a service through a service provider to provide the necessary software and hardware infrastructure to perform the experiment. This will allow the user to focus only on his core experiment while the secondary tasks of software and hardware updates and maintenance or the task of managing and scaling resources could be taken care by the cloud computing service provider.

Generation and distribution of energy efficiently are important on the background that the energy demand is increasing rapidly. This means that the process of energy generation, distribution, data collection, data analysis, and load profile predictions should take place in real time. This is possible with the development of smart grid.

In a smart grid, data are collected from the consumer and his consumption profile is analyzed. Depending upon the consumption profile of the consumer, generation profile and distribution of energy are planned and implemented. Regenerative energy technologies such as solar or wind systems could be used to generate energy which will be located close to the customer.

This reduces transmission and distribution losses. User can be provided energy as needed depending on the time of the day. Energy consumed at peak period will cost more as compared to that during off-peak period. User will also be provided with the chance to trade excess energy. Presence of smart devices in smart grid facilitates better control of the grid in the event of peak loads. From the above discussion, it is understood that there is large amount of data exchange from the utilities to consumers and vice versa. Large storage space will be required to store this data and later use it for analysis. Powerful computers with high computational capabilities, different software for analysis and monitoring, and some additional infrastructure could be required which can be rented from a cloud computing provider. The purpose of this paper was to introduce the effective use of remote laboratory, cloud computing, and smart grid simulation software such as GridIQ for smart grid education in a cost-effective manner.

5.2 Hybrid Power System and Telelab

To realize the above-described system, it is feasible to visualize the whole system as basic experiment setup, cloud computing platform and smart grid infrastructure. Experiment setup includes the laboratory demonstrator or the hybrid power system consisting of solar panel, wind system, solar charge controller, inverter, current sensor, battery, data logger, and computer for monitoring and control. PLC with required analog and digital i/os will be the mode of interaction between the physical experiment setup and the local computer with monitoring and control software such as WinCC and step 7 installed.

Web server, as in Fig. 5.1, could monitor the Internet traffic accessing the experiment. A camera server hosts the camera images in real time and makes them available to the users. This gives the user a sense of changes taking place in the experiment, and the user can easily study the experiment behavior as per the implemented changes. With authenticated user name and password, user from any location of the world can access this experiment.

5.3 Cloud Computing and Its Significance in Our Application

Cloud computing, as it is familiar, delivers different services to the user depending on application requirement. It provides hosted services over the Internet (Shi et al. 2009). Cloud computing is a paradigm shift based on a collection of many old and

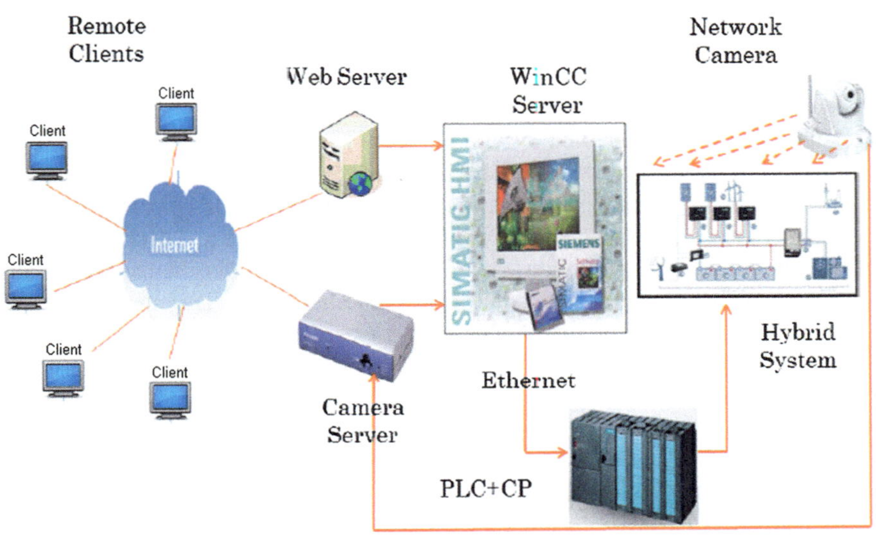

Fig. 5.1 Power system with remote laboratory setup

few new concepts in several domains such as service-oriented architecture (SOA), distributed and grid computing as well as virtualization (Lamia 2009). The services offered are defined as Infrastructure as a Service (IaaS), Platform as a Service (PaaS), and Software as a Service (SaaS). Pay-as-you-go is an important feature of cloud computing which gives user opportunity to use massive computation ability and scalability features. Many companies are using these features of cloud computing such as software applications, programming platforms, data storage, computing infrastructure, and hardware as service offered by various cloud computing providers. It refers to both applications delivered as services over the Internet and the hardware in the data centers that provide those services. A cloud computing platform can dynamically configure and reconfigure computation resources as per the application. These resources can be physical machines or virtual machines with scalable computation resources such as CPU, storages, network equipment, or other devices. This application of cloud computing resides on a large-scale data center or power servers that host the Web services and Web applications. Cloud computing can support grid computing by providing physical or virtual resources on which the grid application can run (Boss 2007). smart grids involve incorporation of renewable energy resources to generate power located close to customer site. This requires lot of data exchange between the nodes and high computational ability. Moreover, smart grids have computerized systems that give efficient and smooth information exchange for monitoring and control of the widely dispersed distributed power resources (Apostolopoulose and Oikonomou 2004).

If we focus on our system under consideration (Fig. 5.1), then it is obvious for instance that WinCC visualization software which is required for monitoring purposes could be provided by a cloud provider under SaaS feature. Both the servers—Web server and camera server—could be provided as hardware services. This means the user authentification task would be taken care by a cloud provider, and even the live images obtained by camera server could be effectively managed by cloud provider. This reduces our hardware requirements and simplifies our tasks of maintaining the software updates required for these servers. All the security arrangements necessary to protect these servers from unauthorized users or from virus threats would be handled by the provider. The bandwidth and storage space required for camera images would be provided by the provider. The scalability feature enables us to offer our experiment to maximum users since the resource management tasks are managed by the provider. It is possible to select for instance off-peak hours in the day or night, where the provider will offer the bandwidth and other services at minimum costs due to less Internet traffic. During these hours, maximum users can log in and benefit from the experiment.

It can be observed in Fig. 5.2 that our basic structure of laboratory demonstrator gets modified due to introduction of Cloud computing provider. The provider makes WinCC software and the two servers as software/hardware services available to users. Even other softwares such as HOMER and the smart grid simulation software GRIDIQ can be made available by the cloud provider. The Telelab or Web site could be maintained in the cloud, and all the data from different nodes of hybrid system could be collected in cloud and stored. When required, specific data can be

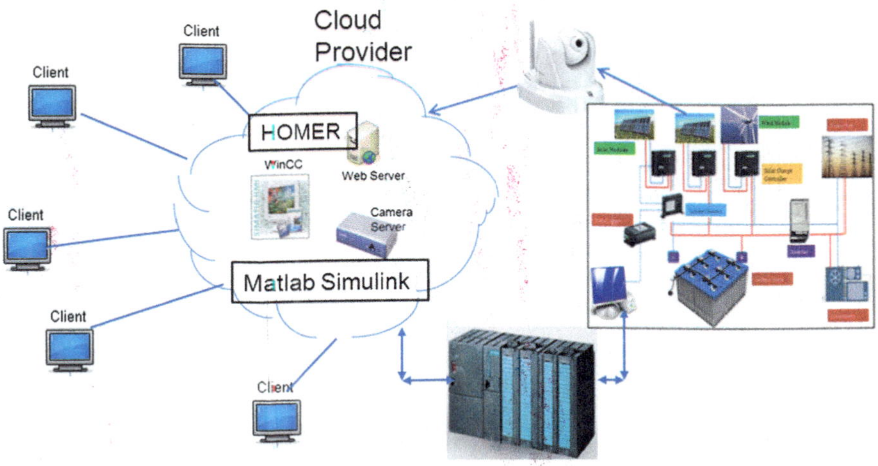

Fig. 5.2 Hybrid system remote laboratory with cloud computing

requested by the user, further analysis on generation profiles, load profiles, and consumption profiles could be done, and based on it, real-time predictions could be possibly obtained.

5.4 GE Smart Grid Package

After discussing hybrid system, remote laboratory concept and cloud computing the smart grid application can be discussed in this section. Smart grid can be simulated with the help of different softwares such as MATLAB/Simulink, Gridsim, Gridlab, GridIQ, EMTP-ATP, NEPLAN, ETAP, and several others. GridIQ solution could be considered for smart grid implementation. GE expertise and knowledge could be used to implement it as end-to-end software solution. Real-time information is provided to optimize consumption of power. Smart grid package consists of metering, advanced metering infrastructure (AMI), outage management, consumer portal, and GIS capability. GE offers implementation service. However, it is possible to carry out installation and commissioning of system by the user. During the remote laboratory, these things could be explained and a know-how could be given to the users performing the experiment. The hosted service of GE enables the user to use the software without requiring to maintain IT infrastructure, or there is no requirement to purchase licenses. The software will be hosted by a GE data hosting center. It can be accessed and utilized like a cloud service. The hardware such as meters is owned by user. One more feature offered by GE provider is to manage meters, software, and IT and deliver the data to the user. This feature is usually preferred if less resources are available.

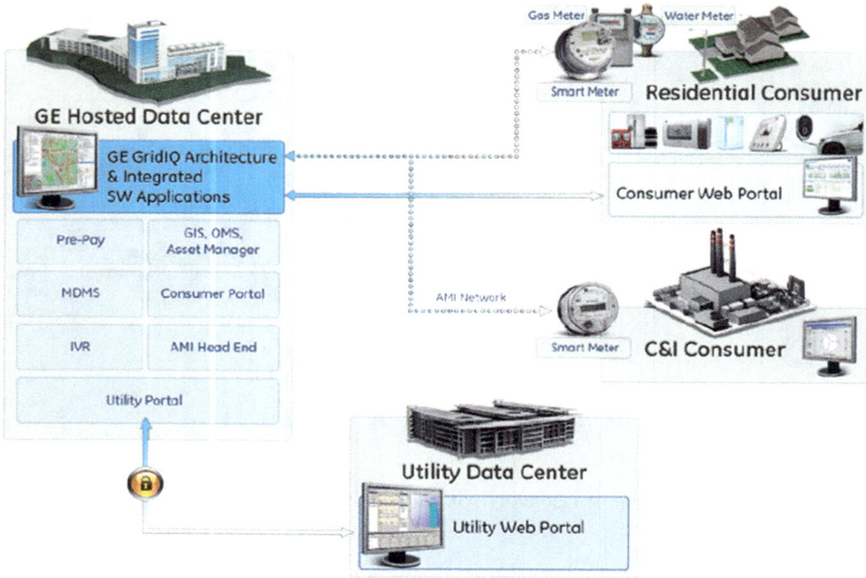

Fig. 5.3 GE services for the user

It can be observed in the Fig. 5.3 above that using AMI feature, data can be obtained from smart meters installed and commissioned at the residential and commercial consumers. Software is hosted at the hosted data center which is provided with high security. Data are protected from other consumers. Between two systems, data are transferred through secured VPN tunnel. The center complies with industry standard cyber security requirements. I210+c is a standard feature of GE. Geospatial information system (GIS) can help the user to view the network exactly the way it is implemented on the field. This helps to exactly locate the metering assets and utilize network resources effectively. Outage management system (OMS) helps in quicker fault detection and increases operational efficiency. Interactive voice response (IVR) is used to communicate with users via text, phone calls, or e-mails to ensure that they are updated on estimated outage duration. Web portal ensures that the users get real-time information on energy usage and price.

This smart grid package from GE combined with Telelab and cloud computing concepts would provide remote users opportunity to understand and educate themselves about the concept of smart grid. The idea of cloud computing for data and software management could be explained to the users with this experiment setup.

5.5 Applications

In this section, different applications related to combining the idea of Telelab for smart grid education using cloud computing will be discussed.

5.5.1 Academic Institutions, Research, and Training

Academic institutions would largely benefit with the implementation of this idea. Telelab makes the complete hardware system available to remote users or remote students or researchers at very low cost. It is possible for companies to train and educate their employees about smart grid concept with the help of this system. Academicians, researchers, and even students can work with this system to understand its functionality and could perform necessary research to improve the system performance in future.

5.5.2 Low-Cost Idea

The users require Internet connection and a computer to get connected to the server. This server will authentify the login data of the user and give him access to the virtual machine located in the university. Through this machine, the user gets access to local hardware and local software. The pathway to get access to the software in cloud goes through this server. As discussed above, a possibility of hosting these servers in cloud could be explored through a service provider. User can understand this whole system at minimum cost. As this would turn out to be a system setup for educational purposes, cheap cloud service providers could be identified. GE could also provide the smart grid package at affordable cost. Since this would be one time investment and the users benefited will be large in number, it is worth to make the investment.

5.5.3 Cloud Computing for Data Management

Smart grids involve large amount of data collection, storage, analysis, and predictions. This task of data management and computation requires efficient computer systems loaded with necessary software. We can take advantage of cloud service provider to provide us all the required software to collect, process, store, and analyze data along with data storage space. The remote user would get a firsthand experience of working with a cloud computing system which would update his knowledge about cloud computing platform.

5.5.4 Smart Grid Education

The increased cost of electricity at generation, transmission, and distribution stages presents challenges to engineers and researchers to device methods to reduce these costs. smart grids analyze the consumption and demand profile of customer and based upon that develop generation profile in run time mode. Distributed renewable energy generation integrated in smart grids saves environment from pollution and minimizes transmission and distribution losses. Smart devices such as smart TV or smart washing machines could be effectively controlled based upon the load demand profile. This would be a good start-up step to educate remote users with the smart grid idea.

5.5.5 Advanced Hardware Technologies

In this experimental setup, advanced and application-specific hardware such as latest sensors, actuators, network cameras, and smart meters would be used. Users would get first hand working knowledge of all these devices and their functionalities.

5.6 Summary

The paper discussed an application system of hybrid power consisting of solar panel, wind turbine, data monitoring, and control components. The significance of cloud computing for data management and computation is also discussed. Education of cloud computing can be served by this experimental setup. The idea to use Telelab to perform remote experiments over an integrated hybrid power system, cloud computing platform, and smart grid packet gives user a chance to know these technologies and get a working knowledge about them. The author is confident that further fine-tuning of this idea will lead to its successful implementation and will add a significant educational value to the power engineering field.

References

Apostolopoulose, T. K., & Oikonomou, G. C. (2004). A scalable, extensible framework for grid management. In *22nd IASTED international conference*. Austria.

Bhandari, A., & Shor, M. (1998). Access to an Instructional control laboratory experiment through the world wide web. In *IEEE American Control Conference* (pp. 1319–1325), Philadelphia, USA.

Boss, G., (2007). *Cloud computing*. IBM, New Delhi.

Hahn, H. H., & Spong, M. W. (2000). Remote laboratories for control education. In *39th IEEE Conference on Decision and Control*. Sydney, Australia.

Henry, J. (1998). Engineering lab online. University of Tennessee, Chattanooga.

Lamia, Y. (2009). Towards a unified ontology of cloud computing. University of California, Santa Barbara.

Schmid, C. (1998). The virtual lab VCLAB for education on the Web. In *IEEE American Control Conference* (pp. 1314–1318), Philadelphia, USA.

Shi, P., Wand, H., Jiang, J., & Lu, K., (2009). Cloud computing for research and implementation of network platform. *Computer Engineering and Science*, *31*, 249–252.

Chapter 6
A Decomposition Analysis of Energy-Related CO_2 Emissions: The Top 10 Emitting Countries

Aylin Çiğdem Köne and Tayfun Büke

Abstract Climate change, caused by greenhouse gas (GHG) emissions, is one of the hot topics all around the world. Carbon dioxide (CO_2) emissions from fossil fuel combustion account for more than half of the total anthropogenic GHG emissions. The top 10 emitting countries accounted 65.36 % of the world carbon dioxide emissions in 2010. China was the largest emitter and generated 23.84 % of the world total. The objective of this study is to identify factors that contribute to changes in energy-related CO_2 emissions in the top 10 emitting countries for the period 1971–2010. To this aim, a decomposition analysis has been employed. Decomposition analysis is a technique used to identify the contribution of different components of a specific variable. Here, four factors, namely population, per capita income, energy intensity, and carbon intensity, are differentiated. The results show that the economic activity effect and the energy intensity effect are the two biggest contributors to CO_2 emissions for all countries with a few exceptions.

6.1 Introduction

The qualitative dimension of energy use is becoming increasingly important for sustainable development. One important question in this context and in the context of global climate change is how one can achieve the separation of greenhouse gas (GHG) emissions. Among six kinds of GHG, the largest contribution to the greenhouse effect is carbon dioxide (CO_2), and its share of greenhouse effect is about more than 50 % (IPCC 1995; He and Chen 2002).

A.Ç. Köne (✉) · T. Büke
Muğla Sıtkı Koçman University, Muğla, Turkey
e-mail: ckone@mu.edu.tr

T. Büke
e-mail: tbuke@mu.edu.tr

A.N. Bilge et al. (eds.), *Energy Systems and Management*,
Springer Proceedings in Energy, DOI 10.1007/978-3-319-16024-5_6

Table 6.1 CO_2 emission by countries (2010)

Rank	Country	CO_2 emission (Mt)	% of total
1	China	7217.1	23.84
2	USA	5368.6	17.73
3	India	1625.8	5.37
4	Russian Federation	1581.4	5.22
5	Japan	1143.1	3.78
6	Germany	761.6	2.52
7	South Korea	563.1	1.86
8	Canada	536.6	1.77
9	Islamic Republic of Iran	509.0	1.68
10	United Kingdom	483.5	1.60
	World total	30,276	100

The top 10 CO_2-emitting countries accounted for 65.36 % of the world CO_2 emissions in 2010. China and the USA were the two highest emitters and generated 23.84 and 17.73 % of the world total, respectively (Table 6.1) (IEA 2012).

Decomposition analysis is a technique used to identify the contribution of different components of a specific variable. It is an effective tool which is used in various disciplines. In economics and environmental sciences, it has been applied to investigate the main factors contributing to the CO_2 emissions and the mechanisms influencing energy consumption.

Its application to policy formulation is generally used to improve sustainability management, to reduce the economic impacts on the environment, to promote energy and technological efficiency, and to design decoupling strategies (Subhes and Arjaree 2004; Diakoulaki et al. 2006; Diakoulaki and Mandaraka 2007; McCollum and Yang 2009).

This work aims to identify the factors that contribute to the changes in CO_2 emissions in the top 10 CO_2-emitting countries for the period of 1971–2010 by the refined Laspeyres method (Steckel et al. 2011; Kumbaroğlu 2011; Andreoni and Galmarini 2012). Population, per capita income, energy intensity, and carbon intensity were the four effects that were investigated.

6.2 Materials and Methods

6.2.1 The Decomposition Analysis

The CO_2 emission can be expressed as an extended Kaya identity (Xiangzhao and Ji 2008; Girod et al. 2009; Linyun and Hongwu 2011) which is a useful tool to decompose the total carbon emission as a product of four variables as shown in Eq. (6.1).

$$(CO_2) = (P)(GDP/P)(TPES/GDP)(CO_2/TPES) \tag{6.1}$$

The right-hand side of Eq. (6.1) refers to the population (P), income per capita $G = (GDP/P)$, energy intensity of economic activity $E = (TPES/GDP)$, and carbon intensity of energy use $C = (CO_2/TPES)$.

The change of CO_2 emission between a base year (t) and a target year ($\Delta t + t$), denoted by (ΔCO_2), can be defined as a function of four variables, namely the change in the population effect, the change in the economic activity effect, the change in the energy intensity effect, and the change in the carbon intensity effect, as shown in Eq. (6.2).

$$(\Delta CO_2) = (CO_2)^{t+\Delta t} - (CO_2)^t = P_{\text{effect}} + G_{\text{effect}} + E_{\text{effect}} + C_{\text{effect}} \tag{6.2}$$

where superscripts (t) and ($\Delta t + t$) denote a base year and a target year, respectively.

According to the complete decomposition model given by refined Laspeyres method, each effect in the right-hand side of Eq. (6.2) can be computed as follows: Equation (6.3) calculates the population effect:

$$\begin{aligned} P_{\text{effect}} = {} & (\Delta P)G^tE^tC^t + \frac{1}{2}(\Delta P)[(\Delta G)E^tC^t + G^t(\Delta E)C^t + G^tE^t(\Delta C)] \\ & + \frac{1}{3}(\Delta P)[(\Delta G)(\Delta E)C^t + (\Delta G)E^t(\Delta C) + G^t(\Delta E)(\Delta C)] \\ & + \frac{1}{4}(\Delta P)(\Delta G)(\Delta E)(\Delta C) \end{aligned} \tag{6.3}$$

Equation (6.4) calculates the economic activity effect:

$$\begin{aligned} G_{\text{effect}} = {} & (\Delta G)P^tE^tC^t + \frac{1}{2}(\Delta G)[(\Delta P)E^tC^t + P^t(\Delta E)C^t + P^tE^t(\Delta C)] \\ & + \frac{1}{3}(\Delta G)[(\Delta P)(\Delta E)C^t + (\Delta P)E^t(\Delta C) + P^t(\Delta E)(\Delta C)] \\ & + \frac{1}{4}(\Delta P)(\Delta G)(\Delta E)(\Delta C) \end{aligned} \tag{6.4}$$

Equation (6.5) calculates the energy intensity effect:

$$\begin{aligned} E_{\text{effect}} = {} & (\Delta E)P^tG^tC^t + \frac{1}{2}(\Delta E)[(\Delta P)G^tC^t + P^t(\Delta G)C^t + P^tG^t(\Delta C)] \\ & + \frac{1}{3}(\Delta E)[(\Delta P)(\Delta G)C^t + (\Delta P)G^t(\Delta C) + P^t(\Delta G)(\Delta C)] \\ & + \frac{1}{4}(\Delta P)(\Delta G)(\Delta E)(\Delta C) \end{aligned} \tag{6.5}$$

Equation (6.6) calculates the carbon intensity effect:

$$C_{\text{effect}} = (\varDelta C)P^tG^tE^t + \frac{1}{2}(\varDelta C)[(\varDelta P)G^tE^t + P^t(\varDelta G)E^t + P^tG^t(\varDelta E)] + \frac{1}{3}(\varDelta C)[(\varDelta P)(\varDelta G)E^t + (\varDelta P)G^t(\varDelta E) + P^t(\varDelta G)(\varDelta E)] + \frac{1}{4}(\varDelta P)(\varDelta G)(\varDelta E)(\varDelta C) \quad (6.6)$$

The first parts of Eqs. (6.3–6.6) can be interpreted as the partial effect of the population, partial effect of the economic activity, partial effect of the energy intensity, and partial effect of the carbon intensity components on the change of (ΔCO_2) emissions between time step ($\Delta t + t$) and the preceding step (t). The following parts of Eqs. (6.3–6.6) capture the interactions between the remaining variables and the residual terms.

It is necessary to make clear that different factors caused the changes in CO_2 emission. The population change effect is used to control the population size. The economic activity effect reflects the economic development. Energy consumption is mainly related to some variables, such as economic structures, the efficiency of the energy systems, energy utilization technologies, energy prices, energy conservation, and energy-saving investments, which are composed of energy intensity effect. And the carbon intensity effect is used to evaluate fuel quality, fuel substitution, and the installation of abatement technologies.

Equations (6.2–6.6) present the required formulas for the decomposition analysis. A computer code in MATHEMATICA (Wolfram 2004) has been developed to do the calculations in this text.

The data used in the study for top 10 CO_2-emitting countries for the period 1971–2010 have been collected from the International Energy Agency (IEA 2012).

6.2.2 Population Growth

Figure 6.1 shows the development of population by countries in the period 1971–2010 (IEA 2012). As seen from Fig. 6.1, it should be noted that there is no analysis for Russian Federation in the period 1971–1990 because of an abruption in the data. Annual growth rate of population for Russian Federation has decreasing effect representing annual growth rate of −0.25 % in the period 1991–2010. Annual growth rate of population has increasing effect for nine countries in the period 1971–2010. Islamic Republic of Iran has the largest annual growth rate of population has increased from 29.4 million in 1971 to 74 million in 2010, representing an overall annual growth rate of 2.36 % while Germany has the lowest annual growth rate of population has increased from 78.3 million in 1971 to 81.8 million in 2010, representing an overall annual growth rate of 0.11 %.

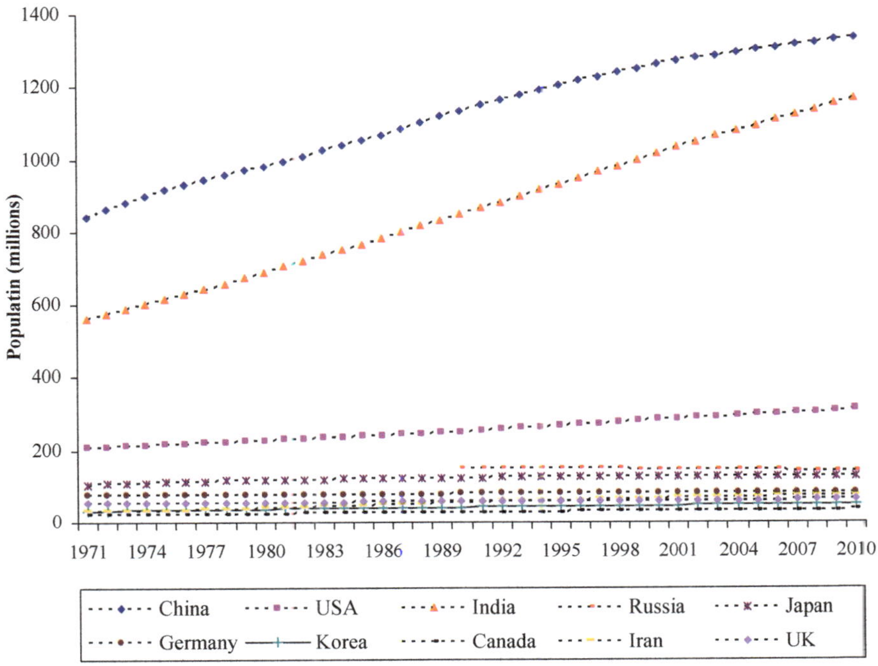

Fig. 6.1 Development of population by countries

6.2.3 Economic Growth

The development of income per capita by countries in the period 1971–2010 is presented in Fig. 6.2 (IEA 2012). Annual growth rate of income per capita has increasing effect for nine countries in the period 1971–2010 (Fig. 6.2). China has the largest annual growth rate of income per capita, which has increased from 358.65 (2005 USD/capita) in 1971 to 6816.29 (2005 USD/capita) in 2010, representing an annual growth rate of 7.55 %, while Islamic Republic of Iran has the lowest annual growth rate of income per capita, which has increased from 7662.35 (2005 USD/capita) in 1971 to 10450.32 (2005 USD/capita) in 2010, representing an annual growth rate of 0.80 %.

Annual growth rate of income per capita for Russian Federation is 0.90 % in the period 1991–2010. At the same time period, China and Japan have the largest and lowest annual growth rates of 9.21 and 0.68 %, respectively.

6.2.4 Energy Intensity

Figure 6.3 shows the development of energy intensity by countries in the period 1971–2010 (IEA 2012). As seen from Fig. 6.3, annual growth rate of energy

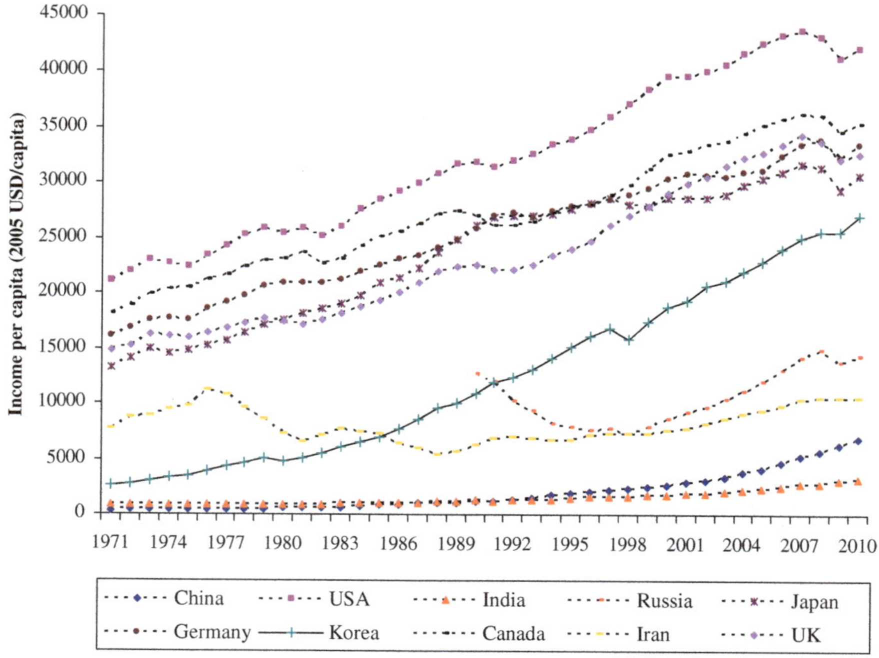

Fig. 6.2 Development of income per capita by countries

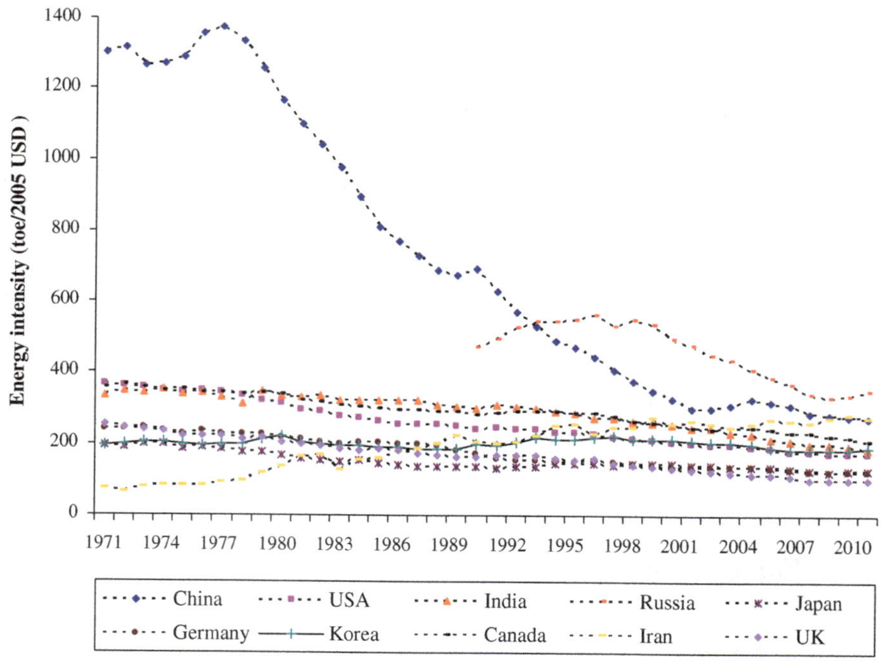

Fig. 6.3 Development of energy intensity by countries

intensity has a decreasing effect for China, the USA, India, Japan, Germany, South Korea, Canada, and the United Kingdom, while Islamic Republic of Iran has increasing effect in the period 1971–2010. China has the largest annual growth rate of energy intensity, which has decreased from 1298.49 (toe/2005 USD) in 1971 to 269.20 (toe/2005 USD) in 2010, representing an annual growth rate of −4.03 %, while Korea has the lowest annual growth rate of energy intensity, which has decreased from 196.00 (toe/2005 USD) in 1971 to 189.27 (toe/2005 USD) in 2010, representing an annual growth rate of −0.09 %.

Annual growth rate of energy intensity for Russian Federation is −1.79 % in the period 1991–2010. At the same time period, China and Korea have the largest and lowest annual growth rates of −4.46 and −0.15 %, respectively.

Annual growth rates of energy intensity for Islamic Republic of Iran are 1.55 and 3.33 % for the periods 1991–2010 and 1971–2010, respectively.

6.2.5 Carbon Intensity

The development of carbon intensity by countries in the period 1971–2010 is presented in Fig. 6.4 (IEA 2012). Annual growth rate of carbon intensity has a decreasing effect for the USA, Japan, Germany, South Korea, Canada, Islamic

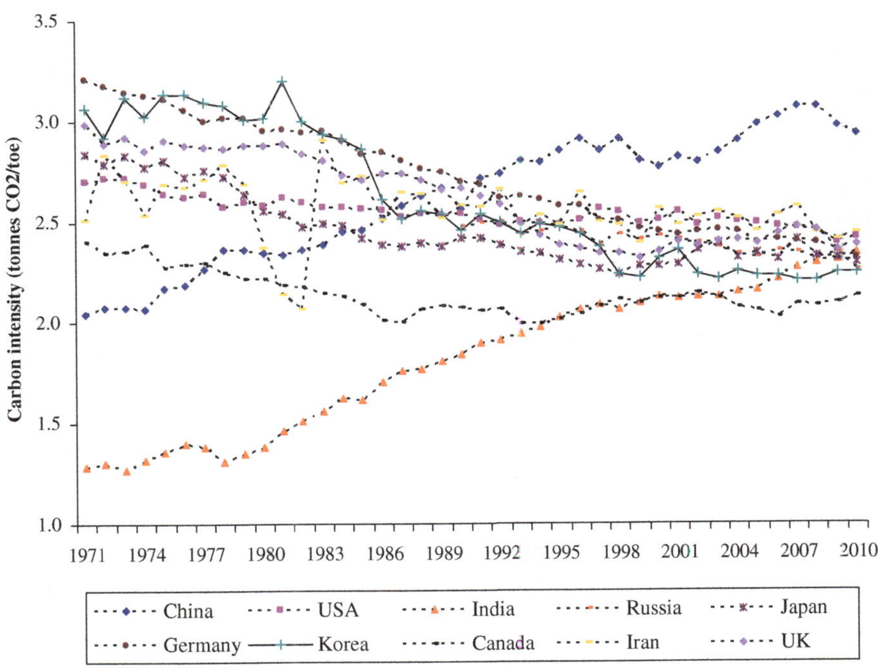

Fig. 6.4 Development of carbon intensity by countries

Republic of Iran, and the United Kingdom, while annual growth rate of carbon intensity has increasing effect for China and India in the period 1971–2010 (see Fig. 6.4). China has the largest annual growth rate of carbon intensity, which has increased from 2.04 (tones of CO_2/toe) in 1971 to 2.94 (tones of CO_2/toe) in 2010, representing an annual growth rate of 0.93 %, while Germany has the highest annual growth rate of carbon intensity, which has decreased from 3.21 (tones of CO_2/toe) in 1971 to 2.33 (tones of CO_2/toe) in 2010, representing an annual growth rate of −0.82 %.

Annual growth rate of carbon intensity for Russian Federation has decreasing effect (−0.52 %) in the period 1991–2010. At the same time period, annual growth rate of carbon intensity for China has increasing effect (0.42 %), while annual growth rate of carbon intensity for Germany has decreasing effect (−0.762 %).

6.3 Results and Discussion

The results of the decomposition analysis of CO_2 emission related to the energy consumption of the top 10 emitting countries for the period 1971–2010 divided into five-year time intervals are presented in Table 6.2. The central columns report the decomposition in the four explanatory variables (P_{effect}, G_{effect}, E_{effect}, C_{effect}).

Table 6.2 Decomposition of CO_2 emission by countries (Mt)

Time period	P_{effect}	G_{effect}	E_{effect}	C_{effect}	ΔCO_2
China					
1971–1975	78.9	122.1	−6.4	56.2	250.8
1976–1980	66.2	346.2	−190.8	90.9	312.4
1981–1985	108.2	610.2	−484.4	79.5	313.5
1986–1990	125.0	465.7	−218.8	33.5	405.4
1991–1995	124.2	1191.4	−788.1	133.6	661.0
1996–2000	113.9	886.8	−969.7	−154.5	−123.5
2001–2005	99.6	1435.9	222.9	220.7	1979.1
2006–2010	133.2	2506.6	−845.1	−180.5	1614.2
USA					
1971–1975	169.6	280.7	−274.0	−106.8	69.5
1976–1980	200.4	396.0	−496.2	−66.8	33.5
1981–1985	166.2	447.4	−563.7	−100.2	−50.2
1986–1990	181.8	404.1	−219.5	−19.7	346.7
1991–1995	251.4	383.6	−294.2	−37.2	303.7
1996–2000	253.6	713.6	−565.3	−7.4	394.4
2001–2005	213.5	400.3	−393.4	−53.9	166.5
2006–2010	199.7	−149.0	−247.5	−119.5	−316.3

(continued)

Table 6.2 (continued)

Time period	P_{effect}	G_{effect}	E_{effect}	C_{effect}	ΔCO_2
India					
1971–1975	20.0	7.9	0.1	13.1	41.0
1976–1980	24.6	12.5	−9.2	−2.7	25.2
1981–1985	30.7	39.0	−10.1	36.9	96.5
1986–1990	42.6	81.9	−31.8	40.1	132.8
1991–1995	51.1	114.6	−58.5	45.9	153.1
1996–2000	61.1	127.2	−61.6	27.7	154.4
2001–2005	62.9	244.7	−147.7	20.8	180.7
2006–2010	77.3	372.8	−163.1	82.5	369.5
Russian Federation					
1991–1995	−6.4	−784.8	209.4	−12.2	−594.0
1996–2000	−14.6	193.3	−205.1	−14.6	−40.9
2001–2005	−29.6	408.0	−317.1	−53.2	8.2
2006–2010	−7.8	152.2	−73.3	−69.4	1.6
Japan					
1971–1975	50.66	91.5	−36.07	−8.59	97.5
1976–1980	31.57	123.91	−102.22	−57.76	−4.5
1981–1985	27.58	118.87	−82.12	−42.83	21.5
1986–1990	15.04	193.91	−34.98	13.33	187.3
1991–1995	13.34	27.84	82.22	−48.5	74.9
1996–2000	9.28	16.07	1.86	−6.42	20.8
2001–2005	4.69	70.47	−52.69	28.43	50.9
2006–2010	−3.68	−13.22	−36.22	−8.78	−61.9
Germany					
1971–1975	5.0	82.0	−60.1	−30.0	−3.1
1976–1980	0.0	122.6	−64.8	−34.3	23.4
1981–1985	−9.1	72.8	−28.3	−43.1	−7.7
1986–1990	21.3	116.5	−153.3	−51.2	−66.6
1991–1995	18.9	25.9	−65.3	−36.5	−57
1996–2000	3.2	69.6	−100.8	−43.4	−71.5
2001–2005	2.0	10.2	−32.4	−14.0	−34.3
2006–2010	−6.7	27.8	−52.5	−27.8	−59.3
South Korea					
1971–1975	4.5	18.5	0.3	1.5	24.7
1976–1980	6.5	21.6	15.0	−4.1	39.0
1981–1985	7.5	43.7	−11.8	−15.5	23.9
1986–1990	7.9	65.4	7.7	−11.3	69.6
1991–1995	12.4	74.5	25.8	−8.4	104.3

(continued)

Table 6.2 (continued)

Time period	P_{effect}	G_{effect}	E_{effect}	C_{effect}	ΔCO_2
Canada					
1971–1975	17.5	50.0	−19.5	−10.3	37.8
1976–1980	17.1	36.8	−5.1	−13.3	35.4
1981–1985	16.1	25.1	−30.0	−19.1	−8.0
1986–1990	24.6	24.2	−23.9	14.1	39.0
1991–1995	20.3	27.2	−1.5	−7.5	38.5
1996–2000	18.6	76.1	−62.9	20.4	52.2
2001–2005	20.6	38.0	−7.8	−17.5	33.3
2006–2010	24.3	−7.0	−51.7	26.8	−7.5
Islamic Republic of Iran					
1971–1975	6.1	13.3	6.7	3.8	29.8)
1976–1980	11.8	−39.3	47.5	−10.7	9.3
1981–1985	17.6	4.8	4.1	28.2	54.7
1986–1990	24.7	−4.4	37.3	4.4	62.0
1991–1995	18.0	−7.2	50.6	−7.8	53.6
1996–2000	20.0	16.3	28.5	−20.8	44.0
2001–2005	18.8	69.7	−31.3	33.9	91.1
2006–2010	6.1	13.3	6.7	3.8	54.0
United Kingdom					
1971–1975	3.2	47.3	−51.4	−43.1	−44.0
1976–1980	1.1	35.8	−51.7	54.0	39.2
1981–1985	2.0	64.3	−42.8	−34.3	−10.7
1986–1990	4.9	64.0	−63.8	−15.1	−9.9
1991–1995	5.6	45.8	−42.9	−52.3	−43.7
1996–2000	6.4	88.1	−86.5	−19.2	−11.2
2001–2005	9.9	49.4	−62.7	−0.8	−4.2
2006–2010	13.3	−13.8	−39.3	−11.4	−51.2

The last column shows the cumulated changes that are calculated as the aggregation of these variables. The percentage change of the four different effects of the top 10 emitting countries for the first (1971–1975) and the last (2006–2010) time periods is also presented in Fig. 6.5 except Russian Federation. Due to lack of the data for the first (1971–1975) time period for Russian Federation, Russian Federation is not included in Fig. 6.5.

Table 6.2 shows that the economic activity effect (G_{effect}) and the energy intensity effect (E_{effect}) are the two biggest contributors to CO_2 emission for all countries with a few exceptions. The population effect (P_{effect}) for the sub-periods (1971–1975) and (1976–1980) in India, the carbon intensity effect (C_{effect}) for the sub-period (1981–1985) in Islamic Republic of Iran, and for the sub-periods (1976–1980) and (1991–1995) in the UK are the biggest contributors to CO_2 emission.

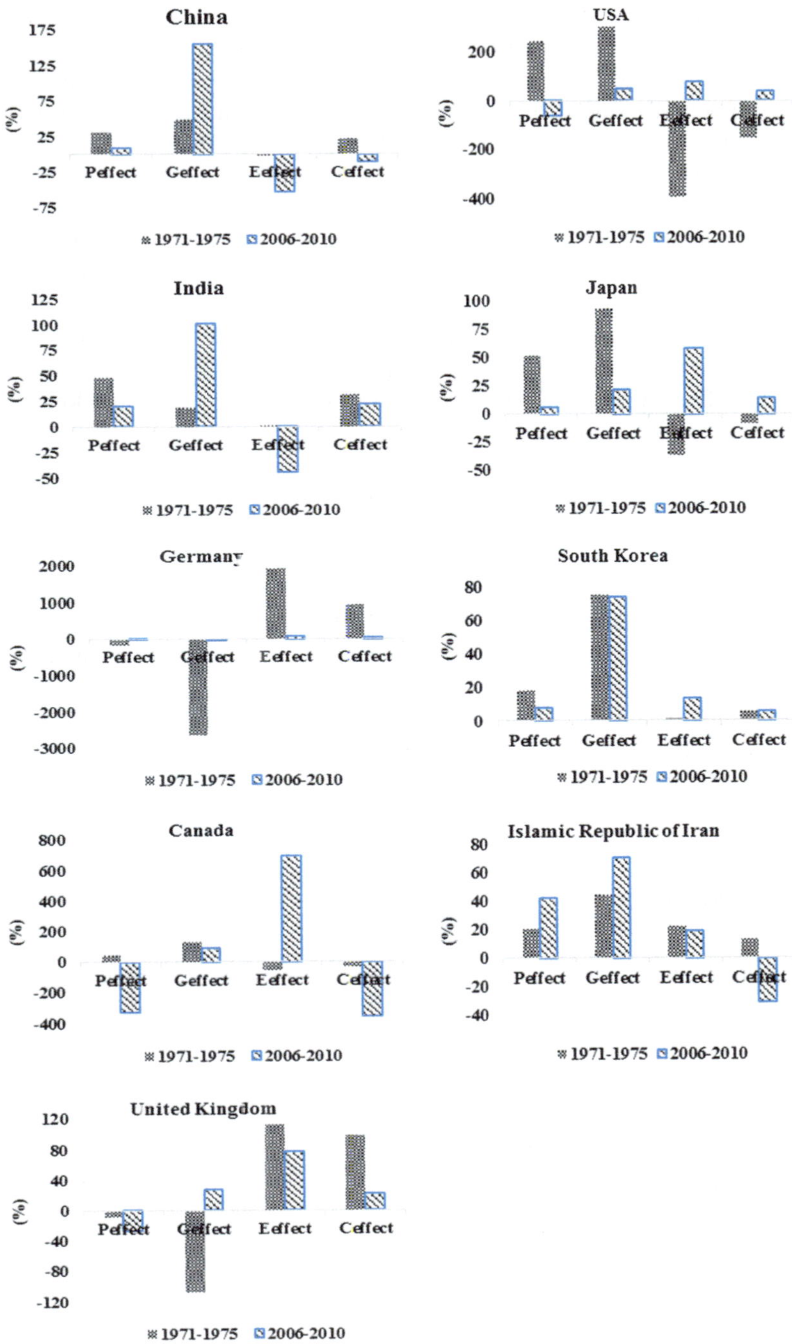

Fig. 6.5 Decomposition of CO_2 emission by countries (%)

In general, the population effect accelerated the increase in CO_2 emission for all countries in the entire sub-periods except Russian Federation. Russian Federation was the only country that the population reduced the increase in CO_2 emission for all the sub-periods. This effect also reduced the increase in CO_2 emissions in Japan and Germany for one and two sub-periods, respectively (Table 6.2).

The economic activity effect accelerated the increase in CO_2 emission for all countries in most of the sub-periods (Table 6.2). This effect reduced the increase in CO_2 emission in the developed countries such as the USA, Japan, Canada, and the United Kingdom for the sub-period 2005–2010 when the economic recession occurred.

The energy intensity effect reduced the increase in CO_2 emission for all the countries in most of the sub-periods except South Korea and Islamic Republic of Iran. This effect accelerated the increase in CO_2 emission in Islamic Republic of Iran in most of the sub-periods (Table 6.2).

The carbon intensity effect reduced the increase in CO_2 emission in the USA, Russian Federation, Japan, Germany, South Korea, Canada, and the United Kingdom, while it accelerated the increase in CO_2 emissions in China, India, and Islamic Republic of Iran (Table 6.2).

The percentage change in the economic activity effect of the top 10 emitting countries is quite different for the first (1971–1975) and the last (2006–2010) time periods except the countries South Korea, Canada, and Islamic Republic of Iran (Fig. 6.5).

The results obtained in this study are consistent with the previous studies for China (Zhang et al. 2009) and the USA (Tol et al. 2009), those of which are the top two CO_2-emitting countries.

6.4 Conclusion

The results show that the economic activity effect and the energy intensity effect are the two biggest contributors to CO_2 emissions for all the countries with a few exceptions. The economic activity caused an increase in CO_2 emission, while the energy intensity contributed a decrease in CO_2 emission.

References

Andreoni, V., & Galmarini, S. (2012). Decoupling economic growth from carbon dioxide emissions: A decomposition analysis of Italian energy consumption. *Energy, 44*, 682–691.

Diakoulaki, D., Mavrotas, G., Orkopoulus, D., & Papayannakis, L. (2006). A bottom-up decomposition analysis of energy-related CO_2 emissions in Greece. *Energy, 31*, 2638–2651.

Diakoulaki, D., & Mandaraka, M. (2007). Decomposition analysis for assessing the progress in decoupling industrial growth from CO_2 emissions in the EU manufacturing sector. *Energy Economics, 29*, 636–664.

Girod, B., Wiek, A., Mieg, H., & Hulme, M. (2009). The evolution of the IPCC's emissions scenarios. *Environmental Science and Policy, 12*, 103–118.

He, B., & Chen, B. C. (2002). Energy ecological efficiency of coal fired plant in China. *Energy Conversion and Management, 43*, 2553–2567.

IEA. (2012). *CO_2 emissions from fuel combustion: Annual historical series (1971–2010)*. Paris, France: International Energy Agency.

IPCC. (1995). *Greenhouse gas inventory: IPCC guidelines for national greenhouse gas inventories*. Bracknell, England: United Kingdom Meteorological Office.

Kumbaroğlu, G. (2011). A sectoral decomposition analysis of Turkish CO_2 emissions over 1990–2007. *Energy, 36*, 2419–2433.

Linyun, S., & Hongwu, Z. (2011). Factor analysis of CO_2 emission changes in China. *Energy Procedia, 5*, 79–84.

McCollum, D., & Yang, C. (2009). Achieving deep reductions in US transport greenhouse gas emissions: Scenario analysis and policy implications. *Energy Policy, 37*, 5580–5596.

Steckel, J. C., Jakob, M., Marschinski, R., & Luderer, G. (2011). From carbonization to decarbonization?-Past trends and future scenarios for China's CO_2 emissions. *Energy Policy, 39*, 3443–3455.

Subhes, C. B., & Arjaree, U. (2004). Decomposition of energy and CO_2 intensities of Thai industry between 1981 and 2000. *Energy Economics, 26*, 765–781.

Tol, R. S. J., Pacala, S. W., & Socolow, R. H. (2009). Understanding long-term energy use and carbon dioxide emissions in the USA. *Journal of Policy Modelling, 31*, 425–445.

Wolfram, S. (2004). Mathematica 5.1. Champaign, USA: Wolfram Research Inc.

Xiangzhao, F., & Ji, Z. (2008). Economic analysis of CO_2 emission trends in China. *China Population Resources and Environment, 18*, 43–47.

Zhang, M., Mu, H., Ning, Y., & Song, Y. (2009). Decomposition of energy-related CO_2 emission over 1991–2006 in China. *Ecological Economics, 68*, 2122–2128.

Chapter 7
Turkey's Electric Energy Needs: Sustainability Challenges and Opportunities

Washington J. Braida

Abstract In order to satisfy its electric energy demand for the next 20 years (440–484 TWh projected demand for year 2020), Turkey has embarked on a series of major investment programs involving energy generation and distribution. A wide variety of energy generation projects are being implemented or will be executed in the near future involving nuclear, coal-, and natural gas-fired thermoelectric plants, combined cycle plants, hydroelectric dams, geothermal plants, and wind and solar energy farms. The engineering and scientific communities along with decision makers at the technical, financial, and political level are facing both, huge challenges (e.g., reduce energy dependence, financial feasibility, environmental protection, social acceptance, and resources management) and a once in a lifetime opportunity for improvement of the Turkey's social welfare and the environment for several generations. This paper presents a view of some of these challenges and opportunities along with a review of the energy–water nexus from a holistic life cycle perspective. Furthermore, it explores different scenarios of technology integration in order to improve the sustainability of the electric energy generation matrix by the sustainable use of available resources and minimization of the carbon and environmental footprint of energy generation.

7.1 Introduction

It has been argued, not without reason, that as long as people have been living on Earth, they have striven to improve their living standards and energy has been the foundation of social organization from hunter-gatherer societies to our modern fossil fuel-based industrialized society (Schlor et al. 2012). Turkey is not an exception to this argument. The economic development, population increase, and improved living standards have resulted in a 6 % compound annual growth rate in energy

W.J. Braida (✉)
Stevens Institute of Technology, Hoboken, USA
e-mail: wbraida@stevens.edu

A.N. Bilge et al. (eds.), *Energy Systems and Management*,
Springer Proceedings in Energy, DOI 10.1007/978-3-319-16024-5_7

generation which has been addressed through a series of major investment programs aimed to expand electric energy generation and distribution. The energy demand for the year 2020 has been estimated between 440 and 484 TWh, depending upon the scenario being considered (Baris and Kucukali 2012). A wide variety of energy generation projects are being executed or will be realized in the near future involving nuclear, coal- and natural gas-fired thermoelectric plants, combined cycle plants, hydroelectric dams, geothermal plants, and wind and solar energy farms. The engineering and scientific communities along with decision makers at the technical, financial, and political level are facing both, huge challenges (e.g., reduce energy dependence, financial feasibility, environmental protection, social acceptance, resources management) and a once in a lifetime opportunity to improve Turkey's social welfare and the environment for several generations. This study addresses the energy–water nexus from a holistic life cycle perspective and the potential implications of climate change in electricity generation. Furthermore, it summarily explores potential scenarios to improve the sustainability of the electric energy generation matrix.

7.2 Current Situation and Trends

After 2008, the installed capacity for electricity generation has steadily increased as shown in Fig. 7.1. The net electricity generation increased from 23.3 to 228.1 TWh between 1980 and 2012 (U.S. EIA 2014). The electricity generation installed capacity relies heavily on fossil fuels as shown in Fig. 7.2, although an increasing trend in the renewable-based installed capacity can be noticed. Due to the low-capacity factor of renewable power plants, this fossil fuel dominance is even higher when electricity generation is considered, representing as much as 82.5 % in 2008 (Yuksel 2013). Among fossil fuels, in the last few years there has been a shift toward natural gas and indigenous coal sources. Based on TEIAS data (TEIAS 2014), in 2012 coal (hard coal and lignite) accounted for 28.4 % of the electricity

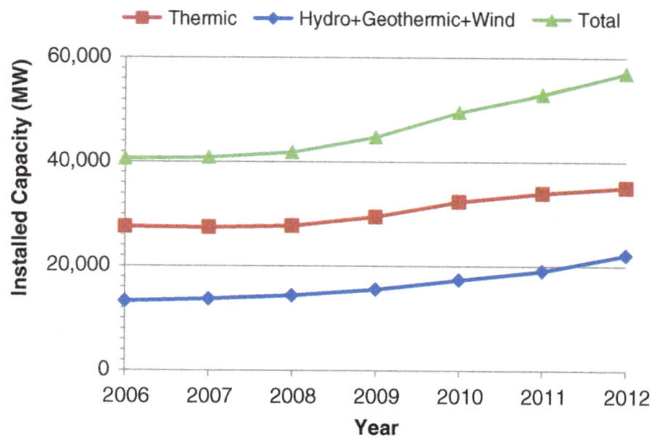

Fig. 7.1 Electric energy generation installed capacity (MW) between 2005 and 2013 (*source* TEIAS 2014)

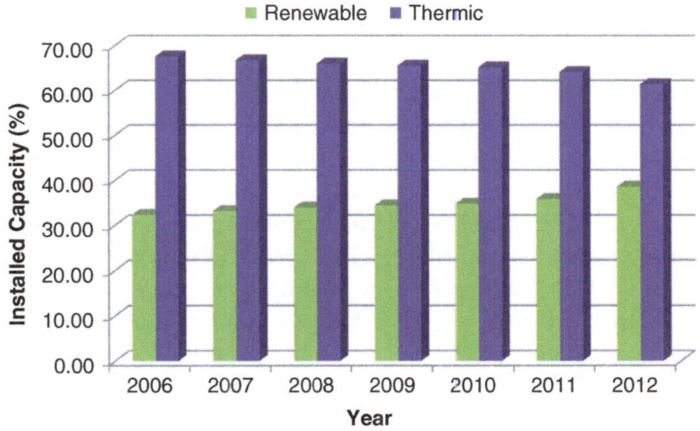

Fig. 7.2 Evolution of installed renewable and thermic electricity generation capacity (*source* TEIAS 2014)

Fig. 7.3 Relative distribution of installed electricity generation capacity in 2012 (*source* TEIAS 2014)

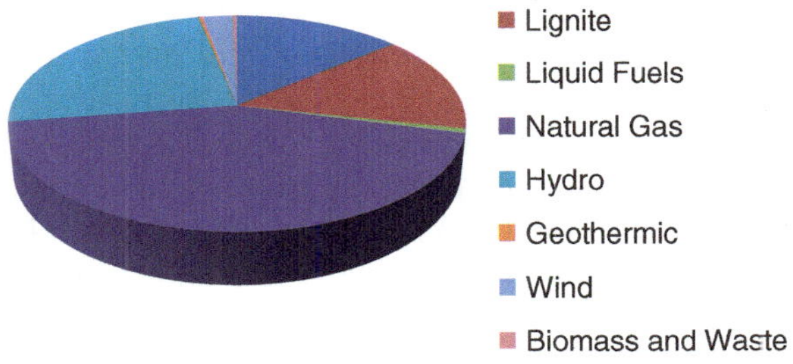

generated, natural gas for 48.63 %, and hydro for 24.16 % (Fig. 7.3). It is expected that by 2020 natural gas-generated electricity will account for 42.7 % of the total generation, hydropower for 23 %, and lignite for 16.7 %. The remaining fraction will be satisfied by hard coal and renewable sources (TEIAS 2009).

Moreover, although a substantial reduction in the CO_2 emissions per kWh generated (from 568 to 472 g/kWh) between 1990 and 2011 had been achieved, the total emissions of greenhouse gases expressed in mega tons of CO_2 equivalents had increased by 125 % during the same period (IEA 2013). These figures clearly show that technological improvements are not enough to deal with emissions resulting from population growth and economic development.

For the year 2020 and depending upon the scenario being considered, the electricity demand is projected between 440 and 484 TWh (Baris and Kucukali 2012) which implies an increase of 30,000–40,000 MW of the installed generation capacity by 2020.

7.3 The Challenging Future

A wide variety of energy generation projects are currently being implemented or will be implemented in the near future. These projects involve two nuclear, several coal- and natural gas-fired thermoelectric plants, combined cycle plants along with renewable-based hydroelectric dams, geothermal plants, and wind and solar energy farms. Evrendilek and Ertekin (2003) estimated that Turkey's economically feasible renewable energy potential exceeds 495 TWh/year as follows: 197 TWh/year from biomass energy, 125 TWh/year from hydropower (large and small plants), 102 TWh/year from solar energy, 50 TWH/year from wind energy, and 22 TWh/year from geothermal energy.

In order to assist in the fulfillment of Turkey's electricity needs, the Turkish Ministry of Energy and Natural Resources strategic plan for the period 2009–2014 sets as 2023 goals the installation of 2000 MW of wind energy, 600 MW of geothermal power, and the generation of 5 % of electricity from nuclear-powered plants. Overall, a series of steps in the correct direction have been taken to deal with the increase of energy demands of the country. However, energy generation needs to be addressed from a system's view perspective by exploring the interactions among the elements of the systems and being aware of system's potential emerging properties. The underlined objective of modern electricity generation systems is "sustainability." Several definitions has been advanced for sustainability; however, effective action on sustainability has been proposed to address three pillars: environment (ecological dimension), society, and economy. On a more realistic view, Spangenberg et al. (2010) proposed a fourth dimension, the institutional one, as part of the sustainability umbrella to be used to address problems using a holistic approach (Fig. 7.4).

Addressing the four dimensions of sustainability is not an easy task, especially for large complex systems like electricity generation. It will be impossible within the scope of this paper to explore all potential interactions; thus, this report focuses on exploring the energy–water nexus and the potential implications of climate change in electricity generation.

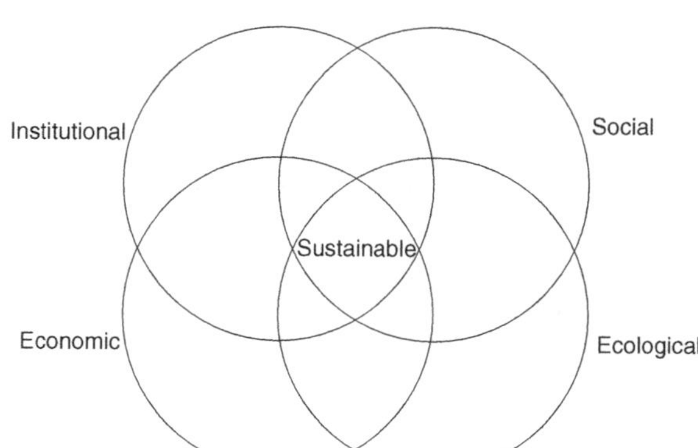

Fig. 7.4 Sustainability definition (adapted from Spangenberg et al. 2010)

In general, national water policies are based on agricultural practices (irrigation) and availability of drinking water. As such, the importance of the water–energy nexus has been downplayed creating vulnerabilities in both systems (energy and water). With a sufficient clean energy supply the world's water problems will be solved (desalination, long-haul transportation, and depth well extraction). Similarly, availability of sufficient quantities of proper quality water will aid in solving our energy problems (e.g., hydro, biofuels, and improved efficiency of thermal plants). This linkage between the water and energy systems/infrastructure also results in the potential for cascade vulnerabilities. In other words, energy scarcity results in water scarcity and vice versa). In the USA, the energy sector is the largest water withdrawal sector accounting for 48 % of the total water withdrawn. Water withdrawal ranges between 0.75 and 160 l/kWh of electricity generated. Water consumption ranges from 0.4 to 3 l/kWh. All those values depend upon the power cycle, the type of fuel used in the power plant, and the cooling system. Moreover, hydroelectric generation requires the creation of water reservoirs which affect water quality, ecological systems and increase losses by evaporation. The importance of water scarcity/availability/conflict issues will become larger as increased competitive water needs such as mining, fracking, agriculture, biofuel production, and human consumption are taken into consideration.

Regardless of the ongoing discussion regarding its causes, climate change is an indisputable reality which needs to be taken into consideration when planning electricity generation systems. Severe weather patterns are seen more and more often. Droughts, freezes, heat waves, and floods will likely impact the performance of electricity generation plants (hydro, thermal). During the heat wave that affected Europe in 2003, nuclear power plants have dialed back electricity generation due to less cooling capacity (higher water temperatures) and rejection water temperature limits. On July 2009, France purchased electricity from England because of the shutting down of one-third of its nuclear reactors to avoid exceeding thermal discharge limits. On August 2012, one of two reactors at Millstone Power Station near New London, Connecticut, was shut down when temperatures in Long Island Sound, the source of the facility's cooling water, reached their highest sustained levels since the facility began monitoring in 1971. Several coal-fired power plants were shut down due to freezing temperatures in Texas in 2011. The grid lost 7 MW of installed capacity, and rolling blackouts hit the state of Texas for more than 8 h. During the 2014 winter, Niagara Falls froze over two times posing a risk to the hydroelectric generation plants in the Niagara River. Pine Flat Power Plant a hydroelectric power plant localized in California (USA) is not producing any electricity due to very low water levels in its reservoir. As a result of droughts, Texas hydropower generation decreased by 24 % from 2012 to 2013. Of the 104 nuclear plants in the USA, 24 are localized in areas experiencing severe droughts and may face shutdowns during the summer.

As shown in the previous paragraphs, plans for electricity generation infrastructure should take into account the global trends in climate change and should be adaptable and resilient to deal with these newly developed weather patterns.

A 2009 Stanford University study (Jacobson 2009) ranked electricity generating systems according to their impacts on global warming, pollution, water supply, land use, wildlife, and other concerns. The very best options were wind, solar (photovoltaic—PV), geothermal, tidal, and hydroelectric power. Nuclear power, coal with carbon capture, oil and natural gas were all poorer options. However, it is recommended that specific technological applications should be evaluated from a life cycle perspective on a case by case basis. According to this, study, the proliferation of thermic power plants in Turkey, although justified on current strategic and financial grounds, appears to be going against sustainability goals. Moreover, as thermal generation facilities require long construction periods and have even longer operational life, the risk of getting locked into a less than desirable situation is substantial. Wind and PV electricity generation potentially appear to be more desirable options. Nevertheless, implementation of these approaches should be done under a life cycle perspective that includes the availability and environmental impacts of raw materials (tellurium, neodymium, indium…), environmental impacts during the deployment and operation (including electricity's distribution infrastructure), and end of the life scenarios (disposal/recycle of wind turbines and solar panels).

Carbon capture and sequestration (CCS), in saline deposits, oil and/or gas reservoirs, offers a mitigation alternative for the release of greenhouse gases into the atmosphere. There are currently 12 operational large-scale CCS projects around the world, which have the capacity to prevent 25 million tons a year (Mtpa) of CO_2 from reaching the atmosphere. Nine other projects are under construction and another 39 are in various stages of planning or development, six of which may make a final investment decision during 2014 (Global CCS Institute 2014). Interestingly, although regulations for gaseous (e.g., SO_2) emissions from power plants have become more stringent in Turkey, we are not aware of any regulation regarding CO_2 emissions neither projects contemplating CCS. The reason for that is likely rooted in the study of Davis et al. (2013) which states that the cost of the CCS technology, lack of a price signal such as a carbon fee, long-term liability risks of carbon dioxide storage underground, and lack of a regulatory regime are the premier concerns of carbon capture experts in government, private companies, and academic institutions. Perhaps CCS with valorization in the form of microalgae growth for biodiesel production is an option to be investigated. Turkey has a high solar radiation potential which can be linked with CO_2 capture in order to develop a sustainable indigenous bio-based fuel production infrastructure.

7.4 Conclusions

This study investigates the current and future electric energy generation in Turkey. In order to fulfill the increasing electricity demand, a large number of projects are being implemented. Most of them are based on thermal plants and despite the increase in the installation of renewable electricity sources and the reduction in the emission of CO_2 per kWh of energy generated, Turkey's CO_2 emissions had increased over 125 % in between 1990 and 2011. The increased dependence on thermal facilities increases the vulnerabilities of the system due to the energy–water nexus and climate change-associated issues. An increase in the wind-, solar-, and geothermal-based installed capacity appears to be a suitable path to follow to address electricity needs. The exploration of tidal and wave sources along with CCS with valorization should also be accelerated. CCS with valorization will likely reduce the carbon footprint of the currently installed thermal capacity (where retrofit is possible) and also will provide biodiesel as a valuable by-product, which will reduce fuel imports.

References

Baris, K., & Kucukali, S. (2012). Availability of renewable energy sources in Turkey: Current situation, potential, government policies, and the EU perspective. *Energy Policy, 42*, 377–391.

Davis, L. L., Uchitel, K., & Ruple, J. (2013). Understanding barriers to commercial-scale carbon capture and sequestration in the United States: An empirical assessment. *Energy Policy, 59*, 61–745.

Evrendilek, F., & Ertekin, C. (2003). Assessing the potential of renewable energy sources in Turkey. *Renewable Energy, 28*, 2303–2315.

IEA. (2013). *CO2 emissions from fossil fuel combustion. Highlights*. Paris, France: OECD/International Energy Agency.

Global CCS Institute. (2014). The Global Status of CCS. February 2014. Docklands VIC Australia.

Jacobson, M. (2009). Review of solutions to global warming, air pollution, and energy security. *Energy and Environmental Science, 2*, 148–173.

Schlor, H., Fisher, W., & Hake, J.-F. (2012). The meaning of energy systems for the genesis of the concept of sustainable development. *Applied Energy, 97*, 192–200.

Spangenberg, J. H., Fuad-Luke, A., & Blincoe, K. (2010). Design for sustainability (DFS): The interface of sustainable production and consumption. *Journal of Cleaner Production, 18*, 1485–1493.

TEIAS. (2009). Turkish electrical energy; 10-year generation capacity projection (2009–2018). Turkish Electricity Transmission Company, www.teias.gov.tr. Accessed May 2014.

TEIAS. (2014). Turkish electricity generation. Transmission statistics 2012. Turkish Electricity Transmission Company, www.teias.gov.tr. Accessed May 2014.

U.S. EIA. (2014). Energy Information Administration; Turkey, http://www.eia.gov/countries/cab.cfm?fips=TU. Accessed May 2014.

Yuksel, I. (2013). Renewable energy status of electricity generation and future prospect hydropower for Turkey. Renewable Energy, 50, 1037–1043.

Chapter 8
Shale Gas: A Solution to Turkey's Energy Hunger?

Ilknur Yenidede Kozçaz

Abstract The aim of this short analysis is to answer whether shale gas can be a sustainable solution to Turkey's long-term energy needs. Turkey, with no significant hydrocarbon reserves of her own, is vulnerable to the risks and challenges associated with energy import dependency. Having a fast-growing natural gas demand has caused Turkey to undertake many gas import contracts. In 2013, 98 % of the natural gas consumption is imported. Globally increasing natural gas prices and volatile Turkish Lira/US Dollar exchange rate have a series of ramifications, including a substantial burden on national budget and balance of payments. It is crucial for Turkey to reduce the share of imports in energy and to develop domestic resources in order to avoid exposure to relevant risks. In short, Turkey needs gas supply security. However, conventional natural gas reserves of Turkey are far from meeting its needs. Shale gas, in this frame, emerges as a buoyant potential for secure future gas deliveries. Given the example of unconventional gas frenzy in the USA, Turkey is now discussed as a long-term candidate for shale gas production. This possibility triggers high hopes, as well as unsupported expectations. Shale gas production has a long list of requirements: distinct geological formations, concordant conditions in surrounding area, advanced exploration and production technology, and capital-intense investments. Even if these conditions are fulfilled, environmental challenges of this production method are yet to be addressed and tackled diligently. Turkey is still on exploration phase of shale gas experience. It will take Turkey at least another decade to meet the requirements for tangible results and to name the shale gas as an answer to its energy hunger.

I.Y. Kozçaz (✉)
İstanbul, Turkey
e-mail: ilknur.yenidede@gmail.com

A.N. Bilge et al. (eds.), *Energy Systems and Management*,
Springer Proceedings in Energy, DOI 10.1007/978-3-319-16024-5_8

8.1 Introduction

Rapid industrialization in twentieth century had one crucial element: uninterrupted access to energy. While pioneers of industry and technology were thriving for secure and affordable energy supplies, uneven distribution of energy sources[1] around the globe created a trade scheme, from producers toward customer, who are not necessarily located in proximity.

As shown in Fig. 8.1, the Middle East is the forerunner in crude oil production, whereas the majority of the crude oil is consumed in North America, Europe, and industrialized Asia-Pacific countries. In case of natural gas, the Middle East is replaced by Eurasian region. Former Soviet Republics hold the lion's share of the current natural gas production in the world, while the majority of the gas is transported to Europe through large-capacity pipelines.

In the last decade, North America has shifted its position from an importer, toward a self-sufficient point) and potentially an exporter. This radical change has its roots in the very own properties of the American oil and gas industry. Yet many countries around the globe have the ambition to create their own shale gas revolution. Turkey is one of the most assertive countries on the list. Understanding whether and how Turkey can make this dream come true needs a careful analysis of available assets and coercive measures which the shale gas business ground rules dictate.

In this study, an outlook on Turkey's energy market is presented with focus on natural gas market. Turkey's shale gas ambition stems from the country's shortage of conventional gas resources. Therefore, a close look on the gas supply and demand balance would set the tone for understanding Turkish enthusiasm on shale gas. In the next section, a brief chapter on shale gas is presented in order to introduce hallmarks of the exploration and production techniques, which have borned the term "unconventional." In the following section, a list of peculiarities of shale gas production is provided, with their corresponding challenges for Turkey under each headline. After having laid the fundamentals of shale gas development in Turkey, the conclusion section intends to answer the question whether Turkey holds a strong case with regard to being next shale revolution in near future.

8.2 Energy of Turkey

Turkey has a fast-growing population and has gone through a robust industrialization since the establishment of the Republic. Economic growth, urbanization, and population increase have boosted the need for hydrocarbon sources (Aydın 2010). Increase in energy need has become even higher in the last two decades. As seen in

---

[1] In this work, "energy sources" refer only to crude oil and natural gas.

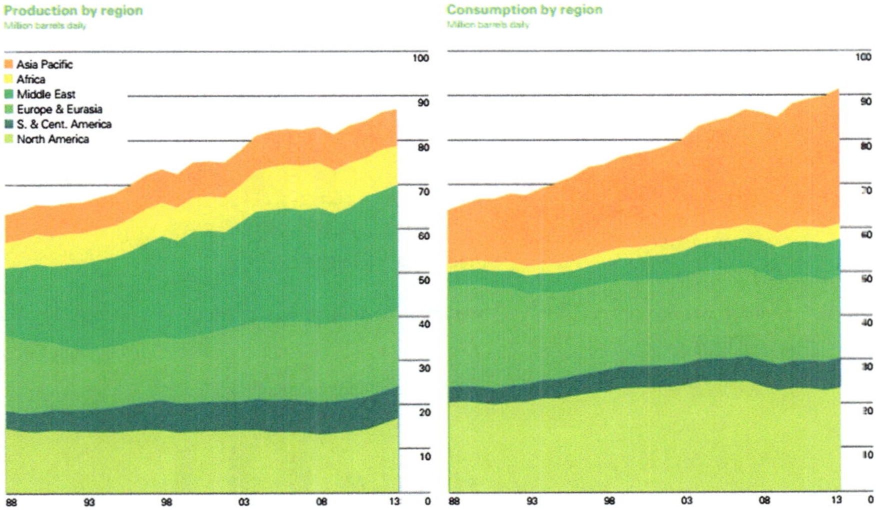

Fig. 8.1 Comparison of oil-producing and oil-consuming regions (BP Statistical Energy Review 2013)

the Fig. 8.2, displaying primary energy consumption has been in steady increase since 1990. Annual energy demand is expected to increase by 5.9 % every year until 2025 (Lise and Van Montfort 2007).

Turkey's indigenous energy production, however, is far from meeting this increasing need. According to the historical data provided by the Ministry of Energy and Natural Resources for the same time period, Turkey's primary energy production corresponds only to one-third of the consumption. Although majority of the energy need is met by crude oil and natural gas, none of these two sources has sufficient exposure in Turkey. In 2011, only 9.5 % of the crude oil demand is answered by domestic sources. In natural gas, share of indigenous production in domestic gas demand is only 2 % (TPAO 2011) (Fig. 8.3).

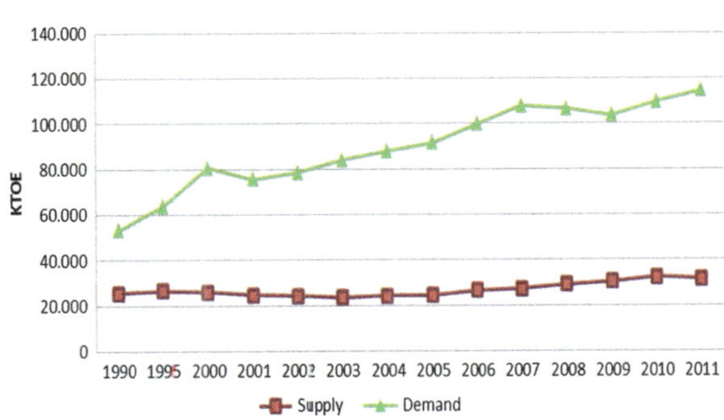

Fig. 8.2 Primary energy consumption of Turkey between 1990 and 2011, in thousand tons petroleum equivalent (World Energy Council—Turkish National Committee 2013)

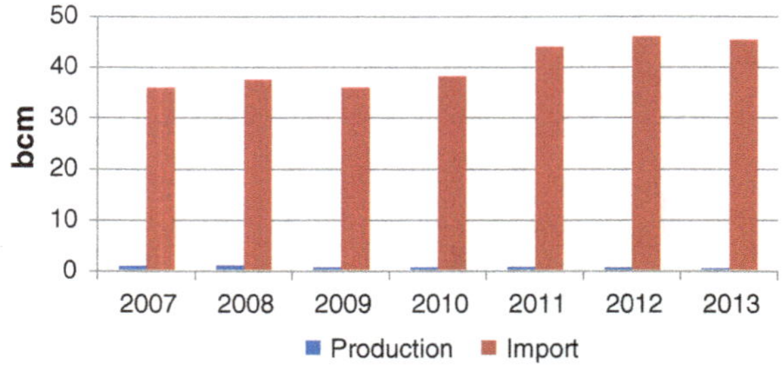

Fig. 8.3 Natural gas supply to Turkey; imports and indigenous production (Energy Market Regulatory Authority EMRA 2013)

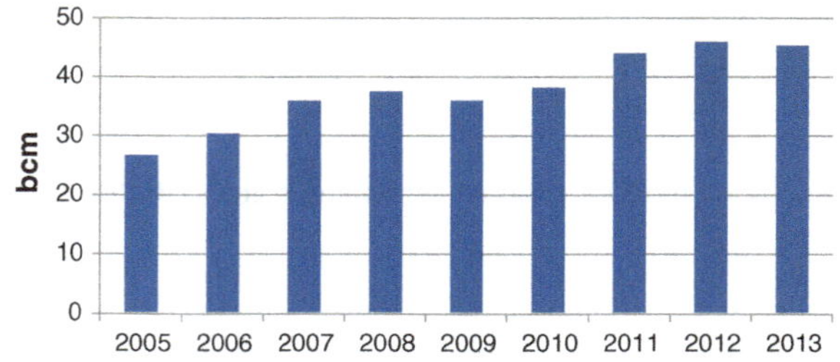

Fig. 8.4 Turkey's annual natural gas consumption between 2005 and 2013 (Energy Market Regulatory Authority EMRA 2013)

Natural gas use in Turkey has started in 1976, with the discovery of Hamitabat gas reserve in Thrace area. Although the only customer happened to be the local industry for a decade, robust population increase and air pollution in big cities triggered natural gas consumption in larger sense. To meet the increasing demand, Turkey has signed its first natural gas import contract with the Soviet Union in 1984.

Deliveries started in 1986, after a pipeline had been constructed for the gas to flow via Bulgaria and the Thrace region of Turkey. In the following decades, natural gas demand has shown a steep increase, particularly in 2000s, due to expansion of the natural gas transmission network and investments made in gas-fired power plants (Fig. 8.4).

However, Turkey has very limited reserves of her own to answer the increasing gas demand. According to Turkish Petroleum Corporation (TPAO)'s data, Turkey has produced 0.53 bcm natural gas in 2013, while Turkey's natural gas consumption has reached 44.1 bcm in the same year. In order to fill the gap in supplies, Turkey has signed gas import deals for further volumes, with the Soviet Union (later Russian Federation) and other suppliers (Table 8.1).

Turkey's dependence on imports of natural gas raises concerns for a list of reasons. The first and the most important one is the supply security. In case of a disruption in gas deliveries, Turkish energy balance is not able to cover the lack of natural gas. Since Turkey does not have a substantial storage capacity, it is crucial

Table 8.1 Turkey's natural gas import contracts (Energy Market Regulatory Authority EMRA 2013)

Supplier	Volume(bcma)	Contract date	Duration	Delivery date
Russia—West 1	6	1986/2012	25	1987
Russia—West 2	8	1998	23	1998
Russia—blue stream	16	1997	25	2003
Iran	10	1996	25	2001
Azerbaijan—SD stage 1	6, 6	2001	15	2007
Azerbaijan—SD stage 2	6	2013	20	2018
Algeria	4	1988	23	1998
Nigeria	1, 2	1995	22	1999

to guarantee uninterrupted flow of gas from contracted suppliers. Lacking of gas storage capacity in Turkish natural gas infrastructure, Turkey is vulnerable to any externalities in natural gas imports (Deloitte 2012).

Second important aspect of import dependency on energy resources is the effect of the payments on national budget. Typically, long-term gas sales and purchase agreements use USD as payment currency. Importing natural gas in USD and selling to domestic market in Turkish Lira leaves Turkish energy authorities with no choice, but to take the exchange rate risk.

Thirdly, Turkish energy market has been heavily subsidized by the Turkish State. BOTAS, State Natural Gas Corporation, has not made any domestic price increases in two years (as of September 2014), despite the increasing import cost of natural gas. In 2012, Turkish economy had a current account deficit of USD 84 billion and energy imports constituted 62.3 % of this deficit.

Combining the need for supply security and easing off the burden on economy, Turkish energy authorities welcomed the possibility of shale gas development in Turkey. Given Turkey's very limited natural gas production with conventional methods, "shale gas boom" in the USA has promoted high expectations on the possibility of implementing the same experience in Turkey.

8.3 Shale Gas

The term "shale gas" presents a new technique of extracting the natural gas trapped in a certain formation of rocks, which is not possible to exploit with conventional methods. Due to the new production method, the gas produced is also referred to as "unconventional gas." This new exploration and production method requires a more advanced technology than the conventional ways of production. Unconventional gas production entails high-pressure water spraying with chemicals into the cracks in the rock formation to produce a "fracture." Fracturing or hydraulic fracking method allows the producer to access the natural gas trapped in the formation, which would be impossible to access otherwise. The cost is also higher compared to the conventional method.

Unconventional natural gas production has been developed and successfully implemented in the USA in the last decade. Once a large natural gas importer, the USA rapidly turned itself into a self-sustaining gas producer, with anticipation of being an exporter in coming years. Many gas importer countries around the world have taken the USA as example and have developed an expectation for duplicating the same experience in their respective gas markets to eliminate the challenges of being gas importer (Geny 2010). Turkey is one of the countries where wishful thinking and reality collide and not necessarily coincide.

Accessing shale gas has a list of requirements and environmental risks, which are widely criticized in countries where production has already started. Majority of these challenges are directly applicable in Turkey. Since Turkey is still in the exploratory phase in shale gas development, it is in best interest of Turkey to consider these issues early in the process and take necessary measures to tackle these challenges.

8.3.1 Use of Extensive Amount of Clean Water

The major element of the unconventional gas production is the hydraulic fracturing technique. Fracking involves drilling a well deep underground and then pumping water, sand, and chemicals down at high pressure to fracture the rocks and enable the gas trapped within them to flow out. The technique requires access to vast water resources in the vicinity of the wells.

Two main shale gas basins in Turkey—Thrace and Dadaslar basins—are located in areas where water resources are limited. Thrace region has gradually lost its natural agricultural character due to rapid industrialization, which led to pollution of existing water resources. At Diyarbakır province, where Dadaslar basin is located, river Euphrates has been lately become the source of all economic activity. There are five dams actively working on the river Euphrates, together with the agriculture zones supplied by these dams. If and when Dadaslar basin is proven fully recoverable, access to vast amount of water would be one of the main concerns.

8.3.2 Disposal of Flow-Back Fracking Fluids Containing Residual Chemicals

The water is contaminated during the process of hydraulic fracturing. After the contaminated water is taken up on the ground, measures will have to be taken to ensure that leaking is prevented. Otherwise, any possible toxic water leaks pose high risks to pollute the clean underground water sources (Howarth et al. 2011).

Toxic water can be stored and purified in isolated tanks or pools before it is safe enough to release it to nature. However, such outdoor cleansing facilities do not necessarily alleviate the risk of the mixture of toxic water with the environment and endangering the residential communities nearby.

The well integrity poses additional challenges to avoid possible damages of poisonous chemicals. Inadequate well structure can lead to gas or toxic water leakage and eventually threaten the nature and the people.

8.3.3 Possibility to Trigger Seismic Activities

A less visible, yet more vigilant risk of unconventional gas production is the potential seismic effect of hydraulic fracturing. Potential risks to trigger tectonic activities have been argued by some seismologists and environmental institutions, since the fracking takes place deep underground.[2,3] In few European Union member states, this claim is taken seriously and led political authorities to cancel the licenses and to put shale gas production projects on a shelf.[4] There are conflicting views on the causal effect of fracking on triggering an earthquake. However, in countries with frequent tectonic activities, even the slightest possibility has to be ruled out before stepping forward in unconventional production.

8.3.4 Need for Large and Unhabituated Areas for Fracturing

One way to avoid the risks mentioned above would be to carry out unconventional gas production activities in unhabituated rural regions. This way, even if a natural hazard takes place, it would not directly threaten the human life, livestock, agriculture, or the environment. In the US experience, thanks to the size of the continent, the risk of endangering the life and environment with shale production is relatively lower.

8.3.5 Gas Prices and Cost of Shale Gas

The last concern for shale gas production in Turkey is the cost element. TPAO has been in cooperation with international companies to share the know-how and the financial burden of the explorations. Since the explorations are in early phase, it is

[2]http://www.ldeo.columbia.edu/news-events/seismologists-link-ohio-earthquakes-waste-disposal-wells.

[3]https://www.gov.uk/government/uploads/system/uploads/attachment_data/file/283837/Seismic_v3.pdf.

[4]http://www.businessweek.com/news/2011-10-04/france-to-keep-fracking-ban-to-protect-environment-sarkozy-says.html, 2.

very difficult to predict the future project economics. Additionally, natural gas prices are very much affected by the policies and politics of international suppliers. Therefore, there is always a risk of investing in the shale gas production in long run and not being able to present it to the regional markets with a competitive price.

8.4 Shale Gas in Turkey

The first and the foremost requirement for Turkey to turn itself into a gas sufficient country is the physical existence of substantial amount of recoverable shale gas. Only existing data on the magnitude of potential shale gas reserves in Turkey belong to American Energy Information Agency (EIA). According to EIA, Turkey has two promising basins with a total reserve of approximately 450 bcm of natural gas recoverable with unconventional methods.[5] In the EIA analysis, two regions, Thrace and Dadaslar basin in South East Turkey, are marked as two regions with potential. Although some unanimous experts are referred to in media for stating 13 tcm of total gas reserves in Turkey, these figures remain untested and speculative.

Other than data provided by EIA, size of shale gas reserves located in Turkey is not known. TPAO has been active in exploring the regions referred to in the EIA report and confirmed findings on potential rock formations. TPAO has engaged in cooperation with international oil companies that consider the above-mentioned two basins attractive. First agreement is signed in 2010 with Transatlantic and Valeura for exploratory studies in both Thrace region and South East Turkey in the form of a memorandum of understanding. Following this protocol, TPAO entered into another agreement with Shell for cooperation in 2011 and exploratory work started in South East Turkey, close to Diyarbakır province.

TPAO has been historically cautious for not revealing any certain figures on the size of the basins which are worked on. International companies which TPAO works with have also been refraining from publishing figures on how much of a gas reserve they have been estimating on the reserves they engage in.

8.5 Conclusions

Turkey, as a net energy importer country, is in immediate need of securing its future supplies. Turkey is located between the markets and suppliers for natural gas, yet overly relying on a single supplier. Cost of importing natural gas is the second reason why Turkey would be very much interested in developing shale gas production. Cost of importing energy in 2013 was around USD 60 billion, and any alternative to reduce this bill is sincerely welcomed in Turkey.

---

[5]http://www.eia.gov/analysis/studies/worldshalegas/pdf/chaptersxx_xxvi.pdf.

Having observed the "game changer" feature of the shale gas boom in the USA, Turkish energy authorities have become quickly interested in shale gas developments. Together with international partners, Turkey is now engaged in two basins on Turkish soil, which are promising areas for shale gas production. With the absence of official figures on the reserve sizes, it has been easy in Turkish media to get carried away and many ambitious calculations are referred to. Beside the uncertainties around the reliability of reserve data stemming from EIA, there are other challenges which make shale gas production in Turkey risky. Given the lack of strict environmental control and distance to international environmental standards, it is more important for Turkey to start on right foot and prevent any danger to livelihood in and around the shale basins.

For the purpose of this paper, only the most immediate and universal concerns are listed. Access to water, risk of contamination, potential effect on earthquakes, and need for vast areas are the major issues which are discussed in EU member countries before Turkey and has created a strong public opinion. In addition, the environmental concerns, project economics, potential cost of production, and future gas prices create uncertainty around how attractive the shale production projects would be for international investors. Lack of specific references in the Turkish legal code on shale gas production would be another unattractive factor for international investors.

Geological, legal, and economic uncertainties, together with environmental concerns, will shape Turkey's shale gas experience in the next decade. Without tangible steps taken on these fronts, shale gas is very unlikely to be the cure for Turkey's energy problem. With a population of 74 million and a fast-growing economy, Turkey's energy needs are far from being tamed anytime soon. Due to the factors listed above, domestic shale gas production is very unlikely to answer Turkey's ever-growing energy need in coming few years.

References

Deloitte. (2012). Turkish Natural Gas Market- Expectations, Developments.

EMRA. (2013). Natural Gas Market Sector Report. Energy Market Regulatory Authority, EMRA.

Geny, F. (2010). Can Unconventional Gas be a game changer in Continental European Gas Markets? The Oxford Institute for Energy Studies.

Howarth, R.W., Santoro, R., & Ingraffea, A. (2011). Methane and the greenhouse-gas footprint of natural gas from shale formations. *Climatic Change, 106*(4), 679–690.

Lise W., & Van Monfort K. (2007). Energy consumption and GDP in turkey: is there a cointegration relationship? *Energy Economics, 29*(6), 1166–1178.

TPAO. (2013). Crude Oil and Natural Gas Sector Report. Turkish Petroleum Corporation (TPAO).

Turkey Energy Outlook. (2013). Presentation by Mr. Oğuz Türkyılmaz, Chairman of Energy Commission of Chamber Of Mechanical Engineers—Member of Executive Board Turkish National Committee Of WEC.

Turkey Energy Outlook. (2013). World Energy Council—Turkish National Committee.

The US Energy Information Administration. (2011). World Shale Gas Resources: An Initial Assessment of 14 Regions Outside the United States.

Further Readings

Energy Outlook in Turkey and in the World. Turkish Ministry of Energy and Natural Resources, 2011.

IEA. (2012). Golden Rules for a Golden Age of Gas. World Energy Outlook Special Report on Unconventional Gas. International Energy Agency (IEA).

IEA. (2014). World Energy Outlook. International Energy Agency (IEA).

Unconventional Gas-Transforming the Global Gas Industry. IHS and International Gas Union Special Report, 2012.

Chapter 9
Assessment of Adsorption Parameter Effectiveness for Radio-Selenium and Radio-Iodine Adsorption on Activated Carbon

A. Beril Tugrul, Nilgun Karatepe, Sevilay Haciyakupoglu, Sema Erenturk, Nesrin Altinsoy, Nilgun Baydogan, Filiz Baytas, Bulent Buyuk and Ertugrul Demir

Abstract Selenium and iodine are found in human body and primarily used in nutrition, and excess or absence of them can lead to diseases. Therefore, their possible dispersion to environment through mining and reprocessing of metals, combustion of coal and fossil fuel, nuclear accidents, or similar activities needs remediation. Adsorption is one of the useful techniques to remove pollutants. In this study, a factorial design is used to determine the effect of pH, concentration

---

A.B. Tugrul (✉) · N. Karatepe · S. Haciyakupoglu · S. Erenturk · N. Altinsoy · N. Baydogan · F. Baytas · B. Buyuk · E. Demir
Istanbul Technical University, Istanbul, Turkey
e-mail: beril@itu.edu.tr

N. Karatepe
e-mail: kmnilgun@itu.edu.tr

S. Haciyakupoglu
e-mail: haciyakup1@itu.edu.tr

S. Erenturk
e-mail: erenturk@itu.edu.tr

N. Altinsoy
e-mail: altinsoy@itu.edu.tr

N. Baydogan
e-mail: dogannil@itu.edu.tr

F. Baytas
e-mail: fbaytas@itu.edu.tr

B. Buyuk
e-mail: buyukbu@itu.edu.tr

E. Demir
e-mail: ertudemir@itu.edu.tr

A.N. Bilge et al. (eds.), *Energy Systems and Management*,
Springer Proceedings in Energy, DOI 10.1007/978-3-319-16024-5_9

of adsorbate, and contact time upon adsorption. Adsorption capacities of radio-selenium and radio-iodine were evaluated for factorial design using activated carbons. The used activated carbon samples were prepared by chemical and physical activation methods. Radioactivity measurements were carried out by using high-resolution gamma spectroscopy system. Results of the research lead to provide useful information about energy generation and management processes by preventing hazardous elements' dispersion to the environment.

9.1 Introduction

Selenium (Se) is an essential trace element for human beings and plays important roles in human health. On the other hand, selenium is toxic at concentrations above the range of that considered a health level in human diet, 1 mg of selenium per kg of body weight. It is an essential nutrient for animals and for human health in the range of 0.8–1.7 mmol/L, but toxic above this value. Normal human dietary intake of Se is about 50–200 µg/day, and Se toxicity may manifest itself at dietary levels of 400 µg/day. Exceeding the tolerable upper intake level can lead to selenosis. Symptoms of selenosis include a garlic odor on the breath, gastrointestinal disorders, hair loss, sloughing of nails, fatigue, irritability, monstrous deformities, and neurological damage. Extreme cases of selenosis can result in cirrhosis of the liver, pulmonary edema, and death. Selenium is also known to be mutagenic and teratogenic. Selenium is introduced in the environment from different sources, both natural and anthropogenic (Hasan et al. 2010; Wang et al. 2012).

Selenium and iodine and also their radioisotopes should be controlled with regard to both toxicity and deficiencies in humans and livestock. The presence of these elements in waste and surface waters is becoming a severe environmental and public health problem. A variety of treatment technologies have been reported for selenium removal from contaminated waters (Tuzen and Sarı 2010; Zhang et al. 2009; El-Shafey 2007a, b; Bronwyn et al. 2004).

The most widely used methods for removing hazardous substances from wastewaters include ion exchange, chemical precipitation, reverse osmosis, evaporation, membrane filtration, adsorption, and biosorption (Tuzen and Sarı 2010; Najafi et al. 2010; Bleiman and Mishael 2010).

Iodine and its compounds are primarily used in nutrition and industrially in the production of acetic acid and certain polymers. Moreover, iodine radioisotopes are also used in medical applications. Some chemical forms may be classified as hazardous materials if the compound is chemically reactive, flammable, or toxic. The radionuclide of greatest concern from a radiation protection viewpoint is radio-iodine which is regularly used in nuclear medicine procedures and it is comparatively radiotoxic. The use of radio-iodine in the treatment of hyperthyroidism and thyroid disease has increased significantly and today represents about 90 % of all therapies in nuclear medicine (UNSCEAR 2000; ICRP 2004).

The success of the therapy depends on the uptake and retention of radio-iodine in the thyroid, remainder tissue, or metastases (Willegaignon et al. 2006). The activity that is not retained in the lesion is excreted through the urinary system to the sewage as radioactive urine effluent. Storage of this liquid waste for sufficient time would ensure the decay of radio-iodine to insignificant levels. But there may be occasions when it may not be practical to allow storage of the liquid waste for long durations.

Adsorption is an important technique in separation and purification processes which is used in water and wastewater industry for the removal of color, odor, and organic pollution. In this study, it is aimed to determine the most effective adsorption parameter of radio-selenium and radio-iodine adsorption on the activated carbon samples. Two activated carbon materials are prepared by chemical and physical activation methods. The adsorption of radio-selenium and radio-iodine is studied by using factorial design for pH, concentration of adsorbate, and contact time on the prepared materials.

9.2 Experimental

9.2.1 Preparation of Activated Carbon

Activated carbon (AC-I) produced from olive stone by a chemical activation method was used for iodine absorption. The olive stone waste material was initially stirred for 1 h at 373 K in a general-purpose oven. Then, the sample was sieved and washed to remove oil residues and dried. Then, they were grounded, and subsequently, the sample was impregnated with a 50 % phosphoric acid solution, dried in air at about 493 K, and then carbonized in a quartz reactor at a temperature of 673 K for 120 min. The carbonization was done in a flow of nitrogen (300 mL/min). After carbonization, the carbon was cooled down to room temperature in a flow of nitrogen. To remove the excess of H_3PO_4, the carbons after carbonization were extensively washed with hot water until neutral pH. Then, the samples were dried in an oven at 283 K.

The other activated carbon (AC-Se) for Se adsorption was prepared by applying physical activation from lignite using carbon dioxide (CO_2) as the activation agent. The original lignite sample was first carbonized at 1073 K for 1 h under nitrogen (N_2) atmosphere and then activated with CO_2 at 1223 K for 3 h.

Characterization of the porous texture of activated carbons is of relevance since many of their properties are determined or strongly influenced by this characteristic. For this purpose, a gas adsorption instrument (Quantochrome Corp., NOVA-2200 Gas Sorption Analyzer) was used. Adsorptions of N_2 and CO_2, as probe species, were performed at 77 and 273 K, respectively. Before any such analysis, the sample

Table 9.1 Surface properties of activated carbon samples

Sample	BET surface area (m^2/g)	Total pore volume ($P/P_0 = 0.95$) (cm^3/g)
AC-I	1373	0.835
AC-Se	505	0.473

was degassed under vacuum at 473 K. The N_2 Brunauer–Emmett–Teller (BET) surface area (S_{BET}) of the activated carbon sample was calculated from the N_2 adsorption isotherms using the BET equation and is given in Table 9.1.

9.2.2 Batch Experiments

All chemicals used in this work were of analytical reagent grade and used without further purification. Sodium iodide and selenium dioxide compounds were irradiated in ITU TRIGA Mark-II Nuclear Research Reactor. Then, they were used to investigate the adsorption performance of activated carbons for iodine and selenium species.

The batch adsorption experiments were performed on a shaker by varying the parameters such as pH (3.0 and 9.0), initial concentration (100 mg/L and 200 mg/L), and contact time (60 and 120 min).

Radioactivity measurements were performed in Low Level Radioactivity Measurement Laboratory in the Istanbul Technical University Energy Institute. HPGe coaxial n-type germanium detector having copper-lined lead shield (10 cm) was used to determine ^{75}Se and ^{128}I activities of the samples. The detector, with integrated digital gamma spectrometer (DSPEC jr. 2.0), has 45.70 % efficiency and 1.84 keV full width at half maximum for 1.3 MeV of ^{60}Co. In the measurements, statistical confidence level and range were adjusted to 1σ and 8192 channels, respectively. Counting times of 4–30 min were applied. Peak areas of ^{75}Se (136 keV) and ^{128}I (443 keV) were determined by using the GAMMA VISION-32 software program (ORTEC 2003; Firestone 1998). ^{152}Eu standard point source was used in energy calibrations of the spectra (DKD-K-36901-000386, 2006).

Relative activities of initial and adsorption solutions were calculated using the following equation:

$$A_0 = A \cdot e^{\lambda \cdot td} = \frac{\left[P \cdot e^{(\lambda \cdot t_d)}\right]}{t_m} \tag{9.1}$$

where A is the counting rate at time t_d based on a radioactive sample that produced counting rate A_0 at time $t = 0$, P denotes the number of counts in the net area of the peak at related energy in the spectrum, t_m symbolizes the counting time, and λ is the decay constant of the nuclide (Loveland et al. 2006).

The amount of adsorbed iodine and selenium was estimated from the difference between their initial and final relative activities. The adsorption capacity $A(\%)$ is computed as follows:

$$A(\%) = \frac{(C_i - C_e)}{C_i} \times 100 \tag{9.2}$$

where C_i is the adsorbate concentration of the initial solution (mg/L) and C_e is the adsorbate concentration of the solution in equilibrium (mg/L).

The statistical design technique was applied by use of a two-level factorial design matrix to interpret the experimental results (Montgomery 1991; Karatepe 2003). A major advantage of the statistical model over the analytical ones is that they do not use rough approximations and allow for a greater number of factors.

9.3 Results and Discussion

The adsorption results of iodine and selenium were calculated and presented in Table 9.2.

9.3.1 Concentration Experiments

In order to determine the effect of initial iodine concentration on the adsorption behavior of iodine, experiments were conducted using iodine solutions with concentration ranging from 100 to 200 mg/L at different contact time and pH values. The change in initial concentration of the iodine solution from 100 to 200 mg/L for pH 3.0 and 9.0 at 25 °C by 60 and 120 min of contact time increased the amount adsorbed on the activated carbon sample (Fig. 9.1).

Table 9.2 The results of radio-selenium and radio-iodine adsorption experiments

Sample no.	Concentration (mg/L)	pH	Time (min)	Adsorption efficiency (%)	
				Radio-selenium	Radio-iodine
1	100	3	60	22.95	87.5
2			120	19.70	83.8
3		9	60	22.35	85.5
4			120	15.06	83.5
5	200	3	60	26.54	94.48
6			120	24.18	94.38
7		9	60	28.03	92.13
8			120	26.08	89.54

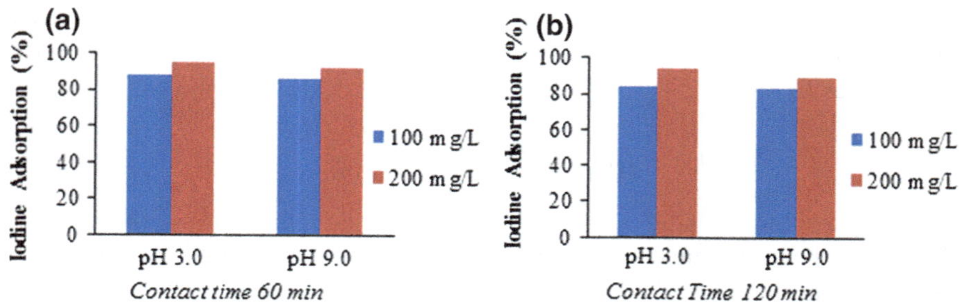

Fig. 9.1 The adsorption of iodine by AC-I versus initial concentration

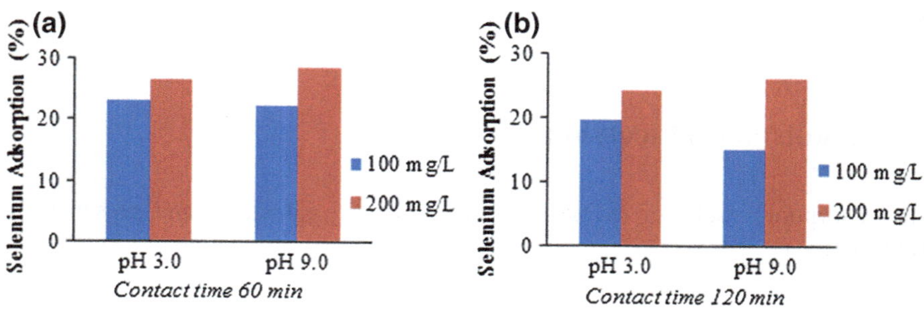

Fig. 9.2 The adsorption of selenium by AC-Se versus initial concentration

To study Se concentration effect in adsorption, selenium solutions containing 0.05–0.5 mg/mL were kept in contact with activated carbon for 2 h at pH 3.0. The results are shown in Fig. 9.2.

9.3.2 PH Experiments

pH is the one of the most important parameters controlling the adsorption behavior. The removal of iodine as a function of pH at different temperatures and initial iodine concentrations is presented in Fig. 9.3.

The removal of iodine decreased with increasing pH from 3.0 to 9.0. This behavior can be explained by considering the nature of the adsorbent at different pH values in iodine adsorption. The cell wall of activated carbon contains a large number of surface functional groups. The pH dependence of iodine adsorption can largely be related to the type and ionic state of these functional groups and also on the iodine chemistry in solution.

Hydrogen and hydroxyl ions have a great impact on the surface charge of adsorbent. What is more, H^+ and OH^- would strongly compete with sorbate during adsorption process. Therefore, pH is one of the most important parameters affecting

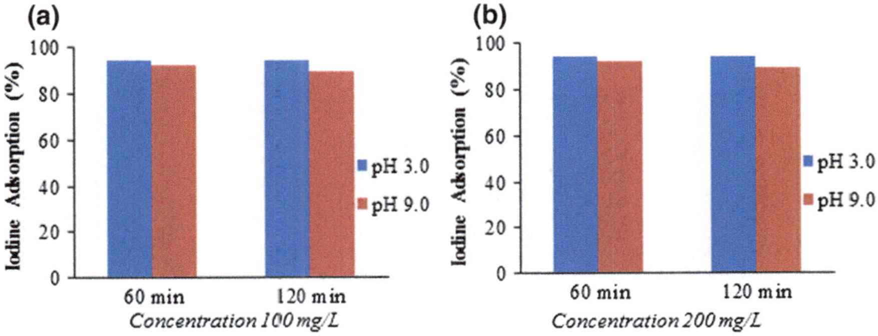

Fig. 9.3 The removal of iodine as a function of pH at different temperatures and initial iodine concentrations 100 mg/L (**a**) and 200 mg/L (**b**) (*Contact time* 60 and 120 min)

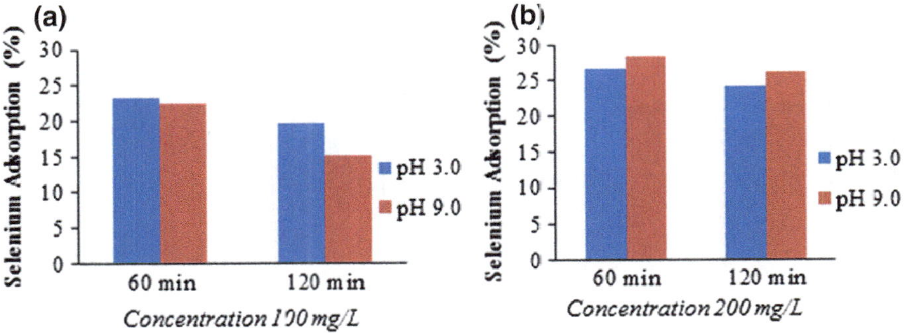

Fig. 9.4 The effect of pH values on the adsorption of Se by activated carbon

ion sorption. Influence of pH values on the adsorption is given in Fig. 9.4 by keeping the conditions of 40 mg activated carbon, 20 °C, and 2 mL of 50 mg/L Se(IV).

To study effect of pH to adsorption, iodine solutions containing 0.05–0.5 mg/mL were kept in contact with activated carbon for 2 h at pH 3.0. The results are shown in Fig. 9.4.

9.3.3 Contact Time Experiments

Figure 9.5 shows the variations of iodine adsorption with contact time. The adsorption decreases with increasing contact time.

The effect of contact time on selenium adsorption was studied at 60 and 120 min. The results are presented in Fig. 9.6.

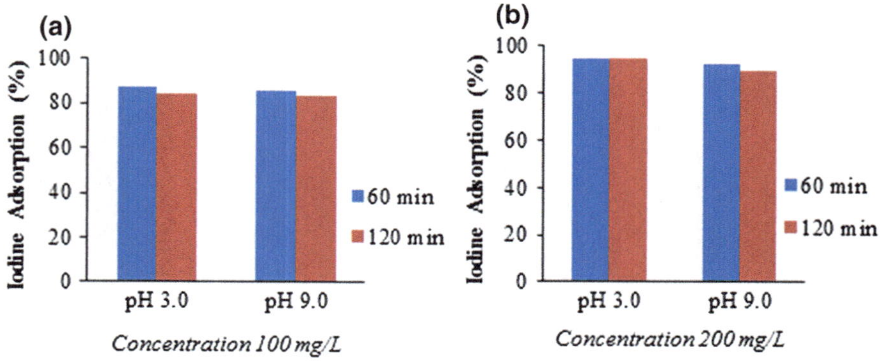

Fig. 9.5 The effect of contact time on the adsorption of I by AC-I

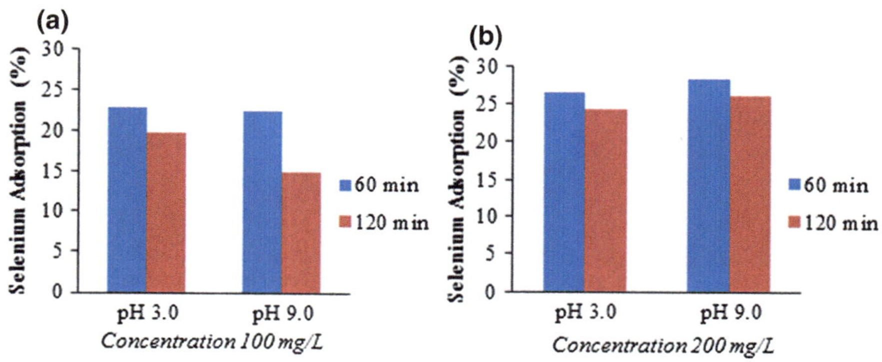

Fig. 9.6 The effect of contact time on the adsorption of Se by AC-Se

9.3.4 Statistical Analysis Results

The results given in Table 9.2 were statistically analyzed to identify and measure the main and interactional effects quantitatively by using analysis of variance, which is the most effective technique in factorial designed experiments (Karatepe et al. 1998). In two-level factorial design experiments, process variables were selected as pH (3.0 and 9.0), initial concentration (100 and 200 mg/L), and contact time (60 and 120 min). The actual and coded values of the variables of experiments are shown in Table 9.3.

In Table 9.4, the design matrix and results of experiments are listed. By using multifactor linear model, a regression equation was developed to predict the adsorption efficiencies of the activated carbons and optimize the process conditions.

Table 9.3 Actual and coded values of the variables

Level	Upper level	Lower level	Base level
Initial concentration (mg/L)	200	100	150
X_1	+1	−1	0
pH	9	3	6
X_2	+1	−1	0
Contact time (min) (t)	120	60	90
X_3	+1	−1	0

Table 9.4 Design matrix and results of experiments

Trial no.	X_1	X_2	X_3	Adsorption (%)	
				Selenium	Iodine
1	−1	−1	−1	22.9	87.5
2	−1	−1	1	19.7	83.5
3	−1	1	−1	22.4	85.5
4	−1	1	1	15.1	83.5
5	1	−1	−1	26.5	94.5
6	1	−1	1	24.2	94.4
7	1	1	−1	28.0	92.1
8	1	1	1	26.1	89.5

$$Y = A_0 + A_1X_1 + A_2X_2 + A_3X_3 + A_4X_1X_2 + A_5X_1X_3 + A_6X_2X_3 + A_7X_1X_2X_3 \tag{9.3}$$

In Eq. (9.3), Y is adsorption efficiencies of activated carbons (%), X_1 is coded value of initial concentration, X_2 is coded value of pH, and X_3 is coded value of contact time.

The results given in Table 9.4 were analyzed statistically to identify and measure the main and interactional effects quantitatively by using analysis of variance which is the most effective technique in factorial designed experiments.

The variance tests of the parameters for the samples demonstrate that some of the variables are not statistically significant. Therefore, their respective terms can be rejected in the following proposed model for iodine (Eq. 9.3) and selenium (Eq. 9.4):

$$Y = 88.706 + 3.632X_1 - 1.039X_2 - 0.902X_3 - 0.464X_1X_2 + 0.524X_{\text{[illegible]}}X_3 - 0.589X_1X_2X_3 \tag{9.4}$$

$$Y = 23.255 + 3.24X_1 - 2.000X_3 + 0.935X_1X_2 - 0.310X_2X_3 + 0.635X_1X_3 + 0.575X_1X_2X_3 \tag{9.5}$$

where Y is the amount of iodine and selenium adsorptions in %, X_1 is coded value of initial iodine concentration, X_2 is coded value of pH, and X_3 is coded value of contact time. The correlation coefficients for Eqs. 9.4 and 9.5 were determined as 0.996 and 0.990, respectively.

The relationship between the coded values (X_k) and actual values can be given as follows:

$$X_1 = \frac{(C-150)}{50} \tag{9.6}$$

$$X_2 = \frac{(\mathrm{pH} - 6)}{3} \tag{9.7}$$

$$X_3 = \frac{(t - 90)}{30} \tag{9.8}$$

where C is the initial iodine concentration in mg/L and t is the contact time in min.

The regression Eqs. 9.4 and 9.5 clearly shows that since the coefficient of initial iodine or selenium concentration is the highest among all the coefficients, the effect of this parameter on the adsorption of the activated carbon sample is dominant. The iodine or selenium adsorptions of activated carbon were affected positively by this variable. Therefore, higher initial concentration causes higher adsorption values (Y).

The adsorption also increased with decreasing the pH value and adsorption contact time. However, the effect of the pH value on the selenium adsorption is lower than that of the contact time. Therefore, this term was rejected in the proposed model. It may also concluded from the regression models (Eqs. 9.4 and 9.5) that the interactional effects such as (initial iodine concentration × pH) (X_1X_2), (pH × contact time) (X_2X_3), (initial iodine concentration × contact time) (X_1X_3), and (initial iodine concentration × pH × contact time) ($X_1X_2X_3$) influence the iodine and selenium adsorptions positively and negatively, at 99.6 and 99 % confidence level, respectively. In other words, if one of the variables is changed with respect to another one, it will have a considerable effect on the iodine adsorption of the activated carbon.

9.4 Conclusion

Activated carbon that was prepared with chemical activation method is an effective adsorbent for the removal of radio-iodine from aqueous solutions. In adsorption process, it was found that increasing the pH and contact time causes a decrease in iodine I(I) adsorption. On the contrary, increasing the initial iodine concentration induces an increase in iodine adsorption. Activated carbon prepared by physical activation method has been demonstrated to be generally low effective for the removal of the Se(IV) ions from aqueous solutions than the iodine.

The results provide information for the effective adsorption of I(I) and Se(IV) in natural surface and groundwater by the means of factorial design. The regression equations of iodine and selenium adsorptions show that the initial concentration has dominant effect in adsorption.

Coal burning in energy power plants or mining activities for energy generation would cause increasing selenium concentration in aqueous media. Furthermore, iodine concentration can rise in the abnormal conditions for nuclear power plants. Consequently, further research is in progress and also necessary to support management of energy generation by preventing hazardous elements dispersion to the environment.

Acknowledgments Authors are grateful to the ITU TRIGA Mark-II Training and Research Reactor group for their help during the irradiation of selenium.

References

Bleiman, N., & Mishael, Y. G. (2010). Selenium removal from drinking water by adsorption to chitosan–clay composites and oxides: Batch and columns tests. *Journal of Hazardous Materials, 183*, 590–595.

El-Shafey, E. I. (2007a). Removal of Se(IV) from aqueous solution using sulphuric acid-treated peanut shell. *Journal of Environmental Management, 84*, 620–627.

El-Shafey, E. I. (2007b). Sorption of Cd(II) and Se(IV) from aqueous solution using modified rice husk. *Journal of Hazardous Materials, 147*, 546–555.

Firestone R. B. (1998). Table of Isotopes [electronic resource]. In Chu F (Ed.), *CD-ROM*, 8th ed. (update). New York: Wiley.

Hasan, S. H., Ranjan, D., & Talat, M. (2010). Agro-industrial waste ‘wheat bran’ for the biosorptive remediation of selenium through continuous up-flow fixed-bed column. *Journal of Hazardous Materials, 181*, 1134–1142.

In Loveland, W. D., Morrissey, D. J., & Seaborg, G. T. (2006). *Modern nuclear chemistry*. New Jersey: Wiley.

International Commission on Radiological Protection (ICRP). (2004). Release of Patients after Therapy with Unsealed Radionuclides, ICRP Publication 94, Elsevier.

Karatepe, N. (2003). Adsorption of a non-ionic dispersant on lignite particle surfaces. *Energy Conversion and Management, 44*(8), 1275–1284.

Karatepe, N., Mericboyu, A. E., & Kucukbayrak, S. (1998). Preparation of fly Ash-Ca$(OH)_2$ sorbents by pressure hydration for SO_2 removal. *Energy Sources, 20*, 945–953.

Montgomery, D. C. (1991). The 2k factorial design. In *Design and Analysis Experiments* (pp. 270–310). Singapore: Wiley.

Najafi, N. M., Seidi, S., Alizadeh, R., & Tavakoli, H. (2010). Inorganic selenium speciation in environmental samples using selective electrodeposition coupled with electrothermal atomic absorption spectrometry. *Spectrochimica Acta Part B, 65*, 334–339.

ORTEC. (2003). Gamma Vision-32 A66-B32 Software Users Manual.

Tuzen, M., & Sarı, A. (2010). Biosorption of selenium from aqueous solution by green algae (Cladophora hutchinsiae) biomass: Equilibrium, thermodynamic and kinetic studies. *Chemical Engineering Journal, 158*, 200–206.

United Nations Scientific Committee on the Effects of Atomic Radiation (UNSCEAR). (2000). Sources and effects of ionizing radiation, Report to the General Assembly with Scientific Annexes. Vol. I: Sources, United Nations, New York.

Wake, B. D., Bowie, A. R., Butler, E. C. V., & Haddad, P. R. (2004). Modern preconcentration methods for the determination of selenium species in environmental water samples. *Trends in Analytical Chemistry, 23*(7), 491–500.

Wang, Y. D., Wang, X., & Wong, Y. S. (2012). Proteomics analysis reveals multiple regulatory mechanisms in response to selenium in rice. *Journal of Proteomics, 75*, 1849–1866.

Willegaignon, J., Stabin, M. G., Guimaraes, I. C., Malvestiti, L. F., Sapienza, M. T., Maroni, M., & Sordi, A. A. (2006). G-M, Evaluation of the potential absorbed doses from patients based on whole-body 131I clearance in thyroid cancer therapy. *Health Physics Journal, 91*, 123–127.

Zhang, L., Liu, N., Yang, L., & Lin, Q. (2009). Sorption behavior of nano-TiO_2 for the removal of selenium ions from aqueous solution. *Journal of Hazardous Materials, 170*, 1197–1203.

Chapter 10
Assessment of Sustainable Energy Development

A. Beril Tugrul and Selahattin Cimen

Abstract In this study, optimum solutions and action plans for sustainable energy development are discussed. Reduction of CO_2 emission could be realized with increasing nuclear and renewable energy usage, and efficiencies on fuel, power, electricity, and fossil fuels. In here, "ecosystems approach" is vital importance. Worldwide cooperation is the most important with the concepts of **6 Cs** (credibility, capability, continuity, creativity, consistency, and commitment). Therefore, it can be successfully developed on sustainability, sharing with public, strategy and culture, procedures and evaluation together with 6 Cs.

10.1 Introduction

Today, energy has a primary role in every society and creating the welfare of people. Usage and management of energy is an important phenomena and also measuring criterion for the countries in the view of evaluation for developing circumstances in the world.

Supplying the energy needs is difficult and complex matter due to fabulously increased energy demand all over the world. Therefore, in the present time, the energy policies conduct to political events, which could not be undeniable and indispensable effects on the world politics (Tugrul 2011). In fact, oriented factors of energy policies are also lead the world politics effectively.

World primary energy consumption is projected to grow up non-negligible rate in the period 2010–2030. It is assumed that all the power plants, for example, thermal, nuclear, hydro, wind, and solar power plants, will be increased in the next

---

A.B. Tugrul (✉)
Istanbul Technical University, Istanbul, Turkey
e-mail: beril@itu.edu.tr

S. Cimen
Ministry of Energy and Natural Resources (MENR), Ankara, Turkey
e-mail: s.cimen@ttnet.com

A.N. Bilge et al. (eds.), *Energy Systems and Management*,
Springer Proceedings in Energy, DOI 10.1007/978-3-319-16024-5_10

period of twenty years (BP 2012). Furthermore, energy demand estimation shows that the energy demand will also grow up almost all the region in the world (DOE 2009).

World electricity demand is projected to grow more rapidly as total energy over the next 20 years. It seems that the fossil fuels may be major fuel in that period also (Tugrul and Cimen 2013). Then, greenhouse effect will be more affected of course, because it affected on climate change and other harmful effects on the nature. On the other hand, sustainability is important for the liveable earth. Therefore, the interrelationship of natural and man-made elements in the environment is the basis for planning aimed toward improved quality of life. Energy demand and sustainability have paradoxical aspects indeed (Tugrul 2014).

10.2 Sustainable Development

Despite its simplicity, however, sustainability is a hard concept for realization in practice. Many of the people have understood different what sustainability really means.

The Brundtland Report (our common future) defined sustainable development as "**development that meets the needs of the present without compromising the ability of future generations to meet their own needs**" (WCED 1987). Note that the definition says nothing about protecting the environment, but it determines the continuality of the ordinance.

Priority areas for action, identified by UN Secretary-General (Soriano 2012), are as follows:

- **W**ater and sanitation
- **E**nergy
- **H**ealth
- **A**griculture
- **B**iodiversity protection and ecosystem management

In here, it is noticed that energy is one of the argument in WEHAB. Then, sustainable energy is a concept for application of energy systems. Sustainable energy is important for the welfare of the countries. Therefore, sustainable energy related to economic growth, environment, and social equity simultaneously is vital for our earth (Soriano 2012). Figure 10.1 represents energy sustainability in the total sustainability concept

Energy needs stimulate new developments. It should be applied by organizational and systematic activities. Figure 10.2 shows the actions of energy with feedback effects.

In the present time, energy and raw materials are essential input for civilization, and then, heat and high wastes and toxics are outputs of course that is one-way flow (TREN 1990). Relation between energy and materials with one-way flow is shown in Fig. 10.3 schematically.

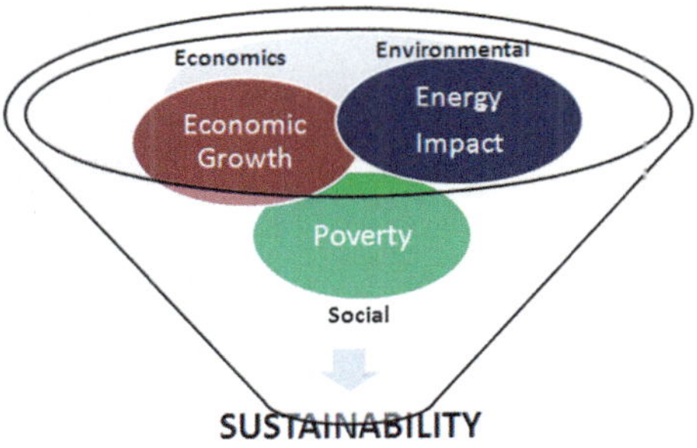

Fig. 10.1 Energy sustainability in the total sustainability concept

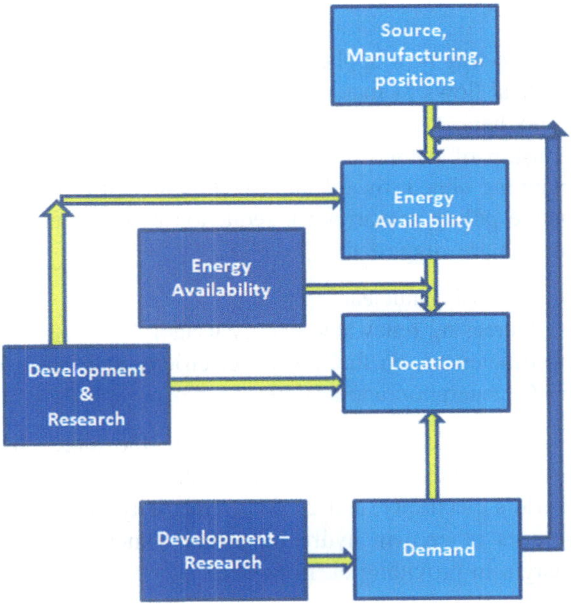

Fig. 10.2 Actions of energy with feedback effects

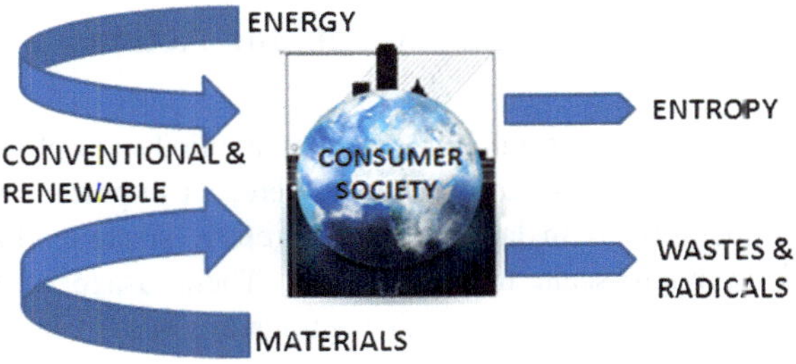

Fig. 10.3 One-way flow energy and materials

After years by years with applying, one-way flow applications have caused the heat index of the world going up and now has reached at the dangerous levels (Url-1 2013).

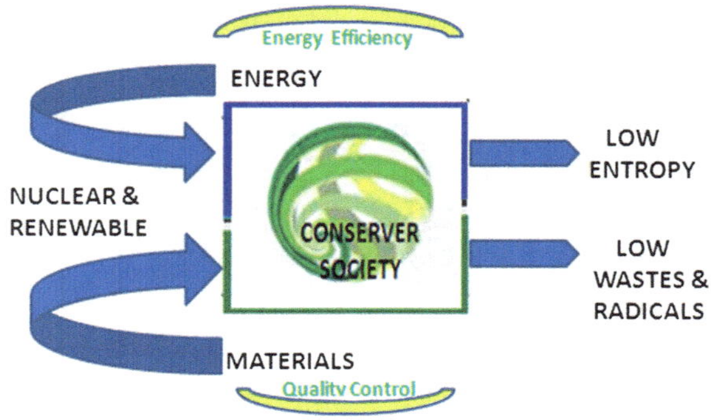

Fig. 10.4 Conserver society

One-way flow applications should be turned as conserver society by applying of cyclical flows of materials and appropriate energy usage (Fig. 10.4).

In here, "ecosystems approach" is vital importance. Therefore, the interrelationship of natural and man-made elements in the environment is the basis for planning aimed toward improved quality of city life. OECD/IEA developed alternative policy scenarios for reduction of CO_2 emission. For this purpose, the following are suggested:

- Increasing nuclear
- Increasing renewable energy usage,
- Efficiencies on fuel power, electricity and fossil fuels
- Cogeneration applications

Renewable and nuclear power are the more environmentally benign way of producing electricity on a large scale. Nuclear power provides about 11 % of the world's electricity, and 21 % of electricity in OECD countries. Renewable energy sources other than hydro have high generating costs, but can be helpful at the margin in providing clean power.

10.3 Major Requirements for Actions

Concepts for sustainability may be applied with physical/geographical, ecological, and jurisdictional arguments. In here, an important argument for appliance of the cooperation is in large scale. Therefore, sustainability concept should be applied from micro-scale to macro-scale. Then, usage of sustainable energy could be possible but more and major actions are needed in global scale.

Worldwide major requirements for actions could be summarized with the concepts of **6 Cs**:

- **Commitment**: Promising on execute reliably for the sustainable energy.
- **Credibility**: Stand up for what is right
- **Capability**: Living with conservation of energy

Fig. 10.5 Important factors for cooperation on sustainability

- **Continuity**: Pursuing on sustainability mission
- **Creativity**: Think beyond boundaries; embrace change and new ways of thinking for sustainability
- **Consistency**: The lack of contradiction for all related activities

Therefore, it can be success for development with sustainability, if it would be sharing with public, strategy and culture/ethics, procedures and evaluation together with 6 Cs (Fig. 10.5).

10.4 Conclusions

The new concepts of world energy require a shift of position in mind and strategic orientation. We are at the edge of a new energy revolution, driven by the world's need for affordable energy and by the real threat of climate change.

The coming decades are likely to bring about huge changes in the world's energy system. Future energy policy will be driven by the triple challenge of achieving substantial reductions in emissions of greenhouse gases while ensuring a secure supply of energy, all at reasonable cost to economies. In order to cope with this challenge, it must be changed the way we use energy. Increasing the energy efficiency of the economies is an absolute necessity. Also, all of us must move rapidly toward a more diverse, sustainable set of energy resources. This move depends on the aggressive development and deployment of more sustainable energy sources.

References

BP. (2012). Energy outlook 2012.

DOE. (2009). Department of energy—EIA, energy information administration (EIA) data.

Soriano, M. L. (2012). Energy-integrated planning for low carbon development in cities, UNDP.

TREN. (1990). Introduction to sustainability, TREN-1F90 http://news.uns.purdue.edu/UNS//images/+2007/diffenbaugh-heat2.jpg. Accessed on 1 Mar 2014.

Tugrul, A. B. (2011). Nuclear energy in the energy expansion of turkey. *Journal of Energy and Power Engineering, 5*(10), 905–910.

Tugrul, A. B., & Cimen, S. (2013). Energy Initiatives for Turkey, International Conference on Economics and Econometrics—ICEE 2013, Dubai-UAE, Proc. (pp. 40–44). 2–3 Dec 2013.

Tugrul, A. B. (2014). Energy, sustainable development and importance of worldwide cooperation, workshop on novel energy in the regenerative built environment, Istanbul. 3–5 Mar 2014.

Url-1. (2013). http://news.uns.purdue.edu/UNS//images/+2007/diffenbaugh-heat2.jpg.

WCED. (1987). *World commission on environment and development, our common future* (p. 27). Oxford: Oxford University Press. ISBN 019282080X.

Chapter 11 Geothermal Energy Sources and Geothermal Power Plant Technologies in Turkey

Fusun Servin Tut Haklidir

Abstract Geothermal energy is used for electric power generation and direct utilization in Turkey. The highest enthalpy geothermal sources are located in Western Anatolia; thus, geothermal power generation projects have also been realized in Western Anatolia since 1984. The present installed gross capacity for electric power generation is 345 MWe from 11 geothermal power plants in 2014, while new 395 MWe of capacity is still under construction or projected at 19 geothermal fields and will be completed in 2016–2017. In Turkey, flash cycle power plants are situated in Kızıldere (Denizli) and Germencik (Aydın) geothermal fields because of over than 230 °C geothermal reservoir temperatures. There are two different geothermal power plants in Kızıldere geothermal field that one is 17.2 MWe single-flash system and the new one consists of 60 MWe triple-flash + 20 MWe binary cycle, as total 80 MWe capacity. In Germencik, 47.4 MWe double-flash geothermal power plant uses power generation. Except from these three geothermal power plants, binary cycle (organic Rankine cycle, ORC) plants use under 200 °C geothermal reservoir temperatures in all installed capacities in Western Anatolia. New geothermal reservoir studies are still under investigation for the eastern part of Turkey. Because of the lower reservoir temperature values at these regions, the possible power generation cycle may be required to binary system (Kalina cycle).

11.1 Introduction

Geothermal energy is one of the most important renewable energy sources such as wind, biomass, solar, and hydro. It is one of the potential energy sources that are sustainable and environmentally friendly and independent from weather conditions. Geothermal sources are derived from the depth of earth's crust by thermal conduction, especially volcanic and tectonically active regions in the world.

F.S. Tut Haklidir (✉)
Istanbul Bilgi University, Istanbul, Turkey
e-mail: fusun.tut@bilgi.edu.tr

A.N. Bilge et al. (eds.), *Energy Systems and Management*,
Springer Proceedings in Energy, DOI 10.1007/978-3-319-16024-5_11

The history of using geothermal energy is to date back from the ancient times. Thermal springs have been used bath, swimming, and underfloor heating purposes at ancient times. In the fourteenth century, the first district heating system has been established in France and heating of greenhouses has been started in USA and Italy at the beginning of 1900. After these direct applications, the first geothermal power plant has been started to produce electricity in Larderello, Italy, in 1911.

Nowadays, geothermal sources can be used at different direct applications such as heating, cooling applications, industrial and agricultural applications, thermal tourism, and power generation around the world. These applications depend on enthalpy of geothermal sources. Low temperatures (20–60 °C) are generally suitable for fish farm, animal husbandry, heating pools, and greenhouse applications. Medium temperatures (70–150 °C) allow district heating and cooling applications. These temperature ranges are also suitable for drying applications for industry. Higher than 150 °C geothermal sources are enabled for indirect applications such as power production in the world. End of the 2009, the 78 countries have been used direct utilization of geothermal sources (Lund et al. 2010). The distribution of geothermal energy is classed by approximately 49.0 % for ground-source heat pumps, 24.9 % for bathing and swimming, 14.4 % for space heating (of which 85 % is for district heating), 5.3 % for greenhouses and open-ground heating, 2.7 % for industrial process heating, 2.6 % for aquaculture, 0.4 % for agricultural drying, 0.5 % for snow melting and cooling, and 0.2 % for other uses (Lund et al. 2010).

The total installed geothermal power capacity is declared as 10.7 GW for 2010 by Bertani (2010), and the biggest geothermal power capacities are located at USA, Mexico, Philippines, and Italy around the world. Geothermal power capacities are increased fastly, and total installed capacity is reached to 1.76 GWe in Europe in 2012 (Dumas 2013).

In Turkey, the potential of direct application is estimated to be nearly 50 GWt, whereas the capacity of installed district heating and heating of greenhouses capacity is 2100 MWt in 2011 (Satman 2013). Beside the direct applications, after the installation of the first power production plant, 15 MWe Kızıldere (Denizli) geothermal power plant in 1984, the power capacity reaches to 350 MWe at the end of the 2013 in Turkey (Tut Haklıdır 2014).

11.2 Geothermal Power Technologies in the World

Geothermal reservoirs are observed as water dominated (two phases system) or steam dominated in geothermal systems.

In steam dominated system, steam has directly sent to turbine-generator system from production well. The largest dry steam fields in the world are the Geysers field in CA-USA and Larderello field in Tuscany, Italy (DiPippo 2005). Nowadays, Larderello geothermal power capacity is reached to net 769 MWe with total 34 power plants (Enel 2013) and geothermal power generation is reached to net 685 MWe with 15 power plants in the Geysers field (Calpine Corporation 2014).

In water-dominated geothermal systems, two main cycles have been used such as flash cycles and binary cycles depend on reservoir temperatures. Flash cycles need higher reservoir temperatures (more than 200 °C) than binary cycles. The basic of flash cycle is based on separation of geothermal fluids as steam + gas and water in separator systems. After the separation, steam directly goes to power generation system and water is sent to reinjection well to feed the reservoir and to protect environment from harmful chemical contents in thermal water. Some emissions released into air depending on the geothermal fluid characteristics in the system. In flash systems, if the reservoir temperatures are higher than 220–230 °C or more, double-flash (Valdimarsson 2011) or triple-flash cycles (Tut Haklıdır and Kındap 2013) can be used to provide energy efficiency in the system. It means after the first separation, high-pressured (hp) steam is sent to hp turbine, and then, the waste thermal water (brine) is sent to second and/or third separation systems to provide more steam in the system.

For using a binary cycle system, geothermal fluid temperature requires less than flash cycles. In this system, geothermal water is sent to a heat exchanger, where the heat is transferred into a second liquid by organic or other special chemicals such as isopentane, isobutane, or NH_3 + H_2O that helps to boil of water at a lower temperature than water (DiPippo 2005). When it is heated, the steam is sent to turbine-generator system and the closed cycle does not cause emission in this type cycle. Binary system can be divided into two systems such as organic Rankine cycle (ORC) and Kalina cycle. The ORC geothermal power plants are very common around the world because of compact system and installation of the system is easier than flash systems. In this system, Rankine cycle is used and generally uses *n*-pentane and *n*-butane or mixing of these chemicals to provide boiling water temperature and produce steam (DiPippo 2005). The Kalina cycle is a modified Clausius–Rankine cycle and uses mixture of NH_3 + H_2O as a chemical. Valdimarsson (2011) noted that NH_3 + H_2O mixture provides both vaporization and condensation of the mixture which happens at different temperatures.

Advanced systems such as flash + binary systems can be used in most of the geothermal fields, to provide energy efficiency and to decrease internal energy consumption in geothermal power generation systems.

11.3 Geothermal Power Generation in Turkey

Turkey is the 7th richest country in geothermal energy potential around the world. Geothermal exploration studies have been started at the beginning of 1960s in Western Anatolia, Turkey. Turkey is located on the Alpine–Himalaya orogenic belt, and it concludes different tectonic zones such as the North Anatolian fault, the Eastern Anatolian fault, and Aegean graben systems. The country has also young volcanics such as Kula (Manisa city) volcanics, Nemrut (Bitlis city) (Bozkurt 2001; Tut Haklıdır 2007). All these natural activities provide to reach geothermal energy potential during these zones. Nowadays, more than 200 geothermal fields are

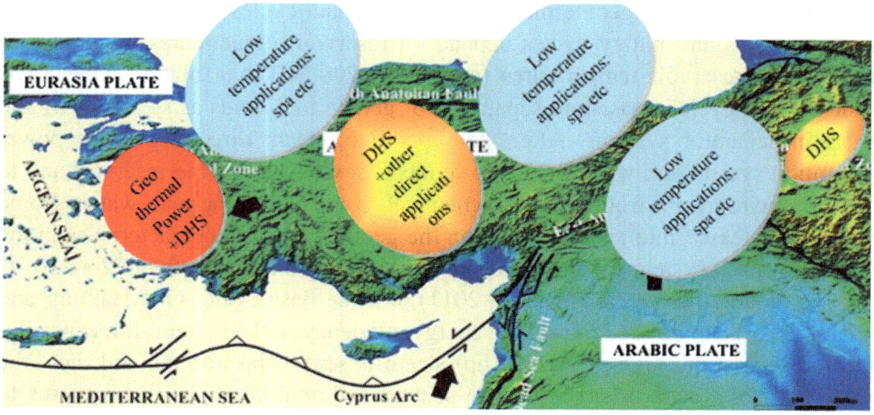

Fig. 11.1 Geothermal applications in Turkey

discovered and geothermal exploration studies still continue in Turkey. The discovered power geothermal power generation capacity is nearly 2000 MWe for Turkey (MTA 2014). Although geothermal fields are discovered in Middle and Eastern Anatolia, the reservoir temperatures are not suitable for power generation such as Western Anatolia (Fig. 11.1). Especially in Eastern Anatolia, the continental crust is very thick and it is not economical for geothermal exploration/drilling operations for present conditions.

In Turkey, the highest temperature (245 °C; Kızıldere, Denizli city) geothermal field, which is suitable for power production, is discovered in Büyük Menderes Graben, Western Anatolia. The first geothermal power plant was established in Kızıldere in 1984, and the net capacity was 15 MWe. After this first important geothermal exploration, the graben has been investigated with detail and other important geothermal fields such as Germencik-Aydın, Salavatlı-Aydın, and Pamukören-Aydın are discovered along the Menderes Graben. In 2010, geothermal reservoirs, which temperature indicates 200 °C, are discovered in Alaşehir (Manisa) Graben in Western Anatolia (Fig. 11.2).

In Turkey, geothermal power generation studies were progressed very slowly between 1984 and 2008. In 2005, "The Law to Use the Renewable Energy Resources for Electricity Production" was enacted by the Grand National Assembly of Turkey and was amended by 2010 and the new electric tariff was determined for different renewable energy sources. This development has been encouraged the investors for new energy investments. In 2007, "5686 numbered Law on Geothermal Resources and Natural Mineral Waters" were put into effect by the government and geothermal investors have been showed great interest to geothermal power generation investments with some sectorial arrangements on using of geothermal energy. These developments show their effects on investors, and end of 2013, the numbers of geothermal power plants have increased to 10, and the total geopower capacity has reached to 255 MWe in Turkey. In 2014, 2 new geothermal

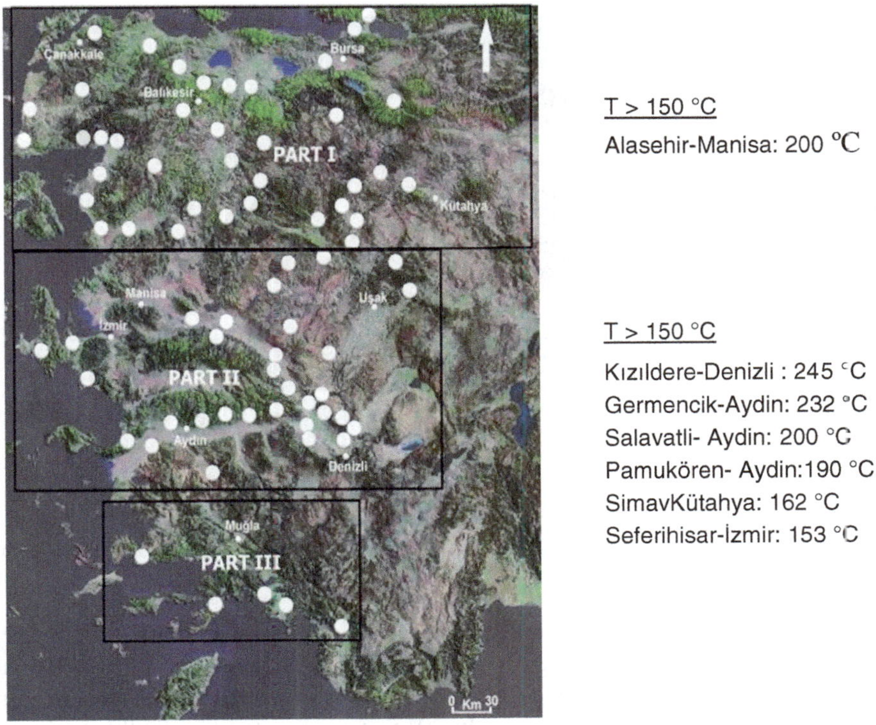

Fig. 11.2 Geothermal sources in Western Anatolia (Tut Haklıdır 2007)

power plants also put into use, the geothermal power capacity increased to 350 and 390 MWe capacities which is still under project status, and the power generation capacity of Turkey will be expected over than 600 MWe in 2016.

11.3.1 High Enthalpy Geothermal Systems in Western Anatolia

The high enthalpy geothermal systems are represented as Part II in Fig. 11.2. This part is strongly affected by Aegean graben systems (Tut Haklıdır 2007). As a result of these graben structures, the crustal thinning is shown besides young volcanism at this part. With this reason, the heat flow is higher than other parts in Western Anatolia (Şalk et al. 2005).

Geothermal power generation is suitable at Part II for Western Anatolia (Fig. 11.3). There are 12 geothermal power plants that are still operated by different energy companies (such as Gürmat, Zorlu Energy, BM Holding, Mege, Çelikler Holding), and more than 10 geothermal power plants are still under construction status along Büyük Menderes Graben.

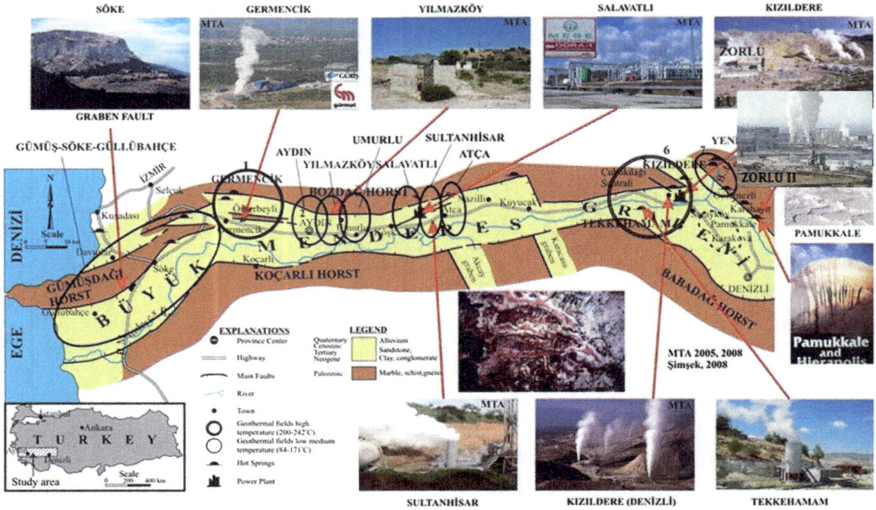

Fig. 11.3 Geothermal sources and power plants in Büyük Menderes Graben (modified from Şimşek 2010)

The second important geothermal zone is Alaşehir Graben at Part II region. The zone is also quite suitable for power generation, and geothermal power generation studies have been continued by different energy companies (such as Zorlu Energy, Türkerler, and Sanko) in this region. In 2015, it is expected that firstly, approximately 70 MWe geothermal power capacity will be put into use at this region.

Geothermal reservoir temperatures change between 180 and 245 °C (Tut Haklıdır et al. 2014), and these reservoir temperatures provide to use different power cycles such as single-, double-, triple-flash, binary and advanced cycles in Büyük Menderes Graben (Tut Haklıdır 2014).

11.3.1.1 Flash Cycle Power Plants in Western Anatolia

There are three flash-type geothermal power plants along Büyük Menderes Graben in 2014. The first one is Kızıldere-I 15 MWe net capacity single-flash power plant that is established in 1984 and still continues to produce power generation (Fig. 11.4a). The reservoir temperatures are around 200 °C, and 8 production wells feed the system to power generation (Tut Haklıdır and Kındap 2013). The second one is Germencik 47.4 gross capacity double-flash geothermal power plant, and it is put into use in 2009 (Haizlip Robinson et al. 2013). There are 8 production and 5 reinjection wells in the system (Fig. 11.4b).The maximum reservoir temperature is 232 °C, and it is suitable for dual separation from geothermal fluids in the region. High- (HP) and low-pressured (LP) separator systems provide more steam to power generation. The third and the biggest flash cycle power plant is Kızıldere-II 80 MWe gross capacity triple-flash + binary system. The system is also advanced and integrated system; thus, there is a 60 MWe capacity flash cycle turbine

Fig. 11.4 **a** Kızıldere-I GPP. **b** Germencik 47.4 GPP. **c** Kızıldere-II 80 MWe GPP

and 2 × 10 MWe binary (ORC) unites beside 50 MWt district heating capacity. Maximum reservoir temperature is 245 °C, and the geothermal fluid has enough enthalpy to 3 times separations in the system (Fig. 11.4c). There are 3 separator systems such as HP, IP, and LP, and some of steam from HP separator is sent also to binary units to produce more energy. The system is put into use in October 2013, and it is feed by 10 production wells and 8 reinjection wells in 2014. Kızıldere-I and Kızıldere-II geothermal power plants are combined systems; thus, they use same reinjection wells in Kızıldere geothermal field.

The new flash cycle geothermal power plant studies continue in Alaşehir geothermal field along Gediz Graben in 2014.

11.3.2 Medium- to Low-enthalpy Geothermal Systems in Western Anatolia

If geothermal reservoir temperatures are lower than 200 °C (160–200 °C), binary cycle must be more efficient to produce steam in geothermal systems. The system is compact, and installation of the equipment is easier than flash-type power plants and requires less area in a region. In Western Anatolia, there is good geothermal potential for power production both in Büyük Menderes and in Gediz Grabens. The first binary system has been started to operate with 8.5 MWe total gross, Dora I plant, in Salavatli, Aydin, in 2006. For the power generation, Ormat binary ORC cycle uses and the reservoir temperature is nearly 170 °C in the region (Karadaş and Akkurt 2014). In the system, an organic working fluid such as *n*-pentane is used to produce superheated steam. Other binary geothermal power plants are located in Pamukören, Salavatlı, and Germencik towns of Aydın along Büyük Menderes Graben. Only one binary plant is located in Sarayköy (Denizli) that it uses some of Kızıldere-II geothermal fluids to produce electricity before district heating of Sarayköy town.

In Salavatlı, there are 4 binary power plants: Dora I, Dora II (11.5 MWe), Dora III (34 MWe), and Dora IV (17 MWe) and last two binary units have been started to produce power 2013–2014 in the region.

In Germencik, there are 3 binary ORC geothermal power plants which are İrem (20 MWe), Sinem (24 MWe), and Deniz (24 MWe) (Parlaktuna et al. 2013).

In Pamukören town, there is a 45 MWe (2 × 22.5) binary ORC geothermal power plant. The geothermal power plant started to operate at the end of 2013.

Except from Part II region in Fig. 11.2, there is only a binary geothermal power plant in Tuzla (Çanakkale) and it is located in Northern Agean in Part I in Fig. 11.2. The reservoir temperature is around 170 °C, and 7 MWe gross capacity geothermal power plant has been generating electricity since 2010 in Tuzla geothermal field (Karadaş and Akkurt 2014). The operation of the system is harder than other binary power plants because of near-the-sea conditions in Western Anatolia.

In Gediz Graben, nearly 34 MWe capacity binary geothermal power plants, studies have been continued and it is expected they will be started to produce steam in 2015–2016 period in the region.

11.3.2.1 Geothermal–Solar Hybrid System in Western Anatolia

The first geothermal-solar hybrid system is built in Gümüşköy, Aydın, in Turkey. The geothermal power plant is 13.2 MWe net capacity ORC binary system, and it is extended by concentrated solar power system (Kuyumcu et al. 2013). The geothermal part is started to operate second part of the 2013, and reservoir temperature is around 165 °C in Gümüşköy. In the system, both ORC and parabolic trough collectors will be used to produce steam to increase efficiency of the system (Fig. 11.5). Kuyumcu et al. (2013) is declared that the heat transfer fluid or water in absorber pipes is firstly heated and after that pumped to the steam turbine-generator system and produce electricity. This system will be a first geothermal hybrid system in Turkey.

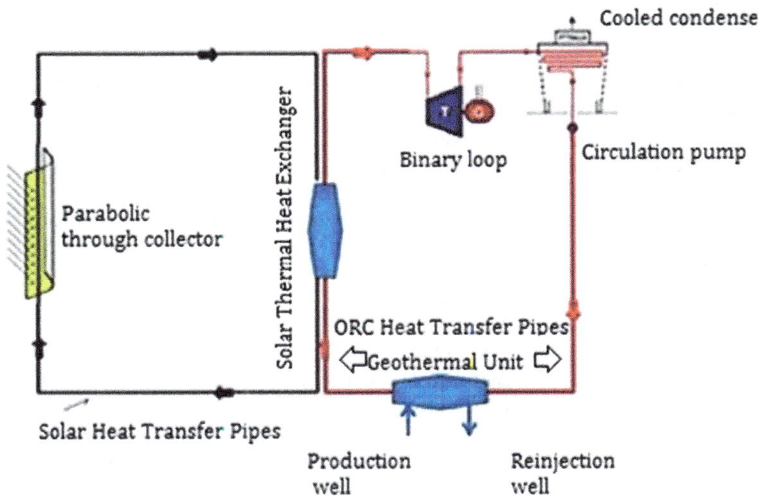

Fig. 11.5 Gümüşköy (Aydın) geothermal–solar hybrid power system (Kuyumcu et al. 2013)

11.4 Future Perspective of Geothermal Power in Turkey

Turkey has substantial geothermal energy capacity because of geological position in the world. Although almost each part of Turkey has geothermal sources, the discovered geothermal reservoir temperatures are higher than eastern part and suitable for power generation in Western Anatolia. Especially Büyük Menderes and Gediz Graben regions have great geothermal potential to produce electricity in Western Anatolia. Depending on the temperatures, different thermodynamic cycles can be used for power generation. While flash cycles which requires more than 200 °C, binary (ORC) system needs less than 200 °C to produce steam effectively. With this reason, there are different geothermal power generation systems such as single-, double- and triple-flash and binary systems.

In 2014, nearly 350 MWe capacity is provided from 12 geothermal fields in Western Anatolia. It is expected that the number will be increased to nearly 600 MWe with completion of new power generation projects in 2016 in Western Anatolia.

Geothermal exploration studies have been gone in both Western and Eastern Anatolia in Turkey. Even if the hottest geothermal reservoirs are discovered in the country, binary cycle technologies (such as ORC and Kalina) will be good options to power generation for new geothermal fields. Power generation from geothermal energy will be quite important because of increasing energy demand and energy prices in future. This energy source will provide great contribution to Turkey's energy demand among the other renewable energy sources with the high energy efficiency and independent of climate conditions of the world of future.

References

Bertani, R. (2010). Geothermal Power Generation in the World 2005–2010 Update Report. World Geothermal Congress Proceedings, April 25–29, 2010, Bali-Indonesia.

Bozkurt, E. (2001). Neotectonics of Turkey—a synthesis. *Geodinamica Acta, 14*, 3–30.

Calpine Corporation. (2014). Geysers by the Numbers (p. 1). The Geysers Geothermal Field 2013 Statistics. Published by Calpine Corporation, CA-USA.

DiPippo, R. (2005). *Geothermal power plants: Principles, applications and case studies* (470 p.). Oxfordshire: Elsevier Advanced Technology.

Dumas, P. (2013). EGEC Geothermal Market Report 2012. European Geothermal Congress Proceedings, June 3–7, 2013, Pisa-Italy.

Enel. (2013). *Geothermal energy* (24 p.). Rome: Enel Green Power.

Haizlip Robinson, J., Tut Haklıdır, F. S., & Garg, S. K. (2013). Comparison of reservoir conditions in high noncondensible gas geothermal systems. In Proceedings of the 38th Stanford Workshop Geothermal Reservoir Engineering, CA-USA.

Karadaş, M., & Akkurt, G. G. (2014). Rapid development of geothermal power generation in Turkey. In: A. Baba, J. Bundschuh, & D. Chandrasekharam (Eds.), *Geothermal systems and energy resources: Turkey and Greece* (pp. 197–224). Boca Raton: CRC Press.

Kuyumcu, Ö. Ç., Solaroğlu, D. U., Serin, O., Atalay, O., & Akar, S. (2013). Gümüşköy geothermal energy power plant: Current status. 19th International Energy and Environment Fair Conference Proceedings (pp. 110–113), Istanbul-Turkey.

Lund, W. L., Freeston, D. H., & Boyd, T. L. (2010). Direct utilization of geothermal energy 2010 worldwide review. World Geothermal Congress Proceedings, April 25–29, 2010, Bali-Indonesia.

Parlaktuna, M., Mertoğlu, O., Şimşek, Ş., Paksoy, H., & Başarır, N. (2013). Geothermal country update report of turkey (2010–2013). European Geothermal Congress Proceedings, June 3–7, 2013, Pisa-Italy.

Şalk, M., Pamukçu, O., & Kaftan, İ. (2005). Determination of the curie point depth an heat flow from Magsat Data of Western Anatolia. *Journal of Balkan Geophysical Society, 8*(4), 149–160.

Satman, A. (2013). Geothermal energy in Turkey and the world. 11th National Sanitary Engineering Proceedings-Geothermal Energy Seminar (pp. 3–20), April 17–20, 2013, Izmir-Turkey.

Şimşek, Ş. (2010). Exploration experiences on geothermal energy in Turkey. Der Geothermiekongress, November 17–19, 2010, Karlsruhe-Germany.

Tut Haklıdır, F. S. (2007). The geothermal geochemistry of Western Turkey. 23rd International Applied Geochemistry Symposium, Proceedings, June 14–19, 2007, Oviedo-Spain.

Tut Haklıdır, F. S. (2014). Geothermal energy sources and geothermal power plant technologies in Turkey. International Conference on Energy and Management (p. 88), Istanbul Bilgi University, Istanbul-Turkey.

Tut Haklıdır, F. S., & Kındap, A. (2013). The first discovered high enthalpy geothermal field in Büyük Menderes Graben: Kızıldere geothermal field with new 80 MWe power plant investment in Western Anatolia, Turkey. Europe Geothermal Conference Proceedings, June 3–8, 2013, Pisa-Italy.

Tut Haklıdır, F. S., Akyüz Dikbaş, A., & Şengün, R. (2014). Comparison of the characteristics of geothermal systems on the Western and Eastern Anatolia. 67th Turkey Geological Congress Proceedings, April 14–18, 2014, MTA, Ankara-Turkey.

Valdimarsson, P. (2011). Geothermal power plant cycles and main components. Short course on geothermal drilling, resource development and power plants (pp. 1–24). Organized by UNU-GTP and LaGeo, in Santa Tecla, El Salvador, January 16–22, 2011.

Chapter 12
Structural Health Monitoring of Multi-MW-Scale Wind Turbines by Non-contact Optical Measurement Techniques: An Application on a 2.5-MW Wind Turbine

Muammer Ozbek and Daniel J. Rixen

Abstract Optical measurement systems utilizing photogrammetry and/or laser interferometry are introduced as cost-efficient alternatives to the conventional wind turbine/farm health-monitoring systems that are currently in use. The proposed techniques are proven to provide an accurate measurement of the dynamic behavior of a 2.5-MW, 80-m-diameter wind turbine. Several measurements are taken on the test turbine by using four CCD cameras and one laser vibrometer, and the response of the turbine is monitored from a distance of 220 m. The results of the infield tests show that photogrammetry (also can be called as computer vision technique) enables the 3-D deformations of the rotor to be measured at 33 different points simultaneously with an average accuracy of ±25 mm while the turbine is rotating. Several important turbine modes can also be extracted from the recorded data. Similarly, laser interferometry (used for the parked turbine) provides very valuable information on the dynamic properties of the turbine structure. Twelve different turbine modes can be identified from the obtained response data. The measurements enable the detection of even very small parameter variations that can be encountered due to the changes in operation conditions. Optical measurement systems are very easily applied on an existing turbine since they do not require any cable installations for power supply and data transfer in the structure. Placement of some reflective stickers on the blades is the only preparation that is necessary and can be completed within a few hours for a large-scale commercial wind turbine. Since all the measurement systems are located on the ground, a possible problem can be detected and solved easily. Optical measurement systems, which consist of several CCD cameras and/or one laser vibrometer, can be used for monitoring several turbines, which enables the monitoring costs of the wind farm to reduce significantly.

M. Ozbek (✉)
Istanbul Bilgi University, Istanbul, Turkey
e-mail: muammer.ozbek@bilgi.edu.tr

D.J. Rixen
Technische Universität München, Munich, Germany
e-mail: rixen@tum.de

A.N. Bilge et al. (eds.), *Energy Systems and Management*,
Springer Proceedings in Energy, DOI 10.1007/978-3-319-16024-5_12

12.1 Introduction

Structural health monitoring can be described as the observation of a system over time by using some dynamic response characteristics (e.g., vibration frequencies, damping ratios, and mode shapes) and other similar indicators for obtaining information on the current state, health, and integrity of the structure. Type, location, and the extent of a possible damage due to aging, extreme loading, or severe operational/environmental conditions can be detected by using the changes observed in these dynamic parameters.

Several types of monitoring applications are currently in use by a wide variety of disciplines and have well-established mathematical theories and analysis algorithms.

Depending on their importance and the complexity of the loads acting on them, most civil engineering structures (e.g., bridges, high-rise buildings, oil platforms, dams, pipelines) as well as mechanical systems (machinery, transportation, or power-generating systems) are continuously or frequently observed for structural health monitoring and damage detection.

Wind turbines with very specific dynamic characteristics and challenging operating conditions are among the structures for which health monitoring plays a crucial role for ensuring safe and reliable operation and increasing the lifetime of the system. However, due to high sensor installation costs and technical difficulties in placing the sensors in existing structures, extensive monitoring is only applied during prototype testing period, which aims at validating the aeroelastic and dynamic stability of the turbine for various wind speeds and load cases.

Due to technical limitations in sensor installations, conventional systems can only be applied at certain locations on an existing turbine. Installations inside the tower are relatively easier, but some parts of the blades, especially last 20–25 m (close to tip), are generally not accessible and therefore cannot be instrumented. In practice, sensors are usually placed at the root regions of the blades. However, some motions such as bending of the rotor axis, small tilt, and yaw motion of the nacelle and teeter cannot be detected by the gauges placed at these locations. Besides, the response measured at the root region only may not be used for detecting a possible damage close to the tip of the blade. Blades are among the most important components of a wind turbine because failure of a blade can damage other blades, the turbine itself, and even other turbines located nearby. Blades can be damaged by moisture absorption, sleet, ultraviolet radiation, atmospheric corrosion, fatigue, wind gusts, or lightning strikes (Yang and Sun 2013). Therefore, it is necessary to perform continuous monitoring of wind turbine blades, for estimation of possible damage at very earlier stages before an irreversible damage occurs or the blade fails catastrophically (Raišutis et al. 2008).

12.2 Conventional Monitoring Technologies

State-of-the-art monitoring systems require different types of sensors to be installed in the turbine. Accelerometers, film sensors, and fiber-optic strain gauges are some of the most important sensors, which are widely used, for condition-based monitoring applications. The main drawbacks and technical limitations in using these measurement systems are discussed in detail in the following sections.

12.2.1 Accelerometers and Film Strain Gauges

Accelerometers and piezoelectric strain gauges are extensively used for measuring the dynamic response of structures and mechanical systems. However, these sensors are sensitive to lightning and electromagnetic fields (e.g., the magnetic field caused by the generator). Besides, some extra installations inside the blades such as placement of cables for power supply and data transfer are required for these applications. The signals from rotating sensors on the blades are transferred to stationary computer via slip rings or by radio/wireless transmission. Accelerometers and strain gauges can be applied on an existing turbine, but in practice, these sensors are only placed at some limited number of locations (tower or root region of the blades) which can be accessed easily. For large commercial turbines, the required installations and preparations (sensor calibration) may be very expensive and time-consuming.

In addition to the practical limitations mentioned above, the complicated nature of wind loads makes the efficient use of accelerometers in condition monitoring of wind turbines very difficult. The deflections under the action of wind loading can be considered as the sum of a static component due to average wind speed and a dynamic component due to turbulence (Tamura et al. 2002). Accelerometers cannot provide very accurate information about the static component. Therefore, several researchers suggest that in wind response measurements, accelerometers should be used together with other systems such as global positioning system (GPS), which are able to detect the static deformations accurately (Nakamura 2000; Breuer et al. 2002; Nickitopoulou et al. 2006). Although GPS–accelerometer combination is widely used to monitor the response of several structures such as high-rise buildings and bridges, it cannot easily be applied to wind turbines because of the technical difficulties in placing GPS sensors in the blade and reduced measurement accuracy due to secondary reflections caused by the rotational effects.

12.2.2 Fiber-Optic Strain Gauges

Fiber-optic strain gauges are proposed to be a promising alternative to accelerometers and conventional strain gauges since optical sensors are not prone to electromagnetic fields or lightning. However, it is reported that some feasibility tests are still needed to ensure the effective and cost-efficient use of this measurement system. The factors affecting the performance of fiber-optic sensors such as sensitivity to humidity and temperature variations and the corresponding error compensation methods should also be investigated further (Schroeder et al. 2006). Similarly, additional long-term durability tests are required to determine whether the bonding between optical fiber and composite blade material deteriorates over time due to repetitive loading and severe environmental factors or not.

Fiber-optic strain gauges are expected to provide a high spatial resolution, but installation costs significantly increase depending on the number of sensors. Besides, high-capacity decoders are needed to be able to acquire data from many sensors simultaneously resulting in a further increase in the hardware costs. Fiber-optic sensors can be applied throughout the blade only if the installation is performed during the manufacturing stage in the factory. The system cannot be easily applied to existing turbines.

12.3 Non-contact Optical Measurement Systems

In this work, two non-contact optical measurement systems (photogrammetry and laser interferometry) are proposed to be promising and cost-efficient alternatives for condition-based health monitoring of wind turbines. Unlike conventional measurement systems (accelerometers, piezoelectric, or fiber-optic strain gauges), optical measurement techniques do not require any sensors to be placed on the turbine. However, some reflective markers should be placed (or painted) on the structure. These markers are made up of a retro-reflective material, which is 1000 times more reflective than the background blade material. Since the markers are in the form of very thin stickers, they do not have any effect on aerodynamic performance of the blades.

The markers, which are used as displacement sensors, can easily be placed on an existing turbine. No extra cable installations for data transfer and power supply are required inside the turbine. Therefore, compared to the conventional sensors, marker installations are very cost-efficient and can be completed within very short periods of time. Retro-reflective paints applied on the turbine components during the manufacturing stage in the factory can substitute these markers, which may result in a further decrease in the installation costs.

During the tests, a total of 55 markers were placed on the turbine (11 markers for each blade and 22 markers on the tower). Placement of the markers on the blade and their final distribution throughout the structure can be seen in Fig. 12.1a, b, respectively. Although the pictures shown in Fig. 12.1 were captured by a handheld digital camera using its flashlight only, the markers can be seen easily. It should be noted that it only took 2 professional people 6 h to place 55 markers on the turbine. Considering the fact that each marker acts as an independent sensor, it can be concluded that it is almost impossible to reach such a high sensor installation speed by using conventional sensor technologies.

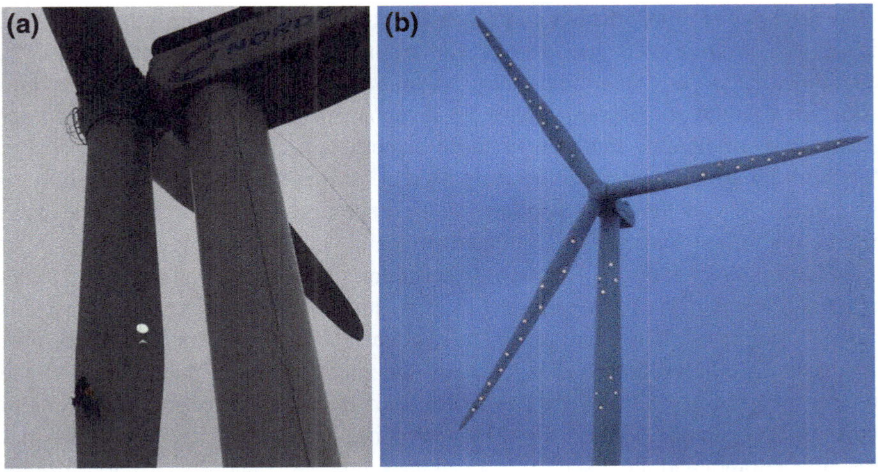

Fig. 12.1 Placement of markers and final layout

These markers are essential for both photogrammetry and laser interferometry, but they are used for different purposes in each method. Photogrammetry is a proven measurement technique based on the determination of the 3-D coordinates of the points on an object by using two or more images taken from different orientations and positions (Mikhail et al. 2001). Although each picture provides 2-D information only, very accurate 3-D information related to the coordinates and/or displacements of the object can be obtained by simultaneous processing of these images as shown in Fig. 12.2. In photogrammetry, markers are used as the targets to be tracked by the camera systems and all the targets can be tracked simultaneously.

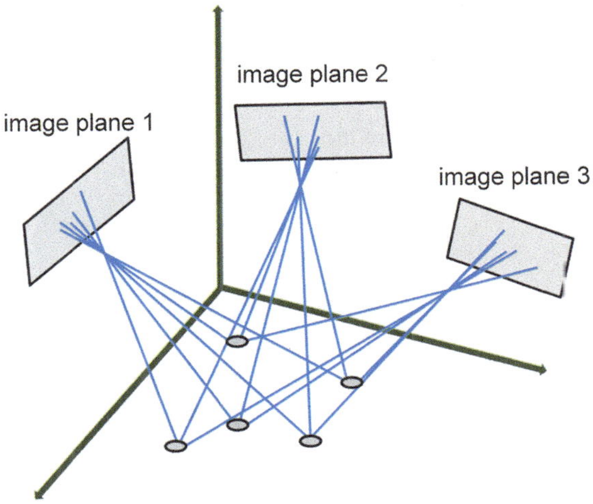

Fig. 12.2 3-D image formation in photogrammetry

Several applications of photogrammetric measurements are currently in use and proven to provide very accurate deformation measurements. The method is sometimes called as "videogrammetry" (which implies that sequences of the pictures are used to monitor the dynamic response of an object) or "stereophotogrammetry" (indicating that two or more cameras are employed simultaneously). Although photogrammetry is efficiently used at smaller scales by a wide variety of disciplines, this method was applied for the very first time to a MW-scale wind turbine within the scope of this research project.

In laser interferometry, a laser vibrometer continuously sends a laser beam to the target and receives the beam reflected from its surface. If the object is moving, this causes a frequency change and phase shift between the sent and the reflected beams. By detecting this frequency change (using Doppler principle), the velocity of the moving object can be found. If the object itself has a reflective surface, no extra retro-reflective markers are needed. However, because the blade material was not reflective enough and the distance between the laser source and the turbine was very long (200 m), high-quality laser signals could only be acquired if the laser was targeted to the markers. Figure 12.3 shows the reflection of the laser beam from the marker on the blade.

Different from photogrammetry, laser vibrometer can only measure the motion of a single point at a time. However, it is still possible to successively measure all the markers distributed throughout the blade. During the tests, the laser interferometry measurements were taken by using a Polytec OFV 505 laser head and OFV 5000 controller with VD06 velocity decoder. These systems were located in the field at a distance of 200 m from the turbine. An super-long-range (SLR) lens, which enables an increased measurement range up to 300 m, was also required to take measurements from this distance. It should be noted that because it is very difficult to keep the laser on the same marker while the turbine is rotating, laser Doppler vibrometer (LDV) was only used for the measurements taken on the parked turbine. Similarly, photogrammetric measurements could not be conducted when the turbine was at parked condition because the low wind speeds could not excite the structure sufficiently resulting in high noise-to-signal ratios.

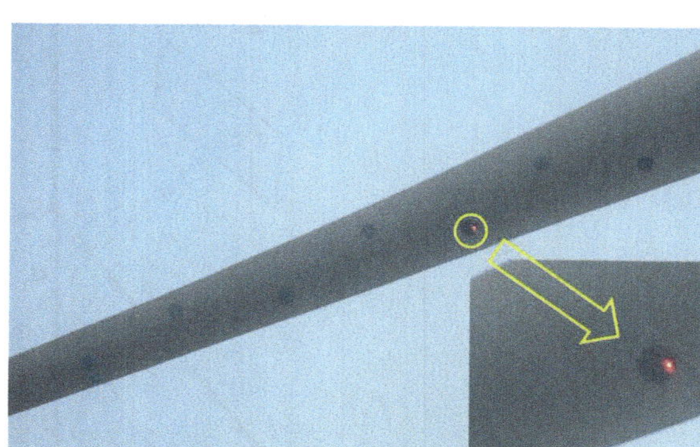

Fig. 12.3 Reflection of the laser beam from the marker on the blade

A third system (used as a reference), which has already been installed in the turbine as a part of a long-term wind load-monitoring campaign, consists of 6 strain gauges placed at the root region of the three blades (2 gauges per blade) and 2 strain gauges located at the tower base. These strain gauges are used to measure flapwise and edgewise vibration of the blades and fore–aft and side-to-side vibration of the tower at a sampling frequency of 32 Hz. All the data recorded by 3 different systems were then synchronized by using a GPS clock whose absolute time accuracy is approximately 10 ms. Considering the fact that frequencies that are expected to dominate the response of the wind turbine are mostly in low-frequency range (0–5 Hz), this accuracy can be considered as quite sufficient.

12.4 Laser Vibrometer Measurements on the Parked Turbine

In order to verify that laser optical sensors (vibrometers) can effectively be used in measuring the vibration response of the structure, it is required to demonstrate that all the vibration modes, which are identified from conventional sensor (strain gauge) data, can also be extracted from laser measurements.

Strain gauges installed in the turbine are placed at some specific locations (rotor, tower, and nacelle) and orientations (flapwise, edgewise) to ensure that all the modes can be observed. Indeed, gauges placed on the rotor may not detect the frequencies related to the tower modes. Similarly, blade sensors oriented in edgewise direction may not provide accurate information about vibration in flapwise direction. Complete description of the dynamic characteristics of the structure can only be obtained by combining the information coming from different sensors. Since LDV can measure the vibration of a single point at a time, this can only be provided by taking measurements at different locations on the turbine.

Table 12.1 summarizes the modal parameters (frequencies) calculated by using strain gauge and LDV measurements. Considering the strain gauge signals, it can be easily seen that some frequencies can be identified either from in-plane strains or from out-of-plane strains only but not from both and that some other frequencies can only be identified from tower signals. It can also be seen that all the frequencies can be identified by analyzing LDV measurements. However, these frequencies cannot be detected from a single data block only. LDV measurement contains a single channel data recorded on a specific marker at a time. The targeted marker may not be at a suitable location to detect some of the modes (or frequencies). Therefore, it is required to try several locations (markers) and time series to identify all the modes. However, the frequencies identified by using two different systems (strain gauge and laser) are always in good coherence.

Table 12.1 Modal parameters calculated for the parked turbine

Mode	Calculated frequency (Hz)	Blade strain out of plane	Blade strain in plane	Tower strain	LDV measurement
First longitudinal tower	0.345	X		X	X
First lateral tower	0.347		X	X	X
First BW flapwise (yaw)	0.902	X		X	X
First FW flapwise (tilt)	0.974	X		X	X
First symmetric flapwise	1.077	X		X	X
First BW edgewise (vertical)	1.834		X	X	X
First FW edgewise (horizontal)	1.855		X	X	X
Second BW flapwise (yaw)	2.430	X	X	X	X
Second FW flapwise (tilt)	2.311	X	X	X	X
Second symmetric flapwise	3.000	X		X	X
Tower torsion	6.154	X			X

12.5 Photogrammetry Measurements on the Rotating Turbine

A typical displacement time history measured in flapwise direction for the tip markers of 3 blades can be seen in Fig. 12.4. It can be seen that the tip of the blade can experience a relative displacement up to 102.4 cm during rotation (Ozbek et al. 2010; Ozbek and Rixen 2013). Frequency-domain analyses of these deformation time histories provide very useful information about the vibration characteristics of the turbine.

Figures 12.5 and 12.6 show the power spectral density (PSD) graphs of edgewise and flapwise direction blade vibration photogrammetric data, respectively. These figures are presented to provide a 3-D frequency distribution that also includes information related to the measurement location. The x-axis represents the frequencies normalized with respect to rotational frequency (P 0.28 Hz); therefore, it is dimensionless. The y-axis corresponds to the marker number. Marker 1 is placed at the blade root, whereas marker 10 is located at the tip of the blade. The z-axis represents the computed PSD amplitude. 1P and 2P components and the first edgewise mode can be recognized from Fig. 12.5.

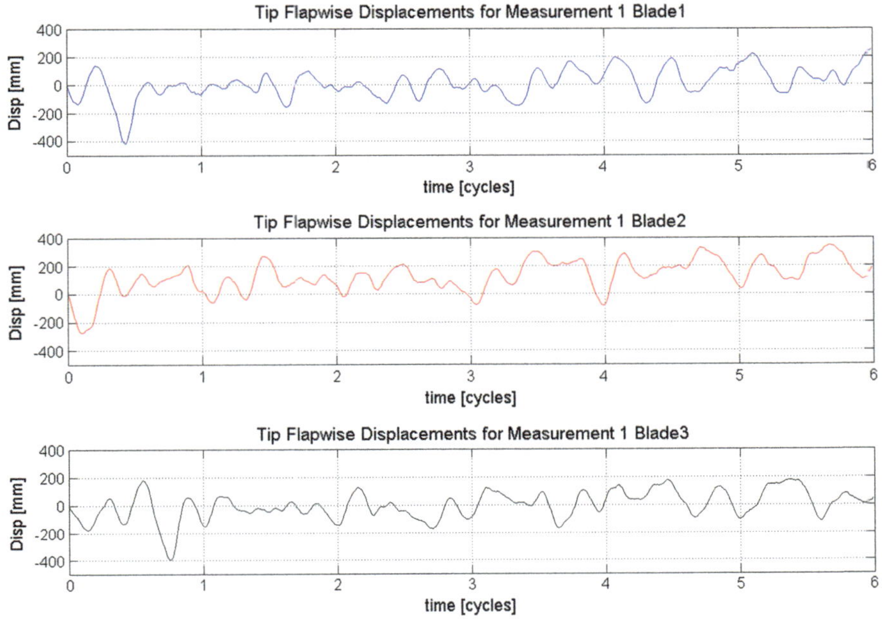

Fig. 12.4 Flapwise displacement time histories recorded for the tip markers

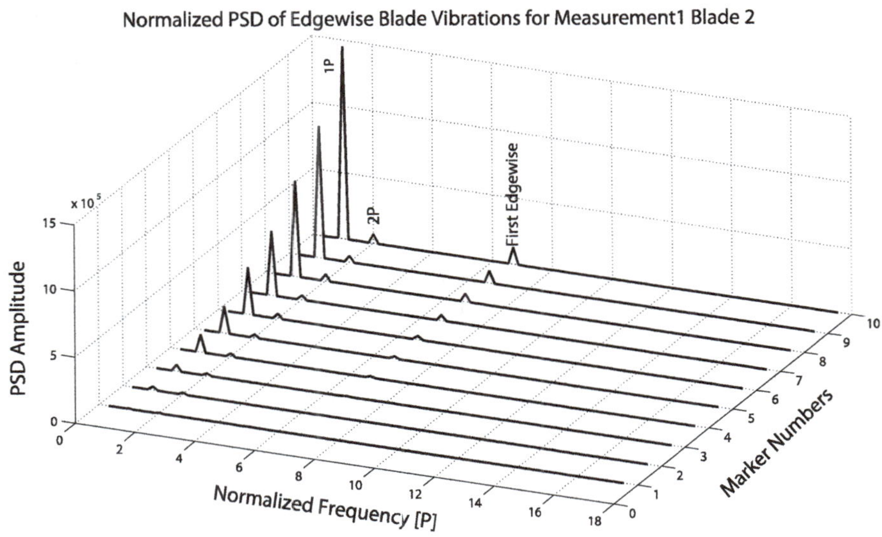

Fig. 12.5 Normalized PSD of edgewise blade vibration Blade 2

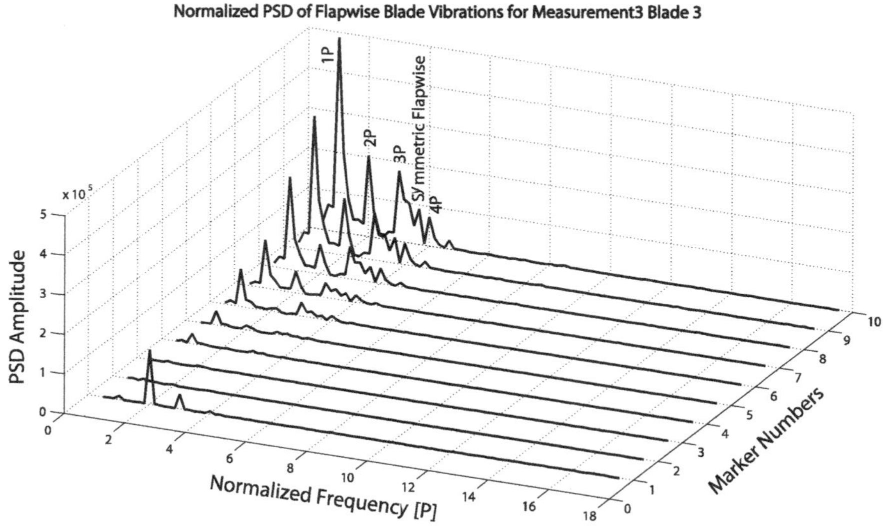

Fig. 12.6 Normalized PSD of flapwise blade vibration for Blade 3

As can be seen in Fig. 12.6, flapwise vibration data enable more frequencies to be identified. Integer multiples of rotational frequency up to 4P can be detected from the corresponding PSD graph. Besides these P components, the first flapwise mode can also be seen in Fig. 12.6. Since the response is mainly dominated by P harmonics, other turbine modes, which have relatively weaker modal participations in the response, cannot be identified easily from PSD plots only.

12.6 Conclusions

Wind turbines have very specific characteristics and challenging operating conditions. Although the optical measurement systems (including both the hardware and image processing software), calibration methods, and utilized operational modal analysis techniques were not specifically designed and optimized to be used for monitoring large wind turbines, the accuracy reached in this feasibility study is very promising. It is believed that this accuracy can easily be increased further by utilizing more specialized hardware and data processing methods (Ozbek et al. 2013).

Photogrammetry enables the deformations on the turbine to be measured with an average accuracy of ±25 mm from a measurement distance of 220 m. The data obtained from photogrammetry appeared to be suitable to identify the 1P–4P harmonics, as well as some of the lower eigenfrequencies of the blades in operation. First edgewise and first flapwise modes can easily be identified from Figs. 12.5 and 12.6, respectively.

Developments over the last decade have resulted in cheaper, higher-resolution, and more sensitive cameras, and in efficient software for photogrammetry, so that we believe that photogrammetry can be a versatile and cost-effective technique for health monitoring and dynamic validation of wind turbines.

The markers, which are used as displacement sensors, can easily be placed on an existing turbine. No extra cable installations for data transfer and power supply are required inside the structure. Therefore, compared to the conventional sensors (accelerometers, piezoelectric, or fiber-optic strain gauges), marker installations are very cost-efficient and can be completed within very short periods of time. Retro-reflective paints applied on the turbine components during manufacturing stage in the factory can substitute these markers, which may result in a further decrease in the installation costs.

A photogrammetric measurement system, which consists of several CCD cameras, flashes, and a central PC, can be reused for monitoring several turbines. If continuous monitoring is not required, all the turbines in a wind farm can be observed by using a single system. The measured data can be stored and used to build a condition-monitoring archive. Since all the measurement systems are located on the ground, a possible technical problem can be detected and solved easily.

Laser optical devices were also observed to provide very useful information about the dynamic characteristics of the turbine. The system parameters obtained by using LDV measurements were always consistent with those obtained by using 8 strain gauges installed on the structure (Ozbek et al. 2009).

All the frequencies that can be identified by using the 8 strain gauges can also be identified by using LDV measurements only. However, since one single point might not be sufficient to detect all the frequencies, different locations on the blade should be measured, which results in longer measurement periods.

Provided that the quality of the laser reflecting from the blade is sufficient, laser vibrometers can reach to very high accuracies (even in micron scale). The tests performed in this study show the high-quality signals can easily be obtained from a distance of 200 m by using some special lenses. The markers required for photogrammetry can also be used for laser measurements; therefore, no additional preparations on the turbine are needed.

Acknowledgments This research project was partly funded by the We@Sea research program, financed by the Dutch Ministry of Economical Affairs. The authors would like to thank Energy Research Center of the Netherlands (ECN) for providing the test turbine and the other technical equipment. The authors also acknowledge the extensive contribution of Pieter Schuer (GOM mbH), Wim Cuypers (GOM mbH), Theo W. Verbruggen (ECN), and Hans J.P. Verhoef (ECN) in organizing and performing the field tests.

References

Breuer, P., Chmielewski, T., Gorski, P., & Konopka, E. (2002). Application of GPS technology to measurements of displacements of high-rise structures due to weak winds. *Journal of Wind Engineering and Industrial Aerodynamics, 90*(3), 223–230.

Mikhail, E. M., Bethel, J. S., & McGlone, J. C. (2001). *Introduction to modern photogrammetry.* New York: Wiley.

Nakamura, S. (2000). GPS measurement of wind-induced suspension bridge girder displacements. *ASCE Journal of Structural Engineering, 126*(12), 1413–1419.

Nickitopoulou, A., Protopsalti, K., & Stiros, S. (2006). Monitoring dynamic and quasi-static deformations of large flexible engineering structures with GPS: Accuracy, limitations and promises. *Engineering Structures, 28*, 1471–1482.

Ozbek, M., & Rixen, D. J. (2013). Operational modal analysis of a 2.5 MW wind turbine using optical measurement techniques and strain gauges. *Wind Energy, 16*, 367–381.

Ozbek, M., Meng, F., & Rixen, D. J. (2013). Challenges in testing and monitoring the in-operation vibration characteristics of wind turbines. *MSSP Mechanical Systems and Signal Processing, 41*, 649–666.

Ozbek, M., Rixen, D. J., Erne, O., & Sanow, G. (2010). Feasibility of monitoring large wind turbines using photogrammetry. *Energy, 35*, 4802–4811.

Ozbek, M., Rixen, D. J., Verbruggen, T. W. (2009). Remote monitoring of wind turbine dynamics by laser interferometry: Phase1. In: Proceedings of the 27th International Modal Analysis Conference, Orlando, Florida.

Raišutis, R., Jasiūnienė, E., Šliteris, R., & Vladišauskas, A. (2008). The review of non-destructive testing techniques suitable for inspection of the wind turbine blades. *Non-Destructive Testing and Condition Monitoring, 63*(1).

Schroeder, K., Ecke, W., Apitz, J., Lembke, E., & Lenschow, G. (2006). A fiber Bragg grating sensor system monitors operational load in a wind turbine rotor blade. *Measurement Science & Technology, 17*, 1167–1172.

Tamura, Y., Matsui, M., Pagnini, L. C., Ishibashi, R., & Yoshida, A. (2002). Measurement of wind-induced response of buildings using RTK-GPS. *Journal of Wind Engineering and Industrial Aerodynamics, 90*, 1783–1793.

Yang, B., & Sun, D. (2013). Testing, inspecting and monitoring technologies for wind turbine blades: A survey. *Renewable and Sustainable Energy Reviews, 22*, 515–526.

Chapter 13
Stability Control of Wind Turbines for Varying Operating Conditions Through Vibration Measurements

Muammer Ozbek and Daniel J. Rixen

Abstract Wind turbines have very specific characteristics and challenging operating conditions. Contemporary MW-scale turbines are usually designed to be operational for wind speeds between 4 and 25 m/s. In order to reach this goal, most turbines utilize active pitch control mechanisms where angle of the blade (pitch angle) is changed as a function of wind speed. Similarly, the whole rotor is rotated toward the effective wind direction by using the yaw mechanism. The ability of the turbine to adapt to the changes in operating conditions plays a crucial role in ensuring maximum energy production and the safety of the structure during extreme wind loads. This, on the other hand, makes it more difficult to investigate the system from dynamic analysis point of view. Unexpected resonance problems due to dynamic interactions among aeroelastic modes and/or excitation forces can always be encountered. Therefore, within the design wind speed interval, for each velocity increment, it has to be proven that there are no risks of resonance problems and that the structure is dynamically stable. This work aims at presenting the results of the dynamic stability analyses performed on a 2.5-MW, 80-m-diameter wind turbine. Within the scope of the research, the system parameters were extracted by using the in-operation vibration data recorded for various wind speeds and operating conditions. The data acquired by 8 strain gauges (2 sensors on each blade and 2 sensors on the tower) installed on the turbine were analyzed by using operational modal analysis (OMA) methods, while several turbine parameters (eigenfrequencies and damping ratios) were extracted. The obtained system parameters were then qualitatively compared with the results presented in a study from the literature, which includes both aeroelastic simulations and in-field measurements performed on a similar size and capacity wind turbine.

M. Ozbek (✉)
Istanbul Bilgi University, Istanbul, Turkey
e-mail: muammer.ozbek@bilgi.edu.tr

D.J. Rixen
Technische Universität München, Munich, Germany
e-mail: rixen@tum.de

A.N. Bilge et al. (eds.), *Energy Systems and Management*,
Springer Proceedings in Energy, DOI 10.1007/978-3-319-16024-5_13

13.1 Introduction

Growing energy demands require wind turbine manufacturers to design more efficient and higher-capacity wind turbines which inevitably results in larger and larger new models to be put into service. However, an important consequence of this increase in size and flexibility of the structure is the complicated dynamic interaction between different parts of the turbine. Motion of the blades interacts with aerodynamic forces, electromagnetic forces in the generator, and the structural dynamics of several turbine components (drive train, nacelle, and tower). Understanding these dynamic interactions, the corresponding structural behavior and response characteristics are essential for optimizing the energy produced, ensuring safe and reliable operation and increasing the lifetime of the system. This requires improving the design methodologies and in-operation control strategies.

Therefore, more attention is paid to developing theoretical models for estimating the behavior of new wind turbines. Contemporary aeroelastic simulation tools coupled with structural dynamics models enable designers to detect, understand, and solve most of the possible problems at very early stages and optimize their designs.

Considering the fact that only the models based on real response measurements are able to represent the complicated interactions among different parts of the structure, several tests have been applied on both parked and rotating turbines. However, conventional dynamic testing techniques based on the exciting structure at several locations with sufficient force amplitudes cannot be easily applied to these challenging structures due to their size and the technical difficulties in providing very large forces that are required to reach sufficient excitation levels.

Standard wind turbine testing includes estimation of the structural frequencies and damping of the turbine modes from manual peak-picking from frequency response spectra of measured response signals, or from the decaying response after exciting the structure through step relaxation or clamping of the brake (Carne et al. 1988; Molenaar 2003; Griffith et al. 2010; Osgood et al. 2010). However, estimations are often performed on turbines at parked condition. Although these estimated modal parameters are mostly related to the turbine structure and do not include aerodynamic effects that dominate the aeroelastic modes of an operating turbine, frequencies and damping ratios of the lower turbine modes are important for tuning and validation of numerical models and for the verification of the prototype design (Carne and James 2010; Hansen et al. 2006).

Carne et al. (1982) extracted natural frequencies and damping ratios of operational turbine modes by applying the step relaxation method on a small (2 m tall) rotating vertical axis wind turbine. The measured input excitation together with the recorded response enabled the authors to calculate frequency response functions (FRFs) and to estimate modal parameters for both parked condition and several rotation speeds.

Carne et al. (1988) also tested a 110-m-tall vertical axis wind turbine at parked condition by using the same step relaxation technique. The step excitation included input forces of 45 and 135 kN applied on one of the blades and the tower, respectively. As in the previous example, the authors calculated FRFs by using the measured input forces and response. However, in this work, Carne et al. also applied an initial approach of operational modal analysis (OMA) where the forces acting on the structure are not required to be measured and modal analyses are solely based on the recorded response. The authors compared the results obtained by using conventional FRFs and OMA approach and reported that they obtained a good coherence between the modal parameters extracted by the two different methods.

In the following years, several researchers also applied step relaxation technique to test turbines at parked condition. Molenaar (2003) performed similar tests on a 750-kW, 50-m-tall horizontal axis turbine by exciting the structure at parked condition with sudden release of a pretensioned cable loaded up to 40 kN.

Griffith et al. (2010) also conducted in-field tests on a 60-kW, 25-m-tall vertical axis wind turbine at parked condition by using impact hammer, step relaxation, and ambient wind excitation and compared the results obtained from various excitation methods.

Osgood et al. (2010) performed similar tests on a 600-kW, 37-m-tall horizontal axis turbine by exciting the structure at parked condition through shakers, which are connected to the tower by cables. Extracted modal parameters were then compared with those obtained by OMA methods while turbine was vibrating under the action of ambient wind forces.

Although step relaxation is successfully applied on wind turbines at parked condition, it is relatively difficult and time-consuming to use the same method for rotating turbines. The system involves specific mechanisms to be installed on the turbine to ensure the sudden release of pretensioned cables. The forces needed to excite a large commercial MW-size turbine with sufficient levels of energy can be very large (even larger than the 135 kN forces mentioned above). Besides, the device has to be reloaded for every input, which means bringing the turbine down to parked condition, reloading the device, and waiting for the turbine to reach a certain rotation speed. If numerous tests are planned to be performed for several wind speeds, this method can be very costly and time-consuming (Carne and James 2010). In fact, it is this time requirement that motivated researchers to look for alternatives to step relaxation, finally resulting in the development of new OMA methods.

Researchers (Hansen et al. 2006; Thomsen 2002) also tried to use different excitation techniques by assuming that a turbine mode can be excited by a harmonic force at its natural frequency, whereby the decaying response after the end of excitation gives an estimate of the damping. Simulations show that turbine vibrations related to several modes can be excited by blade pitch and generator torque variations and eccentric rotating masses placed on the turbine. However, results of the in-field tests performed on wind turbines showed that it is not possible to achieve the required pitch amplitudes to excite the modes with high modal frequency or high damping ratio due to the limited capacity of electrical pitch actuators. On the other hand, excited turbine vibrations are not pure modal vibrations and the estimated damping is therefore not the actual modal damping. In particular, for systems having vibration modes with similar frequencies, but different damping ratios, it is not possible to isolate a certain mode only and aerodynamic damping values cannot be estimated well because of the energy transfer between different modes (Hansen et al. 2006).

OMA tools, a common representation used for several analysis methods which do not require the forces acting on the system to be measured, can be a solution to all these problems. Since estimation of the modal parameters is solely based on the use of measured response signals, these methods can be easily and efficiently used to extract the dynamic properties of these large structures excited by natural environmental inputs (winds). Indeed, early versions of OMA tools were specifically developed to overcome the problems mentioned above and have been in use since early 1990s (James et al. 1992, 1993, 1995, 1996). Some of the researchers (Carne et al. 1988; Griffith et al. 2010; Osgood et al. 2010) mentioned above have also successfully used OMA methods and have reported that they have obtained very good coherence between the modal parameters identified by OMA and the conventional experimental modal analysis techniques. A more comprehensive review of the history and development of this technique can be found in the work by Carne and James (2010).

13.2 Test Turbine

Tests were conducted on a pitch-controlled, variable speed Nordex N80 wind turbine with a rated power of 2.5 MW. The turbine has a rotor diameter and tower height of 80 m. Measurements were taken at Energy Research Center of the Netherlands (ECN) wind turbine test site located in Wieringermeer, the Netherlands. More detailed information about the facilities of the test site can be found at the related Web site (ECNWEB).[1]

The reference turbine used for qualitative comparison, General Electric NM80, is also a pitch-regulated, variable speed wind turbine with a rotor diameter of 80 m. This turbine has a rated power of 2.75 MW and is used as a test case for validation of new aeroelastic stability tools developed within the scope of European Commission-supported STABCON project (Hansen et al. 2006).

13.3 Analysis Results and Identified System Parameters

Researchers (Hansen et al. 2006; Ozbek et al. 2013; Chauhan et al. 2009) agree on the fact that performing modal analysis on a rotating turbine is much more challenging than performing the same analysis on a parked turbine due to the facts that;

---

[1] ECN Energy Research Center of the Netherlands. http://www.ecn.nl/units/wind/wind-turbine-testing/.

- For a rotating wind turbine, some of the important turbine modes have very high aeroelastic damping ratios ranging between 10 and 30 % (in terms of critical damping ratio) which makes them very difficult to be detected by most of the identification algorithms that are currently in use. Aeroelastic damping is a combination of both structural and aerodynamic dampings but mostly dominated by the aerodynamic component caused by rotation of the blades. However, for a parked turbine, the aerodynamic component is small (at low wind speeds); therefore, identified damping is generally considered to be composed of only structural damping which is usually less than 1 %. On the other hand, some exceptions to this are also described in this section.
- For a rotating turbine, integer multiples of rotational frequency (also called P harmonics where P denotes the rotational frequency) always dominate the response of the structure. These frequencies can be effective up to 24P and sometimes coincide with the true eigenfrequencies of the system (Ozbek and Rixen 2013).
- Besides, for rotating turbines, these P harmonics cause violation of steady-state random excitation assumption which is one of the most important requirements of OMA algorithms.
- Another important assumption, time-invariant system requirement, is also difficult to accomplish for rotating wind turbines because of the rotation of the blades and yawing, pitching motion of the turbine. However, for parked turbines, all these motions of the different components are prevented which makes the time-invariant system assumption much easier to fulfill.

13.3.1 Tests on the Parked Turbine

This section summarizes the results of the analyses of measurements taken on the parked turbine. During the measurements, the turbine was kept at a fixed orientation and yawing motion was prevented by application of the yaw brakes. Blade pitch angles were fixed at zero degree where flapwise blade vibration exactly corresponds to motion out of the rotor plane. This is the same as the angle of the blade during rotation below rated wind speeds (<15 m/s for the test turbine). Similarly, the brakes were applied to prevent the movement of the rotor.

Table 13.1 summarizes the calculated modal parameters (frequencies and damping ratios). Frequencies and damping ratios were extracted by using the Natural Excitation Technique (NExT) approach (James et al. 1995, 1996), together with the least square complex exponential (LSCE) time-domain identification method. When the turbine starts rotating, the name of the mode changes to the one indicated in parentheses. The abbreviations FW and BW stand for forward and backward whirling, respectively. Damping ratios are given in terms of critical damping ratio.

These modal parameters are important for tuning and validation of numerical models and for the verification of prototype designs. They can also be used for

Table 13.1 Modal parameters calculated for the parked turbine

Mode	Frequency (Hz)	Damping
First fore–aft tower	0.345	0.003
First side-to-side tower	0.347	0.003–0.009
First yaw (BW flapwise)	0.902	0.010–0.020
First tilt (FW flapwise)	0.974	0.011–0.020
First symmetric flapwise	1.077	0.010–0.020
First vertical edgewise (BW)	1.834	0.004
First horizontal edgewise (FW)	1.855	0.004
Second tilt (FW flapwise)	2.311	0.005
Second yaw (BW flapwise)	2.430	0.004
Second symmetric flapwise	3.00	0.005
Second edgewise	6.36	0.005
Tower torsion mode (needs further verification)	6.154	0.005

health monitoring applications. As can be seen in the table, frequency values are relatively stable and do not change depending on the measurement block analyzed. However, damping values may differ slightly. The damping scatter encountered in the first flapwise and side-to-side tower modes is mostly related to the aerodynamic drag phenomenon (Carne and James 2010). Since the turbine is kept at a fixed orientation during the tests, the relative angle between the effective wind direction and the normal of the rotation plane continuously changes depending on the instantaneous wind direction resulting in a different aerodynamic coupling for each measurement. This also shows that it is not possible to completely eliminate the aerodynamic component of damping even for low wind speeds.

13.3.2 Tests on the Rotating Turbine

This section summarizes the results of the analyses of the measurements taken on the rotating turbine. During the test period, the response of the turbine was continuously measured by strain gauges. Therefore, modal parameters could be extracted for various operating conditions and wind speeds. Calculated modal parameters were then compared with the results presented by Hansen et al. (2006). The work mentioned includes the results of both aeroelastic simulations performed by the stability tool HAWCSstab (Hansen 2004; Riziotis et al. 2004; Marrant and van Holten 2004) and the real measurements taken on a wind turbine which has a similar size and capacity as the test turbine in our work. Therefore, some of the graphs presented below include the parameters extracted from our study and two additional graphs taken from the simulations and the measurements presented in the reference study (Hansen et al. 2006).

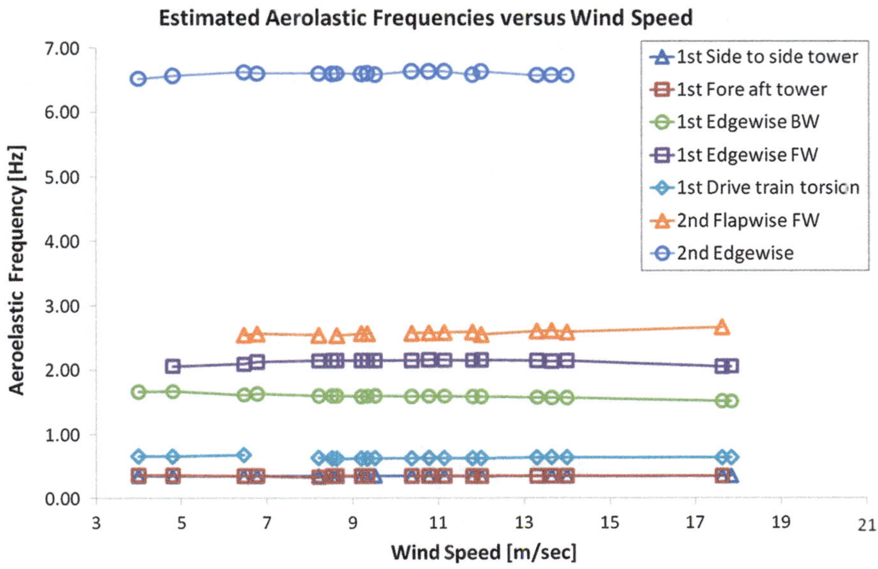

Fig. 13.1 Extracted eigenfrequencies

Figure 13.1 shows the aeroelastic frequencies we identified for different wind speeds. As can be seen in the figure, some of the modes extracted for the parked turbine (shown in Table 13.1) could not be detected for the rotating turbine. The first tilt (FW flapwise), first yaw (BW flapwise), and first symmetric flapwise modes could not be identified due to their very high damping ratios. Hansen et al. 2006 experienced the same problem and reported that these three flapwise modes have too high aerodynamic dampings for identification in response to the excitation by turbulence. Similarly, the second BW flapwise mode, which has a lower damping, could not be observed in the rotation data due to its weak modal participation in the overall motion.

Figure 13.2 shows the change in aeroelastic damping ratio calculated for side-to-side tower mode as a function of wind speed. Identified values are in a very good agreement with the HAWCStab simulation results both in terms of trend and magnitude. The values found are less than 1 % through different operating conditions and wind speeds. It should be noted that damping ratios are given in terms of critical damping ratios.

The same comparison is made for the fore–aft tower mode, and the results are shown in Fig. 13.3. Although the two tower modes have almost the same frequencies, aeroelastic damping calculated for the fore–aft mode is greater due to the motion of the tower in the direction perpendicular to rotor plane. The identified damping ratios are smaller than HAWCStab estimations, but can still be considered as close.

Compared with side-to-side mode, analyzing the fore–aft tower mode and estimating the corresponding modal parameters are more challenging. OMA tools utilized in this work are based on the use of correlation functions, which produce

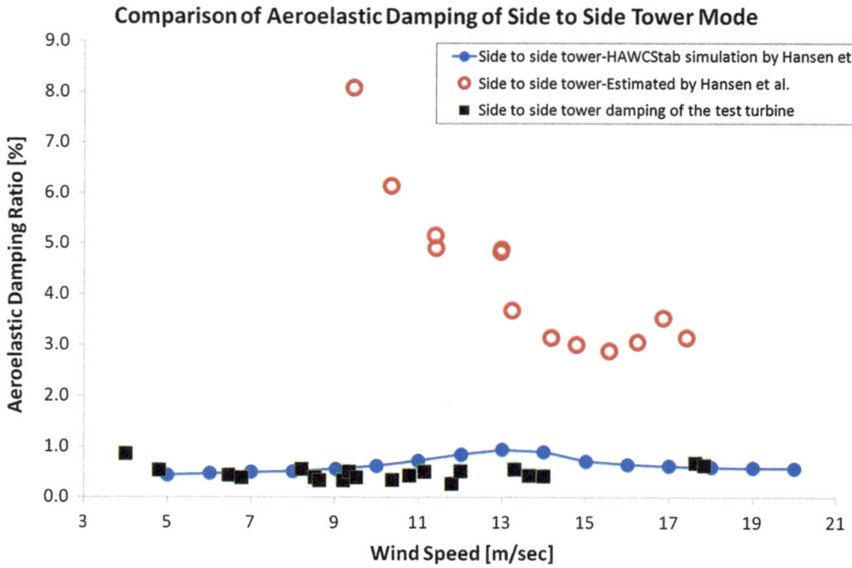

Fig. 13.2 Side-to-side tower-mode damping comparison

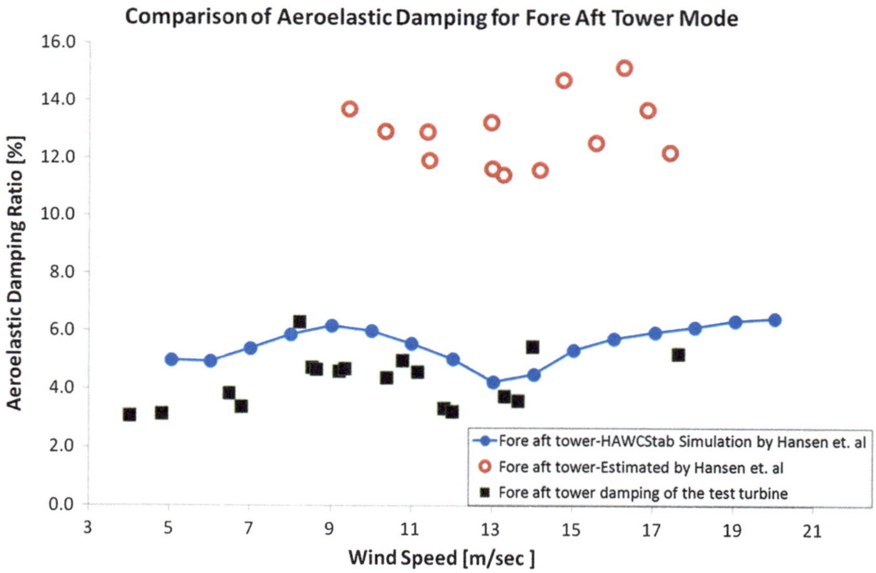

Fig. 13.3 Fore–aft tower-mode damping comparison

very accurate results in case of zero-average steady-state data series. However, changes in mean wind speed cause very slow variations in fore–aft tower movement and bending moment. These quasi-static variations cannot be fully eliminated by

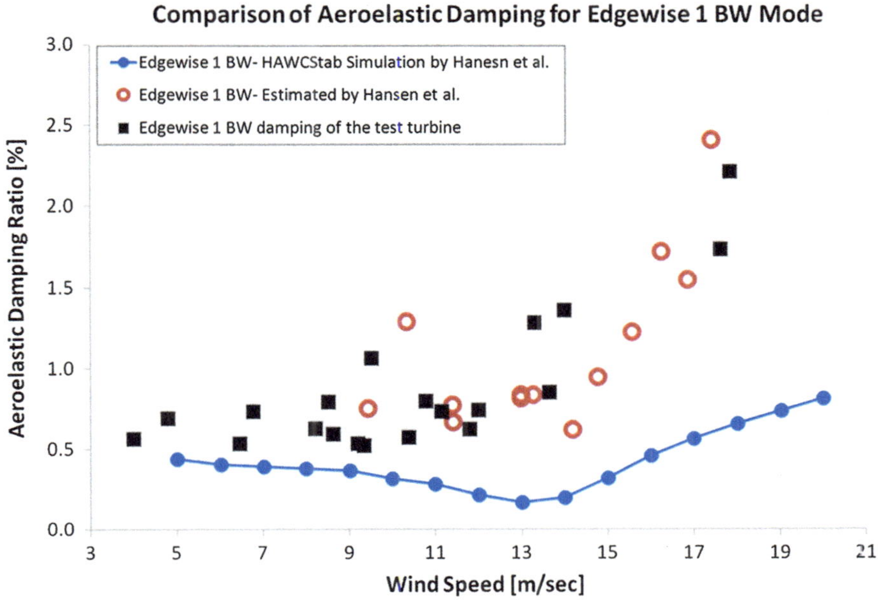

Fig. 13.4 The first edgewise BW-mode damping comparison

using conventional data processing techniques (such as detrending and filtering) and cause scatter in the estimated modal parameters. Therefore, the data series having a nonzero varying average should not be used in the analysis. Although such an approach limits the number of measurements that can be used in the analysis, it significantly increases the reliability of the extracted modal damping.

Comparison of aeroelastic damping ratios found for the first BW edgewise mode is shown in Fig. 13.4. The extracted damping ratios are slightly higher than the HAWCStab results, but are very close to the estimations made by Hansen et al. (2006) using in-field vibration data. Edgewise modes are very straightforward to identify because they have very high modal participation in the overall response of the turbine and low aeroelastic damping.

Unlike tower modes, they can be detected by using the strain gauges placed both on the blades and on the tower. Similarly, Fig. 13.5 displays the same damping comparison for the first edgewise FW mode. Acquired damping ratios are again very close to both simulations and estimations given by Hansen et al. (2006).

It should be noted that during the measurements that were used in modal analysis, operating conditions of the turbine (wind speed, rotor speed, and pitch angle) have to stay constant. Possible variations in these parameters result in a noisy input data and a scatter in identified modal parameters. Therefore, only the measurements corresponding to very low standard deviations of wind speed and rotor speed were selected and analyzed to obtain the results presented in this work. However, it should also be noted that some pitching activity was observed starting

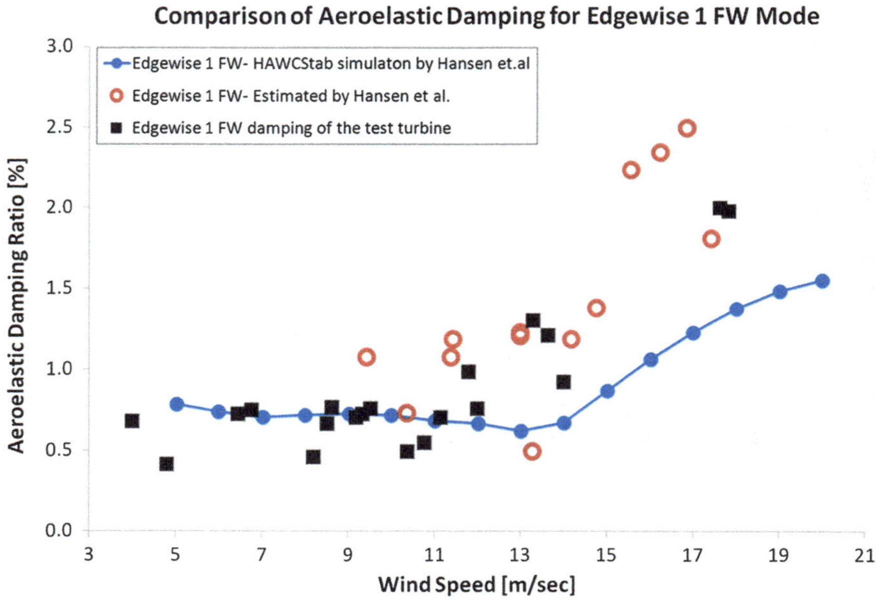

Fig. 13.5 The first edgewise BW-mode damping comparison

from the wind speeds close to the rated speed. As mentioned before, identification of modal parameters for these wind speeds can only be possible by tolerating some pitch activity, which definitely results in some scatter in the obtained values.

13.4 Conclusions

Identification of the modal parameters of wind turbines is very important for optimizing the energy produced, ensuring safe and reliable operation, and increasing the lifetime of the system. However, conventional dynamic testing techniques based on exciting the structure from several locations with sufficient force amplitudes cannot be easily applied to these challenging structures due to their size and the technical difficulties in providing very large forces that are required to reach sufficient excitation levels.

OMA tools, namely the analysis methods that do not require the forces acting on the system to be measured, can be a solution to these problems. Since estimation of the modal parameters is solely based on the use of measured response signals, these methods can easily and efficiently be used to extract the dynamic properties of these large structures excited by natural environmental inputs (winds).

Analyses performed by using OMA methods seem very promising in extracting the modal parameters. Within the scope of the research, twelve different turbine modes were successfully calculated from the measurements taken on the parked turbine.

Similarly, several important turbine modes could be identified from the in-operation measurements. Obtained results are in good coherence with those presented in similar studies in the literature. Performing modal analysis on a rotating turbine is much more challenging than performing the same analysis on a parked turbine due to the high aeroelastic damping of some important modes, rotational P harmonics that dominate the dynamic response, and the difficulties in fulfilling some important system identification assumptions such as time-invariant system and steady-state random excitation.

The turbine modes (first FW, BW, and symmetric flapwise modes), known to have very high aeroelastic damping ratios, could not be detected through the analyses of the rotation data. The second BW flapwise mode, which has a relatively lower damping, could not be extracted due to its low modal participation in the motion.

During the analyses, it was observed that frequency values are more easily identified and the calculated values are mostly stable and reliable. However, significant scatter can be encountered in estimated damping ratios. This scatter can be caused by physical factors such as the change in operating conditions or mathematical uncertainty related to the applied algorithms.

Acknowledgments This research project was partly funded by the We@Sea research program, financed by the Dutch Ministry of Economical Affairs. The authors would like to thank Energy Research Center of the Netherlands (ECN) for providing the test turbine and the other technical equipment. The authors also acknowledge the extensive contribution of Pieter Schuer (GOM mbH), Wim Cuypers (GOM mbH), Theo W. Verbruggen (ECN), and Hans J.P. Verhoef (ECN) in organizing and performing the field tests.

References

Carne, T. G., & James, G. H. (2010). The inception of OMA in the development of modal testing technology for wind turbines. *Mechanical Systems and Signal Processing, 24*, 1213–1226.

Carne, T. G., Lauffer, J. P., & Gomez, A. J. (1988). Modal testing of a very flexible 110 m wind turbine structure. In: Proceedings of the 6th International Modal Analysis Conference, Kissimmee, Florida, USA.

Carne, T. G., Lobitz, D. W., Nord, A. R., & Watson, R. A. (1982). Finite element analysis and modal testing of a rotating wind turbine. In: Proceedings of the AIAA 23rd Structures, Structural Dynamics and Materials Conference, New Orleans, Louisiana, USA.

Chauhan, S., Hansen, M. H., & Tcherniak, D. (2009). Application of operational modal analysis and blind source separation/independent component analysis techniques to wind turbines. In: Proceedings of the IMAC International Modal Analysis Conference, Orlando, FL.

Griffith, D. T., Mayes, R. L., & Hunter, P. S. (2010). Excitation methods for a 60 kW vertical axis wind turbine. In: Proceedings of the 28th International Modal Analysis Conference, Jacksonville, Florida, USA.

Hansen, M. H. (2004). Aeroelastic stability analysis of wind turbines using an eigenvalue approach. *Wind Energy, 7*, 133–143.

Hansen, M. H., Thomsen, K., & Fuglsang, P. (2006). Two methods for estimating aeroelastic damping of operational wind turbine modes from experiments. *Wind Energy, 9*, 179–191.

James, G. H., Carne, T. G., & Lauffer, J. P. (1992). Modal testing using natural excitation. In: Proceedings of the 10th International Modal Analysis Conference, San Diego, California.

James, G. H., Carne, T. G., & Lauffer, J. P. (1993). The natural excitation technique (NExT) for modal parameter extraction from operating wind turbines. Technical Report SAND92-1666, Sandia National Laboratories.

James, G. H., Carne, T. G., & Lauffer, J. P. (1995). The natural excitation technique (NExT) for modal parameter extraction from operating structures. *Journal of Analytical and Experimental Modal Analysis, 10*, 260–277.

James, G. H., Carne, T. G., & Veers, P. S. (1996). Damping measurements using operational data. *ASME Journal of Solar Energy Engineering, 118*, 190–193.

Marrant, B., & van Holten, T. (2004). System identification for the analysis of aeroelastic stability of wind turbine blades. In: Proceedings of European Wind Energy Conference, London.

Molenaar, D. P. (2003). Experimental modal analysis of a 750 kW wind turbine for structural model validation. In: Proceedings of the 41st Aerospace Sciences Meeting and Exhibit, Reno, Nevada.

Osgood, R., Bir, G., Mutha, H., Peeters, B., Luczak, M., & Sablon, G. (2010). Full-scale modal wind turbine tests: comparing shaker excitation with wind excitation. In: Proceedings of the 28th International Modal Analysis Conference, Jacksonville, Florida, USA.

Ozbek, M., Meng, F., & Rixen, D. J. (2013). Challenges in testing and monitoring the in-operation vibration characteristics of wind turbines. *MSSP Mechanical Systems and Signal Processing, 41*, 649–666.

Ozbek, M., & Rixen, D. J. (2013). Operational modal analysis of a 2.5 MW wind turbine using optical measurement techniques and strain gauges. *Wind Energy, 16*, 367–381.

Riziotis, V. A., Voutsinas, S. G., Politis, E. S., & Chaviaropoulos, P. K. (2004). Aeroelastic stability of wind turbines: The problem, the methods and the issues. *Wind Energy, 7*, 373–392.

Thomsen, K. (2002). Determination of damping of blade and tower vibrations. Technical Report, Riso National Laboratory.

Part II
Advanced Energy Materials

Chapter 14
Evaluation of HFO-1234YF as a Replacement for R134A in Frigorific Air Conditioning Systems

Mehmet Direk, Cuneyt Tunckal, Fikret Yuksel and Ozan Menlibar

Abstract The aim of this study was to compare and evaluate the frigorific air conditioning system using HFO-1234yf and R-134a as a refrigerant. For this aim, an experimental frigorific air conditioning system using both refrigerants was developed and refrigerated air was introduced into a refrigerated room. The performance parameters determined were the change of the air temperature in the condenser inlet and time. The performance of frigorific air conditioning system has been evaluated by applying energy analysis. Experiments were conducted for a standard frigorific air conditioning system using the R134a as a refrigerant. Airflow has been introduced to the refrigerated room for 60 min for each performance test. From the result for both refrigerant, the temperature gradient in time was comparable. The HFO-1234yf refrigerant can use the standard frigorific air conditioning system that is currently being used by the R134a refrigerant, without any changes needing to be made.

Nomenclature

COP	Coefficient of performance
h	Enthalpy (kJ kg^{-1})
$\dot{m}$	Mass flow rate (g s^{-1})
$\dot{Q}$	Cooling capacity (W)
$\dot{W}$	Power (W)

Subscripts

comp	Compressor
cond	Condenser
evap	Evaporator
in	Inlet

M. Direk (✉) · C. Tunckal · F. Yuksel · O. Menlibar
Yalova University, Yalova, Turkey
e-mail: mehmetdirek@hotmail.com

O. Menlibar
e-mail: ozanmenlibar@hotmail.com

A.N. Bilge et al. (eds.), *Energy Systems and Management*,
Springer Proceedings in Energy, DOI 10.1007/978-3-319-16024-5_14

out	Exit
r	Refrigerant
ref	Refrigerated room

14.1 Introduction

As stated by the montreal protocol, the use of chlorofluorocarbon (CFC) and hydrochlorofluorocarbon (HFC) refrigerants will be discontinued due to their hazardous nature. Some EU countries have pushed the date back earlier and are working on the design of a new air conditioning system using natural refrigerant. Countries have begun to research more environmentally friendly refrigerants rather than continue to use the harmful CFC and HFC refrigerants.

The European laws specify that beginning on January 1 of 2017, all automobiles equipped with air conditioning systems cannot be manufactured using greenhouse gases that have global warming potentials (GWP) greater than 150. HFO-1234yf could be implemented in existing air conditioning systems using R134a. HFO-1234yf is being considered as a possible replacement for R134a in automotive applications (Meyer 2008).

Meyer (2008) revealed that the cooling capacity and coefficients of performance (COP) of HFO-1234yf could be made relatively equal to the baseline R134a. Spatz and Minor (2008) demonstrated that there remain a couple of issues that are being investigated in order to replace R134a with HFO-1234yf in automotive applications. The effects of using improved components in air conditioning systems were studied experimentally by Petijean and Benouali (2010). They found that the condenser plays important role than the evaporator in the optimization of the HFO-1234yf performance.

Under the conditions for mobile air conditioners, the performance of both HFO-1234yf and R134a was investigated in a heat pump bench tester by Lee and Jung (2012). The results demonstrated that, because of its good environmental properties and reasonable performance, HFO-1234yf could potentially be seen as a long-term environmentally friendly solution for using in mobile air conditioners. They also revealed that energy efficiency devices are needed in all energy conversion area including mobile air conditioning system because of the global energy price rise. Since mobile air conditioning device is rising constantly, some ways such as the optimization of the heat exchanger circuitries should be carefully devised for R1234yf in the long run.

Navarro et al. (2013) compared HFO-1234yf, R134a and R290 for an open piston compressor at different operating conditions. Akram et al. (2013) investigation compared the lubricity of the two refrigerants, the environmentally friendly HFO-1234yf and the traditional R134a, for air conditioning compressor applications. Lee et al.'s (2013) study proposed replacing R134a in applications such as mobile air conditioners with a mixture of R134a and HFO-1234yf. They found that in the HFO-1234yf/R134a mixture when more R134a was added, the properties of flammability decrease, while compositions with R134a above 10 %, the mixture became non-flammable.

In this study, to evaluate the comparative experimental performances, a frigorific air conditioning system using HFO-1234yf and R134a was developed and refrigerated air was introduced into a refrigerated room. The effects of using R134a and HFO-1234yf, as a refrigerant in a frigorific air conditioning system, and their respective performances were determined comparatively.

14.2 Description of the Experimental Set-up

The experimental system is made from original components of a frigorific air conditioning system, as schematically shown in Fig. 14.1. It employs a seven-cylinder fixed-capacity swash-plate SANDEN SD 5750 W compressor at 2200 rpm, a 9500 W parallel-flow micro-channel condenser, a 5750 W laminated-type evaporator and thermostatic expansion valves. An experimental system was connected to a refrigerated room for the test. The room dimensions are 210 cm × 220 cm × 220 cm. This refrigerated room was insulated by 8-cm-thin isolation panels.

All lines in the refrigeration circuit of the system were made from copper tubing and insulated by elastomeric material. The condenser was inserted into separate air ducts of 1.2 m length. In order to provide the required airstreams in the air ducts, an axial fan was placed at the condenser ducts. These ducts also contained electric heaters located upstream of the condenser. The condenser electric heaters could be

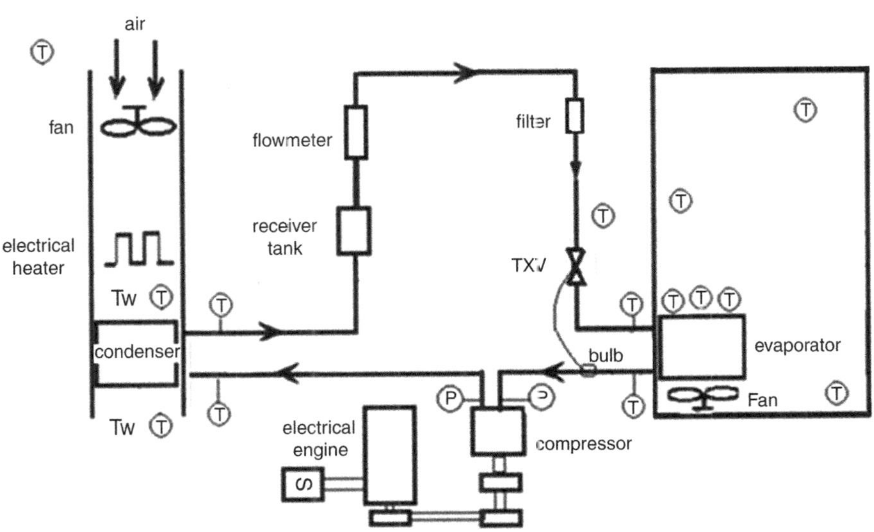

Fig. 14.1 Schematic illustration of the experimental set-up

Table 14.1 Characteristics of the instrumentation

Measured variable	Instrument	Range	Uncertainty
Temperature	K-type thermocouple	−50 with 200 °C	±%0.3
Airflow rate	Anemometer	0.1–15 m s^{-1}	±%3
Refrigerant mass flow rate	Turbine flowmeter	0–350 kg h^{-1}	±%0.1
Refrigerant pressure	Digital manifold	50 bar	0.5 % fs (±1 Digit)

controlled between 0 and 2 kW. To provide the required air temperature at the inlets of the condenser, the refrigeration circuit was charged with 1500 g R134a and HFO-1234yf. In order to gather data for the performance evaluation of the experimental frigorific system, some mechanical measurements were conducted on the system. The instruments and their locations are depicted in Fig. 14.1. The features of the instrumentation can be seen in Table 14.1.

The temperatures of the refrigerant and air at the inlet and exit of each component were measured by K-type thermocouples. The refrigerant pressures at the inlet and exit of the compressor were monitored by digital manifold. The values of most of the measured variables were acquired through a data acquisition system and recorded on a computer. The airstream discharged from the evaporator duct was supplied to the refrigerated room for cooldown tests. In order to measure the compartment temperatures, a thermocouple was located at the exit of the evaporator, while another one was suspended in the air close to the left wall, and the last one was suspended in the air close to the right wall.

The refrigerant flow path in the experimental system is illustrated in Fig. 14.1. The refrigerant passing through the condenser rejects heat to the ambient airstream. Then, the refrigerant condenses and leaves the condenser as sub-cooled liquid. After that, the refrigerant flow reaches the receiver tank, which keeps the unrequired refrigerant when the thermostatic expansion valve decreases the refrigerant flow rate at low-cooling loads reaches TXV located at the inlet of the evaporator. It enters the evaporator, where it rejects cool air taken from the refrigerated room, and leaves the evaporator as low-pressure superheated vapour. After, the refrigerant is directed to the compressor, which receives the low-pressure refrigerant vapour and compresses it to a high pressure. The performance of HFO-1234yf was tested in a standard R134a frigorific air conditioning without alterations being made.

14.3 Thermodynamic Analysis

The performance of the frigorific air conditioning system can be evaluated by using the first law of thermodynamics. The evaporator, the cooling capacity of the experimental frigorific air conditioning system, can be evaluated from

$$\dot{Q}_{evap} = \dot{m}_r \left(h_{evap,in} - h_{evap,out}\right) \tag{14.1}$$

Assuming that the compressor is adiabatic, the power absorbed by the refrigerant during the compression process can be expressed as follows:

$$\dot{W}_{comp} = \dot{m}_r \left(h_{comp,out} - h_{comp,in}\right) \tag{14.2}$$

The most common approach in determining a system's performance is to use COP. The ratios of the cooling capacities to the compressor power give the energetic performances of the system in the frigorific air conditioning. The COP of the frigorific air conditioning system defined as the ratio between cooling capacity and compressor power, i.e., can be determined.

$$COP = \frac{\dot{Q}_{evap}}{\dot{W}_{comp}} \tag{14.3}$$

14.4 Results and Discussion

The effect of the air temperature at the condenser inlet on certain of the steady-state performance parameters of the frigorific air conditioning system using HFO-1234yf and R-134a as a refrigerant is shown in Figs. 14.2 and 14.3. As is shown in Fig. 14.2, the steady-state cooling capacity usually increases on decreasing the temperatures of the airstreams entering the condenser. Higher cooling capacity has been provided with HFO-1234yf compared to the air conditioning systems using R134a. Figure 14.3 indicates that COP drops with air temperature entering into the condenser. This is due to the increasing cooling capacity and decreasing compressor power with decreasing $T_{cond,in}$. Consequently, COP for both refrigerants is closely comparable.

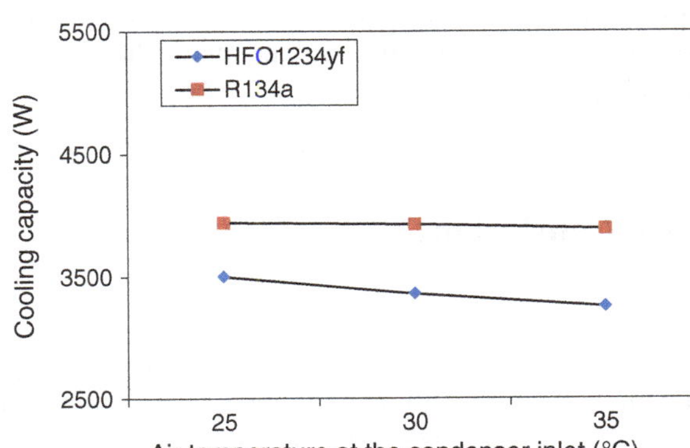

Fig. 14.2 The variation of the cooling capacity as a function of air temperature at the condenser inlet

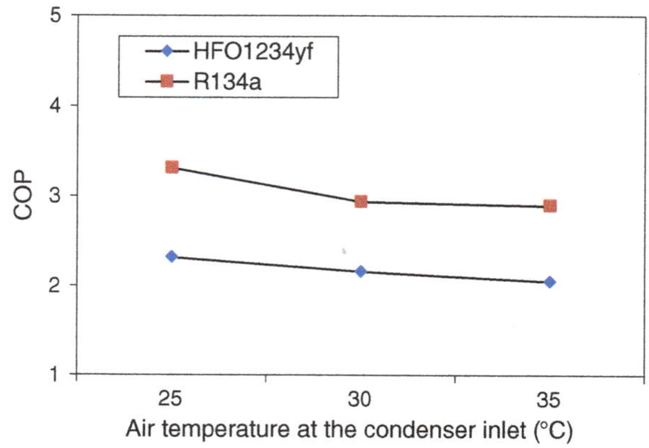

Fig. 14.3 The variation of the COP as a function of air temperature at the condenser inlet

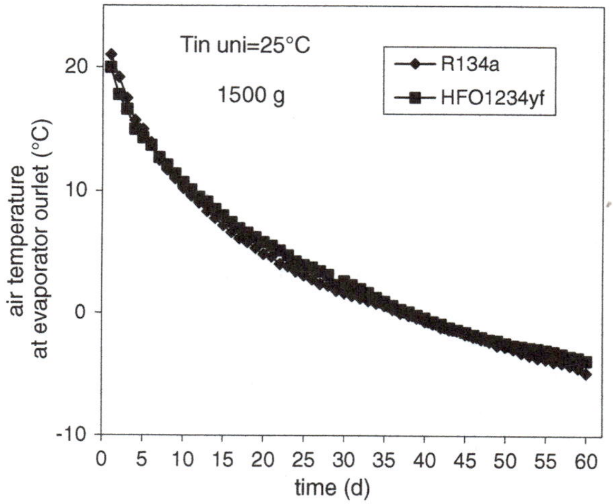

Fig. 14.4 The variation of the air temperature at the evaporator outlet as a function of time

Figure 14.4 indicates a slight alteration in the air temperatures in the first 30 min, the HFO-1234yf refrigerant dropping slightly lower during this period. This change is only temporary, and the graph reveals how these two temperatures reunite following the initial 30 min and maintain the same decline thereafter. The graph reveals promising results.

14.5 Conclusions

The steady-state results demonstrated that the frigorific air conditioning system cooling capacity usually gets higher on decreasing the temperatures of the airstreams travelling through the condenser. The frigorific air conditioning system using R134a provided a higher cooling capacity when compared with the air conditioning systems using HFO-1234yf as a refrigerant. Furthermore, the results determined that the COP for both refrigerants was closely comparable. While additionally, for both refrigerants, the temperature gradient in time was also comparable.

On the other hand, the transient and steady-state performance of an experimental frigorific air conditioning system for the cases of using HFO-1234yf and R134a as a refrigerant as a heat source have been evaluated and compared with each other. Airflow has been introduced to the conditioning room for 60 min for each performance tests. The transient test revealed that the evaporator exit air temperature in the refrigerated room drops with time.

The results of these tests and analysis have shown that the HFO-1234yf refrigerant can use the standard frigorific air conditioning system that is currently being used by the R134a refrigerant, without any changes need to be made. As a result, HFO-1234yf could be viewed as a potential alternative for the frigorific air conditioning system.

Acknowledgments The authors would like to thank the University of Yalova for supporting this study through a Research Project No: 2013/BAP/063.

References

Akram, M. W., Polychronopoulou, K., & Polycarpou, A. (2013). Lubricity of environmentally friendly HFO-1234yf refrigerant. *Tribology International, 57*, 92–100.

Directive 2006/40/EC of The European Parliament and of the Council of 17 May, relating to emissions from air-conditioning systems in motor vehicles and amending Council Directive 70/156/EEC Retrieved online at, http://www.bis.gov.uk/files/file30125.pdf.

Lee, Y., & Jung, D. (2012). A brief performance comparison of HFO-1234yf and R-134a in a bench tester for automobile applications. *Applied Thermal Engineering, 35*, 240–242.

Lee, Y., Kang, D., & Jung, D. (2013). Performance of non-flammable Azeotropic HFO-1234yf/HFC134a mixture for HFC134a applications. *International Journal of Refrigeration, 36*, 1203–1207.

Meyer, J. (2008). HFO-1234yf system enhancements and comparison to R134a. *Proceedings of the SAE alternative refrigerant systems symposium*. Phoenix, AZ, USA.

Navarro, E., Martınez-Galvan, I. O., Nohales, J., & Gonzalvez-Macia, J. (2013). Comparative experimental study of an open piston compressor working with HFO-1234yf, R134a and R-290. *International Journal of Refrigeration, 36*, 768–775.

Petijean, S., & Benouali, J. (July 2010). R-1234yf validation & A/C system energy efficiency improvements. SAE Alternate Refrigerant Symposium. Scottsdale, (AZ): AARS.

Regulation (EC) No 842/2006 of the European Parliament and of the Council of on certain fluorinated greenhouse gases. Official Journal of the European Communities. Retrieved May 17, 2006 from http://eur-lex.europa.eu/LexUriServ/LexUriServ.do?uri=OJ:L:2006:161:0001:0011:EN:PDF.

Spatz, M., & Minor, B. (April 2008). HFO-1234yf. A low GWP refrigeration for MAC. SAE world congress. Detroit. Michigan.

Chapter 15
Biodiesel Production Using Double-Promoted Catalyst CaO/KI/γ-Al_2O_3 in Batch Reactor with Refluxed Methanol

Nyoman Puspa Asri, Bambang Pujojono, Diah Agustina Puspitasari, S. Suprapto and Achmad Roesyadi

Abstract A benign process for biodiesel production has been developed using heterogeneous γ-alumina base as catalyst. This study was conducted using double-promoted catalyst CaO/KI/γ-Al_2O_3 to improve the activity of catalyst and this research was the first one which employs that kind of catalyst for biodiesel production. The preparation of the catalyst was conducted by precipitation and impregnation methods. The effects of reaction temperature, reaction time, and the ratio of oil to methanol on the yield of biodiesel were studied. The reactions were carried out in a batch-type reactor system which consists of three-neck glass flask with 1000-ml capacity equipped with reflux condenser and hot plate stirrer. Results showed that CaO/KI/γ-Al_2O_3 catalyst effectively increased the biodiesel yield about 1.5 times than that the single-promoted catalyst. The optimum condition for the production of biodiesel is as follows: the reaction temperature is 65 °C, the reaction time is 5 h, and the ratio of oil to methanol is 1:42. Under this optimum condition, the highest biodiesel yield of 95 % was obtained.

15.1 Introduction

Nowadays, energy demand in the world has increased rapidly, while the petroleum reserves as the main support of energy needs have significantly decreased. National Energy Board reported that Indonesian crude oil reserves currently are 3.7 billion

N.P. Asri (✉) · B. Pujojono
WR Supratman University, Surabaya, Indonesia
e-mail: nyoman_puspaasri@yahoo.com

S. Suprapto · A. Roesyadi
Sepuluh Nopember Institute of Technology, Surabaya, Indonesia

D.A. Puspitasari
University of Brawijaya, Malang, Indonesia

A.N. Bilge et al. (eds.), *Energy Systems and Management*,
Springer Proceedings in Energy, DOI 10.1007/978-3-319-16024-5_15

barrels (0.3 % of total world oil reserves). The amount is only sufficient for the next few years. Therefore, the Indonesian government actively promotes the use of alternative energy from renewable energy resources. In the last ten years, renewable energy has attracted the attention of many scientists due to its potential to replace fossil fuel (Lam et al. 2010). Vegetable oil is one of the renewable energy sources for diesel engines fuel. The triglycerides as main compound of vegetable oils or animal fats are the main source for biodiesel production, and transesterification of these compounds with simple alcohol preferably methanol in the presence of catalyst will produce biodiesel as the main product and glycerol as the by-product.

Biodiesel is a mixture of monoalkyl esters of long chain fatty acids derived from vegetable oils and animal fats (Asri et al. 2013b). At present, biodiesel has been used as an alternative fuel for substituting diesel oil due to its advantages such as, non toxicity, biodegradability, lack of aromatic compounds and low of SO_x and particulate matter content (Asri et al. 2013a; Kawashima et al. 2008). Moreover, it can be blended with diesel oil and can be used directly in current diesel engines without any modification (Asri et al. 2010; Zabeti et al. 2010). Currently, homogeneous base or acid catalysts are generally used for biodiesel production.

Base catalysts are preferred than acid catalysts due to the higher reaction rate and the lower reaction temperature (Asri et al. 2013a; Stamencovic et al. 2010). However, the use of homogeneous base catalysts possesses at least two main disadvantages.

The first one is the formation of soap leading to decrease of the biodiesel yield (Asri et al. 2013b; Furuta et al. 2006). The second is that the catalyst requires neutralization and separation from the reaction mixture, and it will produce waste water, leading to a series of environmental problems (Dossin et al. 2006; Stamencovic et al. 2010; Asri et al. 2013b). The drawbacks of the homogeneous base catalysts can be overcome by two methods. The first is using supercritical methanol with/without catalysts and the second is introducing heterogeneous base catalysts to the transesterification process as substitute the homogeneous catalyst.

Recently, supercritical methanol has received much attention in the biodiesel production due to its unique properties. At supercritical state, the dielectric constant of methanol is lower than at liquid phase; therefore, the nature two phase of oil/methanol can easily dissolve to form a single phase, leading to a very fast rate of transesterification reaction. Several studies have reported on sub- and supercritical methanol biodiesel production in batch, semi-batch, or continuous processes (Kusdiana and Saka 2001; Song et al. 2008; Joelianingsih et al. 2008; Petchmala et al. 2010; Asri et al. 2013a). However, due to the high temperature and pressure conditions, the security aspect of the process needs to get serious attention in the industrial application. In addition, high energy costs still need to be considered; therefore, further studies are to be continuously developed in order to find an efficient process for the biodiesel production (Joelianingsih et al. 2008; Kouzu et al. 2008). Therefore, the development of heterogeneous catalyst for biodiesel production is still highly desired.

Transesterification with heterogeneous catalysts particularly with refluxing methanol is expected to be an effective biodiesel production process with low cost and minimal environmental impact due to the possibility of simplifying the separation and purification process in the mild condition. Moreover, the catalyst can be recovered and reused, leading to more economic process compared to homogeneous ones. Kouzu et al. (2008) used some kind of catalysts (CaO, Ca $(OH)_2$, and $CaCO_3$) as the solid catalysts in the production of biodiesel from soybean oil with yield biodiesel 93, 12, and 0 %, respectively. Alumina-supported CaO base catalyst has also been used by Zabeti et al. (2010) for the transesterification of palm oil. Optimal condition was obtained on molar ratio of oil to methanol 1:12, amount of catalyst 6 % (wt% to oil) and temperature reaction 65 °C (Zabeti et al. 2010). Asri et al. (2010) had conducted transesterification of palm oil with CaO/γ-alumina catalyst in a batch reactor with refluxing methanol, at 75 °C and time of reaction of 7 h with various amount of catalyst of 4, 6, and 8 % (w% to oil). The highest conversion of palm oil (67 %) was achieved at 6 % of the amount of catalyst. From the viewpoint of the economical advantage, calcium oxide was also used for the study of the noble process for biodiesel production using the solid base catalyst. The experimental results indicate that calcium oxide was quite active in transesterification of soybean oil with refluxing methanol (Kouzu et al. 2008) and in transesterification of palm oil (Zabeti et al. 2009; Asri et al. 2010). However, the activities of the heterogeneous catalysts still lower compared to the homogeneous one. Furthermore, the information regarding double-promoted alumina-supported base catalysts for the biodiesel production in the literature is still scarce. Therefore, in this work, double-promoted catalyst CaO/KI/γ-Al_2O_3 was used for biodiesel production. In order to improve the activity of the catalyst, the addition of various amount of CaO was also conducted. In addition, the effects of reaction variable such as molar ratio of oil to methanol, reaction time, and reaction temperature on biodiesel yield were also investigated.

15.2 Experimental

15.2.1 Materials

The refined palm oil as the raw material was purchased from local market. Meanwhile, waste cooking oil was collected from California Fried Chicken (CFC). Analytical grade γ-alumina (BET surface area of 120–190 m^2/g), analytical grade potassium iodide, and other standard for gas chromatography analyses were purchased from Merck, Germany. Commercial grade of calcium oxide, acetic acid, and methanol were purchased from the local supplier.

15.2.2 Catalyst Preparation

A double-promoted CaO/KI/γ-Al_2O_3 catalyst was synthesized by precipitation and impregnation methods. CaO loadings were varied from 10 to 50 % w to Al_2O_3. Meanwhile, the amounts of KI were kept constant at 35 % w to Al_2O_3. The detail of the procedure can be seen elsewhere (Asri et al. 2013b). The slurry was heated at 100–120 °C in an oven over night in order to remove the water content. The synthesized catalyst was milled into powder and then calcinated at 650 °C in a muffle furnace with continuous flow of air for 4 h. The calcined catalyst was kept in a desiccator in the presence of silica gel in order to avoid the adsorption of water and CO_2 onto the catalyst. The catalysts were characterized by means of nitrogen sorption, X-ray powder diffraction (X-RD), and scanning electron micrograph (SEM). The X-RD gives information of the crystallite structure information. The specific surface area, pore volume, and pore diameter of the prepared catalysts were calculated from nitrogen sorption isotherms using Brunauer–Emmett–Teller (BET) method. SEM obtained on Evo Zeiss-Germany scanning electron microscope was used to identify the morphology and size of CaO /KI/γ-Al_2O_3 catalyst particle (Asri et al. 2013b).

15.2.3 Transesterification Reaction

Transeterification process was conducted in batch-type reactor system which consists of three-neck glass flask fitted with reflux condenser and magnetic hot plate stirrer. A known amount of methanol and 6 % (w% to oil) of CaO/KI/γ-Al_2O_3 catalyst were added into the batch reactor and heated at a certain reaction temperature. Preheated palm oil with a certain molar ratio of oil to methanol was also added into the reactor. After the temperature in the system reached the desire value (35–75 °C), the mixture was agitated using a magnetic stirrer at constant desired speed. After predetermined time (1–7 h), the reaction was stopped by cooling the reactor to room temperature, and the catalyst was separated from the mixture using a vacuum filtering flask. The liquid mixture was discharged into a separator funnel leave it overnight to ensure that the mixture was completely separated into three layers. The top layer is the excess of methanol, the middle is biodiesel, and the bottom layer is glycerol. Biodiesel product was analyzed with gas chromatography; using a GC HP 5890 with HP's first column with carrier gas flow of 28 ml/min, initial temperature of the column was 125 °C with the temperature increment of 15 °C per minute and the final temperature was set at 275 °C. The yield (%) of biodiesel was determined according to the following equation (Asri et al. 2013b):

$$\mathrm{Yield}\,(\%) = \frac{w_{\mathrm{ofbiodieselproduced}} \times w_{\mathrm{ofbiodieselinsample}}}{w_{\mathrm{oilusedintheexperiment}}} \tag{15.1}$$

15.3 Results and Discussion

15.3.1 Characterization of Catalyst

The BET-specific surface area, cumulative pore volume, and mean pore diameter of CaO/KI/γ-Al_2O_3 catalysts are in the range of 15.508–82.769 m^2/g, 0.049–0.135 cc/g, and 34.08–77.904 °A, respectively. The specific surface area of all catalysts (10–50 % of CaO loading) are smaller than that the single-promoted catalyst (83.77 m^2/g) that have been reported by Asri et al. (2010) and Zabeti et al. (2010). The decrease of BET surface area after loaded with CaO possibly due to the covering of surface of catalyst by the double alkaline compounds (CaO and KI). However, the presence of those substances causes the increase in the active sites and causes the increase in the activity of catalyst. These phenomena supported by the X-ray diffraction patterns of CaO/KI/γ-Al_2O_3. Figure 15.1 shows the diffraction pattern of 30 % CaO loading. It can be seen from this figure that more diffraction peaks were appeared on the pattern than the CaO/γ-Al_2O_3 catalyst pattern (figure is not shown). The similar result was reported by Xie and Li (2006). Moreover, as reported by Ilgen and Akin (2008), the activity of the catalyst for transesterification of vegetable oils was also influenced by the basicity of the catalyst. The scanning electron micrograph (SEM) photograph of the sample showed the crystallites structure with the size around 10 μm and similar irregular shape (the figure not shown). These evidences indicate that the catalyst has a mesoporous structure (Asri et al. 2013b).

15.3.2 Effect of CaO Loading on Yield of Biodiesel

The effect of CaO loading on the yield % of biodiesel was studied by varying the CaO loading from 5 to 50 % (w% to alumina). The experiments were conducted in batch glass-type reactor with refluxing methanol at reaction temperature of 65 °C, reaction time of 5 h, and molar ratio of oil to methanol of 1:42. Meanwhile, the amount of catalyst was 6 % (w% to oil) and loading amount of KI was 35 % (w% to alumina). Figure 15.2 shows the effect of CaO loading on the yield of biodiesel. The yield of biodiesel increased with the increasing of % CaO loading. By increasing CaO loading from 10 to 25 %, the yields of biodiesel gradually increased from 79 to 83 %. The increase in the yield of biodiesel possibly due to the well dispersed of CaO in the pores of the catalyst. At 30 % CaO loading, the biodiesel yield drastically increased to 95 %. At 30 % of CaO loading, some of CaO were well dispersed on pore of alumina and some of them were still remaining on surface as an active site. In addition, with 35 % loading of KI on the catalyst, there was an optimum synergy among CaO, KI, and Al_2O_3, due to the interaction between alkali compound and the alumina. Similar case to 30 % of CaO loading, at 35 % loading of KI, some of KI also well dispersed on the alumina pores and another part decomposes into K_2O and still remaining at the catalyst surface as a composite and

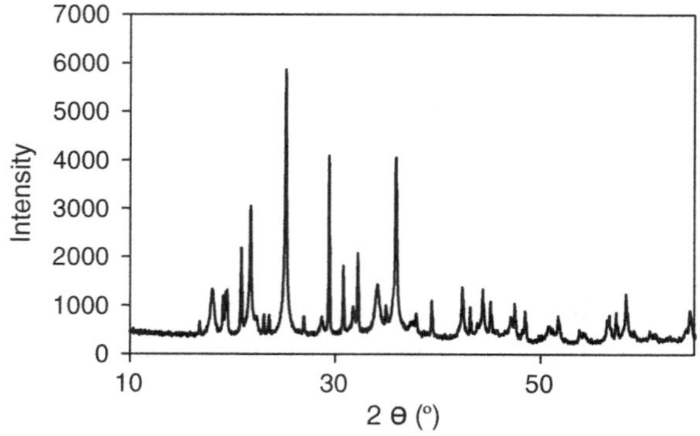

Fig. 15.1 X-RD pattern for CaO/KI/γ-Al_2O_3 with 30 % of CaO loading

eventually increases the basicity and the basic strength of the catalyst. This evidence is supported by the diffractogram of CaO/KI/γ-Al_2O_3 catalyst (Fig. 15.1), where at 30 % of CaO loading, new diffraction peaks which appeared at 2θ of 38.0 and 55.0° were indicated as K_2O species. The presence of this compound can improve the basicity of catalyst (Xie and Li 2006; Noiroj et al. 2009). In contrary, at loading of CaO above 30 % (40, 50 %), there was no significant effect on yields of biodiesel even it tends to slightly decrease, because the excess of CaO began to cover the active site of the catalyst surface, thus decreasing the activity of catalyst. The highest yield of biodiesel (95 %) in this work was higher than that of the previous results reported in the literatures. For example, Xie and Li (2006) studied the transesterification of soybean oil with single-promoted catalyst (KI/alumina), and the highest conversion of 87.4 % was obtained at 6 h of reaction time. Another study on the transesterification of palm oil using single-promoted catalyst CaO/γ-Al_2O_3 (Asri et al. 2010) obtained the highest yield of 65 % at 65 °C, 7 h of reaction time and almost 50 % of CaO loading. These evidences indicate that double-promoted catalyst has significantly higher activity than single-promoted catalysts.

Because the highest yield of 95 % was achieved at 30 % of CaO loading, it can be concluded that 30 % of CaO loading was the optimum one. Therefore, CaO/KI/γ-Al_2O_3 catalyst with 30 % of CaO loading was used to investigate further the effect of molar ratio of oil to methanol, the reaction time, and the reaction temperature on yield of biodiesel.

15.3.3 Effect of Molar Ratio Oil to Methanol and Time of Reaction on Yield of Biodiesel

Transesterification is a displacement process of an alcohol group from an ester by another alcohol, more preferable simple alcohols such as methanol or ethanol (Yacob et al. 2009; Asri et al. 2010). Theoretically, it consists of three consecutive reversible reactions as represented by Eqs. 15.2–15.4, where triglyceride (TG) reacts with methanol (MeOH) to produce diglyceride (DG) which in turn reacts

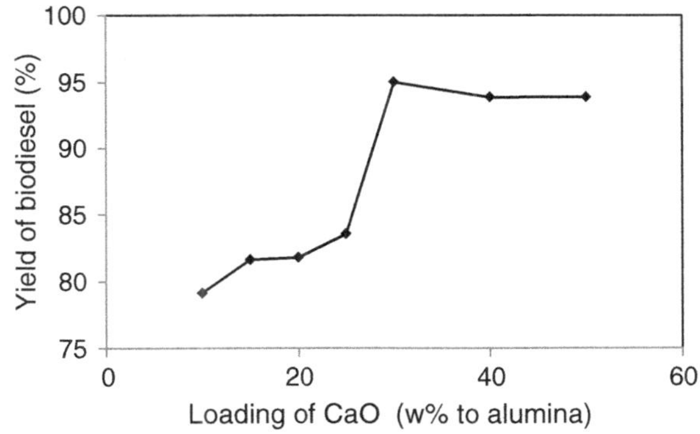

Fig. 15.2 Yield of biodiesel (%) as function of CaO loading (w% to alumina), at 65 °C, reaction time of 5 h, molar ratio of oil to methanol 1:42, and catalyst amount of 6 %

with methanol (MeOH) to produce monoglyceride (MG). Finally, MG reacts with MeOH to produce methyl ester (ME) and glycerol (GL) (Asri et al. 2013; Kusdiana and Saka 2001; Joelianingsih et al. 2009).

$$\text{Triglyceride} + \text{ROH} \leftrightarrow \text{Diglyceride} + \text{RCOOR}_1 \tag{15.2}$$

$$\text{Diglyceride} + \text{ROH} \leftrightarrow \text{Monoglyceride} + \text{RCOOR}_2 \tag{15.3}$$

$$\text{Monoglycride} + \text{ROH} \leftrightarrow \text{Glycerol} + \text{RCOOR}_3 \tag{15.4}$$

Molar ratio of palm oil to methanol is one of the important factors that effect the biodiesel yield of palm oil transesterification. Stoichiometrically, at each step, one mole of TG requires three moles of methanol to produce one mole of methyl ester, but in application, excess of methanol is required in order to drive the reaction to the right side. Generally, molar ratio of oil to methanol of 1:6 was used for homogeneous catalysts. As reported by Encinar et al. (1999) in the conventional process with alkali catalyst, the biodiesel yield increased with increasing molar ratio of Cynara oil—methanol. The optimal conditions obtained at molar ratio of oil to methanol of 1:4–1:6. At molar ratio of oil to methanol higher than 1:6, the glycerol as a by-product becomes difficult to be separated. Conversely, if the ratio is less than 1:4, the reaction is incomplete. However, in the transesterification using heterogeneous catalysts, a higher molar ratio of oil to methanol is required if compared to the homogeneous one (Xie and Li 2006; Kim et al. 2004; Kouzu et al. 2008). Kim et al. (2004) reported that the optimum molar ratio of vegetable oil to methanol with heterogeneous catalysts in refluxing condition was found to be 1:9. They also stated that at the molar ratio more than 1:9, the excess of methanol had no significant effect on the increase in the biodiesel yield. Another study reported by Xie and Li (2006) also deals with transesterification of soybean oil with refluxing methanol and the optimum molar ratio of soybean oil to methanol of 1:15 was obtained. Meanwhile, Kusdiana and Saka (2001) studied the effect of molar ratio of

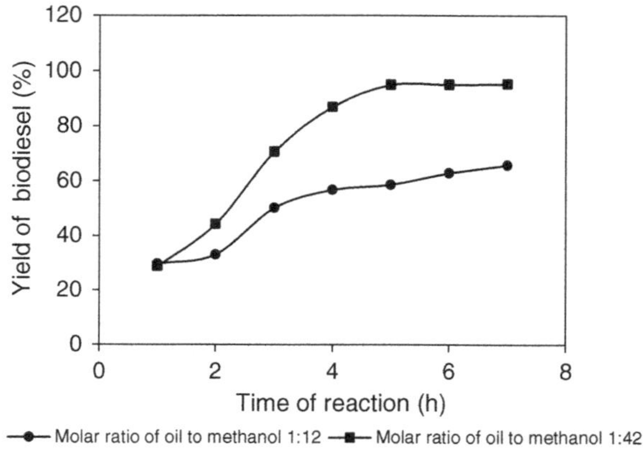

Fig. 15.3 Yield of biodiesel (%) as function of time of reaction at 65 °C, amount of catalyst 6 % (wt% to oil), and 30 % of CaO loading

methanol to rapeseed oil in the range of 3.5–42. They noted that at molar ratio of 42, almost complete conversion was achieved with 95 % yield of biodiesel, whereas for lower molar ratio of 6 or less, incomplete conversion apparent with the lower yield of biodiesel was obtained. Song et al. (2008) conducted the transesterification of RBD palm oil under the same approach proposed by Kusdiana and Saka (350 °C and 40 MPa) for 5–10 min at various mole ratios of RBD palm oil to methanol. They stated that the contents of FAMEs (biodiesel) increased drastically with an increasing mole ratio of RBD palm oil to methanol up to 1:30 and then constant for further increasing mole ratio.

Therefore, in this work, to investigate the effect of molar ratio of palm oil to methanol on yield of biodiesel, the experiments were conducted at two levels of molar ratio 1:12 and 1:42 and time of reaction varies from 1 to 7 h with 1 h of interval, while other conditions are kept constant at reaction time of 65 °C and catalyst amount of 6 % (wt% to oil). Figure 15.3 shows the effect of molar ratio of oil to methanol and time of reaction on the yield of biodiesel. It shows that the yield of biodiesel at molar ratio of oil to methanol 1:42 was significantly higher than that of the other one (molar ratio of 1:12). At molar ratio 1:12, yield of biodiesel gradually increased along with the increasing of the rate of reaction. However, at molar ratio of oil to methanol 1:42, the yield of biodiesel linearly increased from 29 to 95 % with the increasing of reaction time from 1 to 5 h. Beyond 5 h, the exposure of time reaction had no significant effect on the yield of biodiesel. While at a molar ratio of 1:12, the highest yield of 66 % was obtained at reaction time of 7 h. It shows that molar ratio of oil to methanol plays an important role in the transesterification reaction.

15.3.4 Effect of Reaction Temperature on Yield of Biodiesel (%)

In order to investigate the effect of reaction temperature on the yield of biodiesel, the experiments were conducted at various temperatures from 35 to 75 °C with

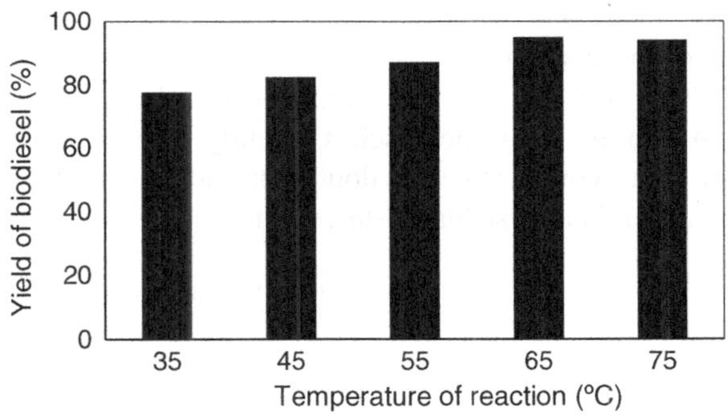

Fig. 15.4 Biodiesel yield (%) as function of temperature reaction (°C) at 5 h of reaction time, 30 % of loading CaO, molar ratio oil to methanol of 1:42, and amount of catalyst 6 %

interval of 10 °C. Meanwhile, the other conditions such as CaO loading, reaction time, amount of catalyst, and molar ratio oil to methanol were kept constant at 30 % (wt% to alumina), 5 h, 6 % (wt% to oil) and 1:42, respectively. Figure 15.4 describes the effect of temperature on the yield of biodiesel. The yields of biodiesel increased with the increase in reaction temperature indicate that the higher temperature is favorable to transesterification triglyceride to biodiesel. This is consistent with the results as mentioned earlier that the reaction temperature has very significant effect on the conversion of palm oil (Asri et al. 2013c). This may also relate to Arrhenius law which stated that rate of reaction constant is a function of reaction temperature. Figure 15.4 shows that the biodiesel yields gradually increase from 78 to 87 % at temperature 35–55 °C. However, at 65 °C, the yield of biodiesel drastically increased to 95 %. This is possible because at 65 °C, methanol completely boils (boiling point of methanol is 64.7 °C) and causes the molecules move faster than at lower temperature, due to the higher kinetics energy of the system, eventually enhancing the frequency factor. The increase in frequency factors resulted in an increased of transesterification reaction rate constant, eventually increase in the yield of biodiesel. Beyond 65 °C, there was no significant effect on yields of biodiesel. Therefore, in this work, 65 °C was decided as the optimum reaction temperature with 95 % yield.

15.3.5 Transesterification of Waste Cooking Oil

Currently, most of the biodiesel plants in Indonesia using refined oils as their main feedstock. Therefore, biodiesel production cost is quite higher compared to diesel oil, because nearly 80 % of the overall biodiesel production cost is contributed by feedstock cost. In order to reduce biodiesel production cost, in this work the synthesized $CaO/KI/\gamma\text{-}Al_2O_3$ catalyst was used to produce biodiesel from waste cooking oil (WCO). WCO collected from California Fried Chicken (CFC) with FFA content of 3.5 % was first filtered to remove solid residues and heated at 105–

110 °C to remove water content. Subsequently, WCO was treated by adding active carbon and heated under stirring for one hour. The FFA content of the homogeneous treated WCO was 0.9 %. The homogeneous WCO was ready for the transesterification process into biodiesel. Currently, the experiments and data analyses are still being conducted. The double-promoted CaO/KI/γ-Al_2O_3 can be potentially used as solid catalyst for waste cooking oil.

15.4 Conclusion

Double-promoted catalyst CaO/KI/γ-Al_2O_3 was synthesized by precipitation and impregnation methods and was employed as the catalyst for the transesterification of palm oil. Double-promoted catalyst CaO/KI/γ-Al_2O_3 was found much more active than the single-promoted catalyst CaO/γ-Al_2O_3. The optimum conditions of CaO loading, reaction temperature, time of reaction, and molar ratio of oil to methanol were 30 %, 65 °C, 5 h, and 1:42, respectively. Under those conditions, the highest yield of 95 % was achieved. Therefore, it can be concluded that the heterogeneous base catalyst CaO/KI/γ-Al_2O_3 has great ability for the production of biodiesel in order to overcome the drawbacks of conventional process.

Acknowledgments The authors gratefully acknowledge Directorate of research and community services, Directorate General of higher education, Education Ministry of the Republic Indonesia, for financial support contained in the memorandum of agreement No: 011/SP2H/P/K7/KM/2014.

References

Asri, N. P., Anisa, A., Rizqi, F., Roesyadi, A., Budikaryono, K., Suprapto, S., & Gunardi, I. (2010). Biodiesel production from palm oil using CaO/Al2O3 as a solid base catalyst. In *Proceeding of the 1st International Seminar on Fundamental and application of Chemical Engineering (ISFACHE), Kuta, Bali.*

Asri, N. P., Hartanto, F., Ramadhan, R., Savitri, S. D., Gunardi, I., Suprapto, S., Budikarjono, K., & Roesyadi, A. (2011). Transesterification of palm oil to methyl ester using γ-alumina supported base catalysts. In *Proceeding of Bali International Seminar on Science and Technology (BISSTECH), Denpasar, Bali* (pp. A IV.5-1–A IV.5-5). ISBN-97998623-1-0.

Asri, N. P., Machmudah, S., Wahyudiono, W., Suprapto, S., Budikarjono, K., Roesyadi, A., & Goto, M. (2013a). Non catalytic transesterification of vegetables oil to biodiesel in sub-and supercritical methanol: A kinetic's study. *Bulletin of Chemical Reaction Engineering & Catalyst, 7*(3), 215–223.

Asri, N. P., Machmudah, S., Wahyudiono, W., Suprapto, S., Budikarjono, K., Roesyadi, A., & Goto, M. (2013b). Palm oil transesterification in sub and supercritial methanol with heterogenious base catalyst. *Chemical Engineering and Processing: Process Intensification, 72*, 63–67.

Asri, N. P., Suprapto, S., Budikarjono, K., & Roesyadi, A. (2013c). Kinetics of palm oil transesterification using double, promoted catalyst cao/ki/γ- Al2O3 in refluxed methanol. In *Proceedings of International Seminar on Biorenewable Resources Utilization for Energy and Chemicals 2013 In conjunction with Chemical Engineering Seminar of SoehadiReksowardojo (BioEnChe)* (pp. 113–128). Bandung: ITB. ISSN 2354-5917.

Dossin, T. F., Reyniers, M. F., & Marin, G. B. (2006). Kinetic of heterogeneously MgO catalyzed transesterification. *Applied Catalysis, B: Environmental, 61*, 35–45.

Encinar, J. M., González, J. F., Sabio, E., & Ramiro, M. J. (1999). Preparation and properties of biodiesel from Cynara cardunculus L. oil. *Industrial and Engineering Chemistry Research, 38*, 2927–2931.

Furuta, S., Matsuhashi, H., & Rata, K. (2006). Green diesel fuel production with solid amorphous-zirconia catalyst in fixed bed reactor. *Journal of Biomass and Bioenergy, 30*, 870–873.

Ilgen, O., & Akin, A. N. (2008). Development of alumina supported alkaline catalyst used for biodiesel production. *Journal of Turkey, 33*, 281–287.

Imahara, H., Minami, E., Hari, S., & Saka, S. (2008). Thermal stability of biodiesel in supercritical methanol. *Fuel, 87*, 1–6.

Kawashima, A., Matsubara, K., & Honda, K. (2008). Development of heterogeneous base catalyst for biodiesel production. *Bioresource Technology, 99*, 3439–3443.

Kim, H. J., Kang, B. S., Kim, M. J., Park, Y. M., Kim, D. K., Lee, J. S., & Lee, K. Y. (2004). Trasnsesterification of vegetable oil to biodiesel using heterogeneous base catalyst. *Catalyst Today, 939*(5), 315–320.

Kouzu, M., Kasuno, T., Tajika, M., Sugimoto, Y., Yamanaka, S., & Hidaka, J. (2008). Calcium oxide as a solid base catalyst for transesterification of soybean oil and its application to biodiesel production. *Fuel, 87*, 2798–2806.

Kusdiana, D., & Saka, S. (2001). Kinetics of transesterificationin rapeseed oil to biodiesel fuel as treated in supercritical methanol. *Fuel, 80*, 693–698.

Lam, M. K., Lee, K. T., & Mohamed, A. R. (2010). Homogeneous, heterogeneous and enzymatic catalysis for transesterification of high free fatty acid oil (waste cooking oil) to biodiesel: A review. *Biotechnology Advances, 28*, 500–518.

Joelianingsih, Maeda, H., Hagiwara, S., Nabetani, H., Sagara, Y., Soerawidjaya, T. H., Tambunan, A. H., & Abdullah, K. (2008). Biodiesel fuels from palm oil via the non catalytic transesterification in a bubble column reactor at atmospheric pressure: A kinetic study. *Renewable Energy, 33*, 1629–1636.

Noiroj, K., Pisitpong, I., & Samai, J. (2009). A comparative study of KOH/Al2O3 dan KOH/NaY catalyst for biodiesel production via transesterification from palm oil. *Journal of Renewable Energy, 34*, 1145–1150.

Petchmala, A., Laosiripojana, N., Jongsomjit, B., Goto, M., Panpranot, J., Mekasuwandumrong, O., & Shotipruk, A. (2010). Transesterification of palm oil fatty acid in near– and super–critical methanol with SO_4–ZrO_2 catalyst. *Fuel, 89*, 2387–2392.

Song, E. S., Lim, J. W., Lee, H. S., & Lee, Y. W. (2008). Transesterification of RBD palm oil supercritical methanol. *Journal of Supercritical Fluid, 44*, 356–363.

Stamencovic, O. S., Lazic, M. L., Todorovic, Z. B., Velkovic, V. B., & Skala, D. U. (2010). The effect agitation intensity on alkali-catalyzed methanolysis on sunflower oil. *Bioresource Technology, 98*, 2688–2699.

Xie, W., & Li, H. (2006). Alumina supported iodide as a heterogeneous catalyst for biodiesel production from soybean oil. *Journal of Molecular Catalyses A: Chemical, 255*, 1–9.

Yacob, A. R., Mustajab, M. K. A., & Samadi, N. S. (2009). Calcination temperature of nano MgO effect on base transesterification of palm oil. *World Academy of Science, Engineering and Technology, 56*, 408–412.

Zabeti, M., Daud, W. H. A. W., & Aroua, M. K. (2009). Activity of solid catalysts for biodiesel production: A review. *Fuel Processing Technology, 90*, 770–777.

Zabeti, M., Daud, W. H. A. W., & Aroua, M. K. (2010). Biodiesel production using alumina—supported cacium oxide: An optimization study. *Fuel Processing Technology, 91*, 243–248.

Chapter 16
I–V Characterization of the Irradiated ZnO:Al Thin Film on P-Si Wafers By Reactor Neutrons

Emrah Gunaydın, Utku Canci Matur, Nilgun Baydogan, A. Beril Tugrul, Huseyin Cimenoglu and Serco Serkis Yesilkaya

Abstract ZnO:Al/p-Si heterojunctions were fabricated by solgel dip coating technique onto p-type Si wafer substrates. Al-doped zinc oxide (ZnO:Al) thin film on p-Si wafer was irradiated by reactor neutrons at ITU TRIGA Mark-II nuclear reactor. Neutron irradiation was performed with neutron/gamma ratio at 1.44×10^4 ($n\ cm^{-2}\ s^{-1}\ mR^{-1}$). The effect of neutron irradiation on the electrical characteristics of the ZnO:Al thin film was evaluated by means of current–voltage (I–V) characteristics for the unirradiated and the irradiated states. For this purpose, the changes of I–V characteristics of the unirradiated ZnO:Al thin films were compared with the irradiated ZnO:Al by reactor neutrons. The irradiated thin ZnO:Al film cell structure is appropriate for the usage of solar cell material which is promising energy material.

E. Gunaydın · N. Baydogan (✉) · A.B. Tugrul · H. Cimenoglu
Istanbul Technical University, Istanbul, Turkey
e-mail: dogannil@itu.edu.tr

E. Gunaydın
e-mail: egunaydin@itu.edu.tr

A.B. Tugrul
e-mail: beril@itu.edu.tr

H. Cimenoglu
e-mail: cimenogluh@itu.edu.tr

U.C. Matur
Gedik University, Istanbul, Turkey
e-mail: utku.canci@gedik.edu.tr

S.S. Yesilkaya
Yildiz Technical University, Istanbul, Turkey
e-mail: serkaya@yildiz.edu.tr

A.N. Bilge et al. (eds.), *Energy Systems and Management*,
Springer Proceedings in Energy, DOI 10.1007/978-3-319-16024-5_16

16.1 Introduction

Photovoltaic (PV) refers to the conversion of light energy into electricity using electronic devices called solar cells. In developing next-generation solar alternatives, a thinner profile is paramount. The majority of solar cells in existence today are made from rigid multi- or single-crystalline silicon (Si) wafers. Typically 150 μm thick, the wafers demand multiple processing steps before they can be integrated into a module. On the contrary, thin-film solar cells utilize only a 1- to 4-μm-thick layer of semiconducting material to produce electricity, thus requiring less processing and fewer materials. These cost-saving alternatives also offer another important advantage as compared to wafer-based modules in that they can be used in a wide range of applications. Thin-film solar cells employ light weight and making them ideal for advanced applications such as building-integrated PV (http://solopower.com/solutions-technology/thin-film-photovoltaics/). The ZnO coating as conducting electrode makes the device as the reliable energy source for the use in transparent energy applications. The conducting electrodes consisting of ZnO:Al thin films have several favorable properties, including good transparency, high electron mobility, wide bandgap, and strong room-temperature luminescence. Hence, these properties of ZnO:Al thin films are used in the emerging applications for transparent electrodes in energy-saving or heat-protecting windows, in liquid crystal displays, and in electronics as thin-film transistors and light-emitting diodes. ZnO:Al/p-Si heterojunctions can be used to obtain clean, abundant, and convenient energy. The researches on ZnO:Al/p-Si heterojunctions attracts an increasing attention for the use of ZnO:Al/p-Si heterojunctions in solar cells as they have a number of advantages, such as economic and simple processing steps and an excellent blue response.

ZnO structures are useful for space and terrestrial applications at strong particle irradiation as ZnO is extremely radiation resistant to radiation damage with respect to other semiconducting materials. High-energy particles can create the point defects (i.e., vacancies and interstitials) with the changes of their optical and electrical properties. Earlier studies have shown that ZnO compounds are fairly resistant to harsh environments and displacement damage resulting from particle irradiation (Ramirez et al. 2013). ZnO structures are useful for space and terrestrial applications for which irradiation hardness is a prerequisite (Elliot 2005). Radiation resistance makes ZnO a suitable candidate for space applications. Hence, ZnO thin films are good candidates for transparent conductive oxide films at aggressive radiation environment and this thin film is a good choice for electronic or optoelectronic applications for the use in solar cells. Their property could be advantageous for electronic and photonic applications in radiation environments (Coskun et al. 2009). For space applications, nuclear researches, extraction uranium ores, and their treatment, semiconducting devices have to operate in harsh radiation conditions. An important point for such applications is the high radiation resistivity of the semiconducting material, providing reliable operation of devices during extended periods of time.

The effect of high-energy neutron irradiation has been reported for ZnO:Al in this study. The effect of reactor neutrons on current voltage characteristics of ZnO is not detailed regarding in literature. In particular, to our knowledge in the literature, data pertaining to the changes of current voltage characteristics in the induced ZnO:Al by neutrons are unknown at different Al concentrations. Hence, this study is initiated to evaluate the neutron effect on ZnO:Al/p-Si heterojunctions. The applications of devices based on ZnO depend essentially on Al concentration and production method. This problem is closely related to the interaction of ZnO with high-energy particles such as neutrons. It leads to a generation of radiation defects in ZnO crystal lattice similar to native defects at thermodynamic equilibrium and at growing thin films by different methods. The types of defects generated by radiation, the processes of their annealing, and the methods for their characterization are considered. The impact of radiation on I–V characteristics of ZnO:Al are compared at different Al concentrations.

16.2 Experimental

ZnO:Al thin films deposited by solgel dip coating technique at four different Al concentrations were annealed in vacuum ambient at 700 °C. The details of the production process of ZnO:Al thin films deposited by solgel dip coating technique were given elsewhere (Baydogan et al. 2013). This paper presents some information about the impact of neutron irradiation on the surface of ZnO:Al thin film. The ZnO:Al thin films were irradiated in the tangential beam tube of the ITU TRIGA Mark-II Training Reactor (Fig. 16.1). ITU TRIGA Mark-II Training Reactor

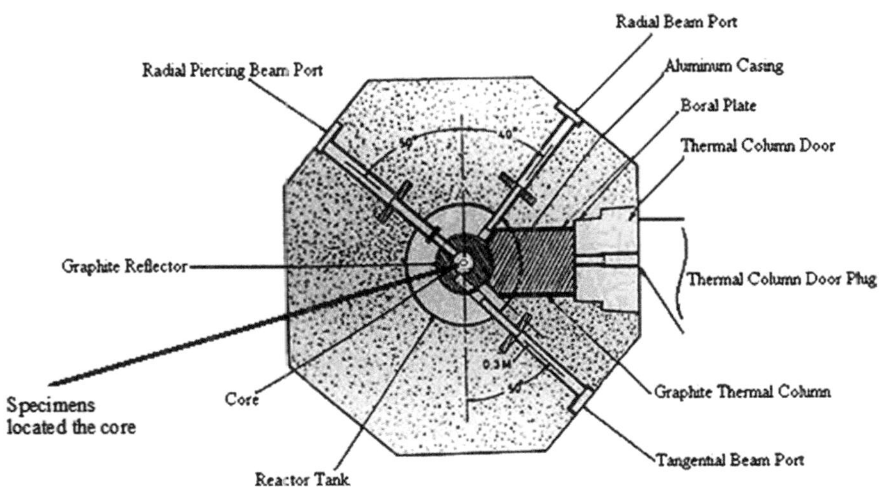

Fig. 16.1 ITU TRIGA Mark-II reactor arrangement with port facility

operates at a maximum power of 250 kW (Yavuz et al. 1993). The reactor was operated at this maximum power during 3 h.

16.3 Results and Discussion

Phase analysis of the ZnO:Al thin films was made by qualitative X-ray diffraction (XRD) using Cu-K_α radiation. For this purpose, a GBC-MMA X-ray diffractometer was operated at 35 kV and 28.5 mA with a constant scan rate of 2°/min. XRD analyses were performed to scan the samples in a 2θ range between 30° and 60°. ZnO peaks from the (100), (002), (101), (110), and (103) planes were identified in the XRD patterns in Fig. 16.2. The changes in electrical resistivity of ZnO:Al thin film with the rise of irradiation time by using reactor neutrons were presented in Fig. 16.2. The decrease in electrical resistivity on the surface of ZnO:Al thin film was determined with the rise of irradiation time by using reactor neutrons in Fig. 16.3.

I–V characteristics of ZnO:Al thin film were investigated for Cu/ZnO:Al/p-Si/Al configuration. I–V characteristics of ZnO:Al thin film annealed at 700 °C in vacuum ambient are presented in Figs. 16.4 (at dark) and 16.5 (at light). The changes of I–V characteristics were determined with the rise of Al concentrations. Current density–voltage characteristics of unirradiated ZnO:Al/p-Si heterojunctions are presented at different Al concentrations in Fig. 16.6. Besides, Current density–voltage characteristics of irradiated ZnO:Al/p-Si heterojunctions were examined by using reactor neutrons and are shown in Fig. 16.7. For this purpose, irradiation process has continued during 3 h.

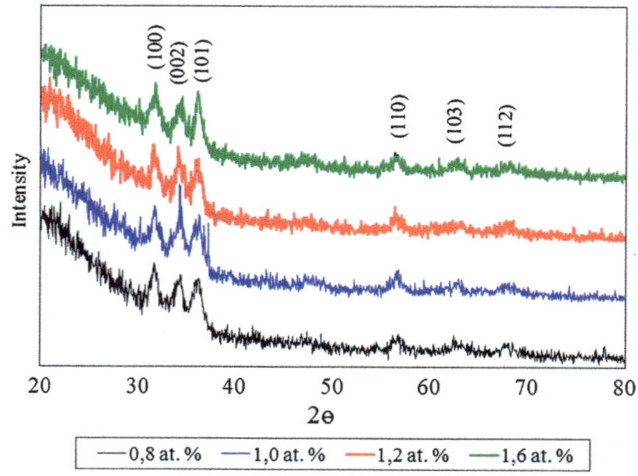

Fig. 16.2 X-ray diffraction patterns of the ZnO:Al films annealed in nitrogen ambient depends on Al concentration

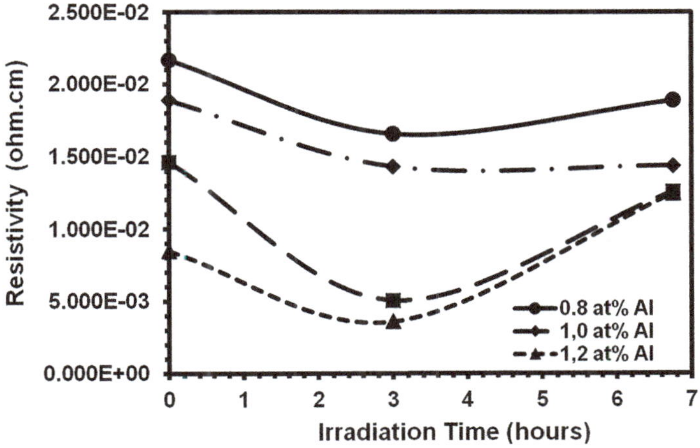

Fig. 16.3 The changes in electrical resistivity of ZnO:Al thin film with the rise of irradiation time by using reactor neutrons

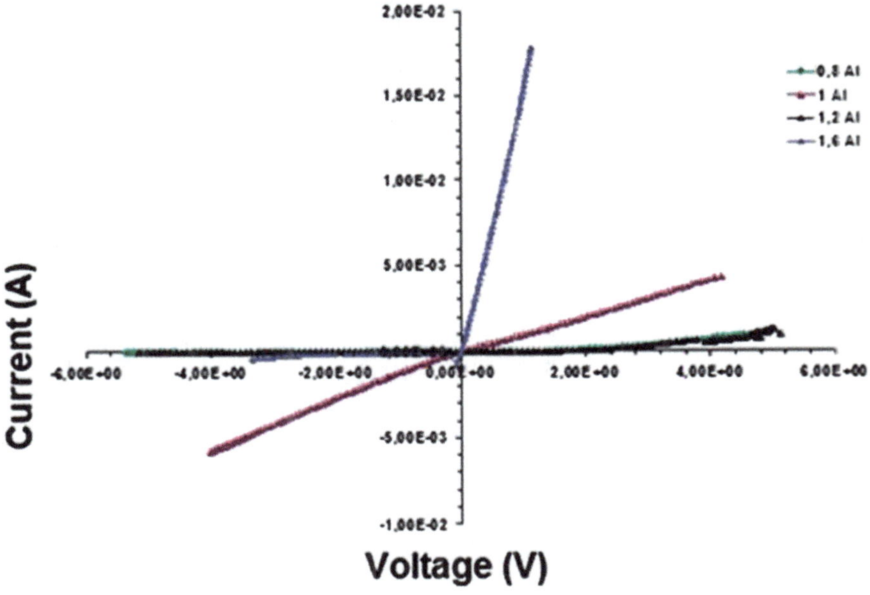

Fig. 16.4 The changes in I–V characteristics of ZnO:Al thin film in dark

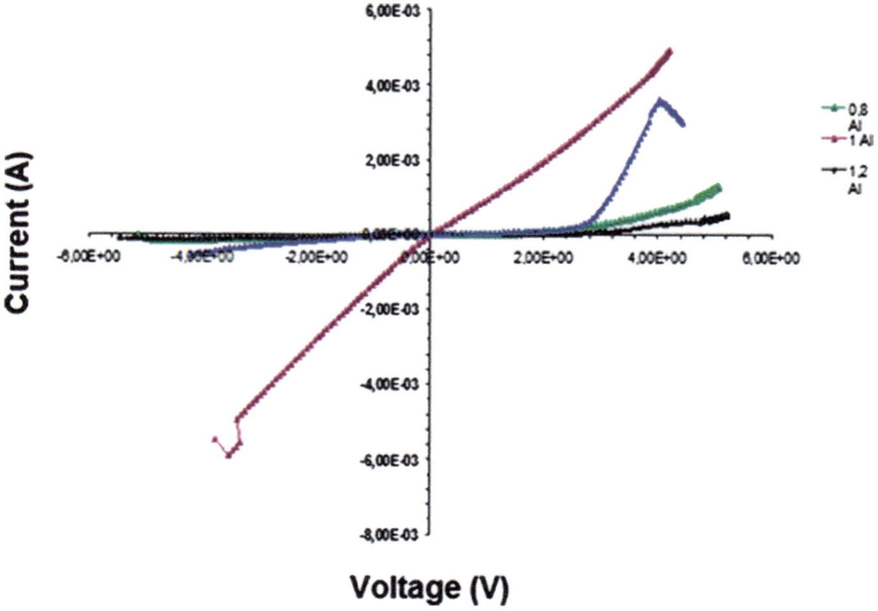

Fig. 16.5 The changes in I–V characteristics of ZnO:Al thin film in light

Fig. 16.6 Current density–voltage characteristics of unirradiated ZnO:Al/p-Si heterojunctions

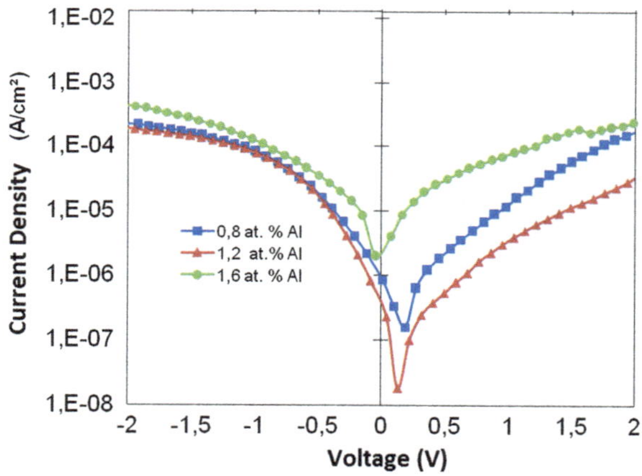

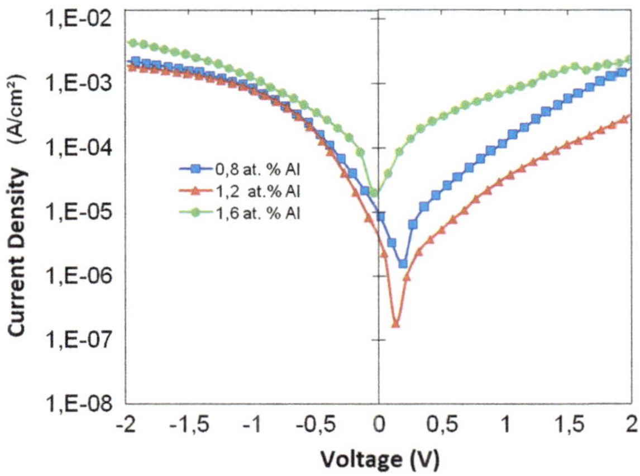

Fig. 16.7 Current density–voltage characteristics of irradiated ZnO:Al/p-Si heterojunctions by reactor neutrons during 3 h

16.4 Conclusions

In Cu/ZnO:Al/p-Si/Al configuration, ZnO:Al/p-Si heterojunctions were examined to evaluate their space and terrestrial solar cell applications at strong neutron irradiation depending on extremely radiation-resistant properties of ZnO structures. In the results of the study, it concluded as below:

- There was an improvement in the electrical properties of the ZnO:Al thin film after the irradiation by reactor neutrons.
- The enhancement of the electrical performance in ZnO:Al/p-Si heterojunctions was determined for all of the Al concentration at Cu/ZnO:Al/p-Si/Al configuration.
- The decrease of electrical resistivity of ZnO:Al thin film deposited by solgel dip coating technique was concluded to have the potential for use in electronic devices.
- Irradiated thin ZnO:Al film modules are special when compared to traditional crystalline products. Therefore, it can be said that the irradiated thin ZnO:Al film cell structure is appropriate for the usage of solar cell material.

References

Baydogan, N., Ozdemir, O., & Cimenoglu, H. (2013). The improvement in the electrical properties of nanospherical ZnO: Al thin film exposed to irradiation using a Co-60 radioisotope. *Radiation Physics and Chemistry, 89*, 20–27.

Coskun, C., Guney, H., Gur, E., & Tuzemen, S. (2009). Effective annealing of ZnO thin films grown by electrochemical deposition technique. *Turkish Journal of Physics, 33*, 49–56.

Elliot, T. B. (2005). *Focus on semiconductor research*. NewYork: Nova.

Ramirez, J., Israel, Y. V., Li, Y. V., Basantani, H., & Jackson, T. N. (2013). *Effects of gamma-ray irradiation and electrical stress on ZnO thin film transistors*. Canada: IEEE. ISBN: 978-1-4799-0814-1/13.

http://solopower.com/solutions-technology/thin-film-photovoltaics/.

Yavuz, H., Bilge, A. N., Unseren, E., Bayulken, A., Tugrul, A. B., Durmayaz, A. (1993). Modernisation of ITU TRIGA Mark-II Reactor Institute for Nuclear Energy. Istanbul: Istanbul Technical University Press.

Chapter 17
The Characteristic Behaviors of Solgel-Derived CIGS Thin Films Exposed to the Specific Environmental Conditions

Utku Canci Matur, Sengul Akyol, Nilgun Baydogan and Huseyin Cimenoglu

Abstract This study was performed to determine the time effect on optical properties of solgel-derived $Cu(In,Ga)Se_2$ (CIGS) thin films at the specific environmental conditions. For this purpose, solgel-derived CIGS thin films were exposed to a variety of environmental conditions at different steps of the production process from the preparation of solution to deposition of substrate. The optical properties of the CIGS thin films changed with the rise of time at the specific environmental conditions such as the increase of the aging time of solgel solution (from 18 to 35 days) and the rise of the annealing time of the thin film (from 15 to 60 min). The CIGS solution was aged with the rise of time to investigate the effect of the aging time on optical properties of the thin film. The films were deposited by aged colloidal solution which kept at −5 °C in a dark environment in order to extend the useful life of solution. The color of the colloidal solution changed slightly with the increase in the elapsed time after the preparation of the solution. The increase of the annealing time has affected the optical behaviors of the CIGS thin films with the changes of the surface morphology.

U.C. Matur
Gedik University and Istanbul Technical University, Istanbul, Turkey
e-mail: utku.canci@gedik.edu.tr

S. Akyol · N. Baydogan (✉) · H. Cimenoglu
Istanbul Technical University, Istanbul, Turkey
e-mail: dogannil@itu.edu.tr

S. Akyol
e-mail: akyolsh@gmail.com

A.N. Bilge et al. (eds.), *Energy Systems and Management*,
Springer Proceedings in Energy, DOI 10.1007/978-3-319-16024-5_17

17.1 Introduction

Solar cells are renewable energy devices that convert sunlight directly into electricity. There are many researches on the application of photovoltaic technology to support the energy efficiency and management nowadays. The promising energy materials containing CIGS thin films are expected to enable photovoltaic systems with inexpensive upscale solution at processing technologies and to become more widely adopted to meet worldwide energy demand (Todorov et al. 2012) $CuIn_{1-x}Ga_xSe_2$ (CIGS) thin film attracts attention nowadays as absorber layer for the solar cells due to their direct band gap, high absorption coefficient, high efficiency, stability, and economic in manufacture (Canci Matur et al. 2013; Baydogan et al. 2014). CIGS thin film solar cells can achieve conversion efficiency of 20.3 % (Jackson et al. 2011). Both in production process of solar cells, and during their performance, the distribution of dopants, impurities, and especially defects cannot be uniform and unpredictable, and they directly influence the processes in the cells (Hall 1981; Hovel 1975). There were many methods to growth CIGS thin films which were sputtered stacked elemental layer, rapid thermal processing, electrodeposited stacked elemental layer, coatings of nanoparticles precursor layer (non-vacuum), selenization with H_2Se, magnetron sputtering in Se atmosphere, screen printing of nanoparticles, co-evaporation, and doctor-blade coating (Brémaud 2009). Co-evaporation of elements and selenization of elemental layers to form $Cu(In,Ga)Se_2$ is the most preferred method (Mickelsen and Chen 1980; Dimmler et al. 1996; Fredric et al. 1993).

But co-evaporation method is not suitable to control the variations in the elemental fluxes. Besides, selenization in H_2Se has toxicity of materials and non-uniformity problems. Because of these problems, new researches are necessary for deposition of CIGS thin films such as the laser annealing of elemental precursors, flash evaporation (Benslim et al. 1992), and solgel method. The solar cells that were made from CIGS absorber layer virtually have no degradation with time (Tarrant et al. 1991). However, this presented study indicated that the investigation on possible degradation of main optical characteristics is necessary at the different production steps of the CIGS thin film. The changes of optical characteristics are required for the evaluation of the elapsed time through the aging of produced solution and the increase in the annealing time of the CIGS thin film for the use in solar cells efficiently.

In our previous works, the electrical properties of the CIGS thin films which derived by solgel technique were investigated at different Se concentrations and pH values. The thin films exhibited $7.82 \times 10^2\ \Omega\ cm^{-1}$ high electrical conductivity with 7.8 at.% Se concentration at pH = 2. This study was initiated to determine the time effect on optical properties of CIGS thin films produced at the specific environmental conditions. The CIGS samples are produced with the same solution which left for aging for 18, 27, and 35 days. Then, each sample annealed in air at temperature of 150, 175, and 200 °C is characterized optically to evaluate optical band gap. Besides, the CIGS thin films were annealed at four different annealing times (such as 15, 30, 45, and 60 min) in air at 180 °C, and then, CIGS thin films were characterized structurally and optically for the tuning of their optical band gap.

17.2 Experimental Procedure

17.2.1 Production of CIGS Thin Films

CIGS thin films were produced by solgel dip-coating technique to apply this technique on large areas such as glass substrates. Prior to deposition, all of the substrates were sonicated in acetone and alcohol for 10 min, rinsed in distilled water for 10 min, and then dried in still air. The solution consisted of absolute ethanol, as the solvent material. High-purity copper nitrate hydrate ($Cu(NO_3)_2{\cdot}3H_2O$; 99.999 %), indium nitrate hydrate ($In(NO_3)_2{\cdot}3H_2O$; 99.999 %), and gallium nitrate hydrate ($Ga(NO_3)_3{\cdot}H_2O$; 99.9 %) were used as the starting material, followed by the addition of an ethanol solution (20 mL) with terpineol (Fruka, 14 g) and ethyl cellulose (Aldrich, 0.75 g) (Park et al. 2011). The SeO_2 was doped in CIG solution which solved in ethanol. CIGS solution was prepared via the solgel route employing diethanolamine and hydrochloric acid for stabilizing pH. The cleaned substrates were immersed in the solution by employing a computer-controlled dip coater (KSV LMX2).

The relative humidity of the laboratory environment was held constant at 40 % by using a dehumidifier while preparing the solution and during the dip-coating process. The cleaned substrates were immersed in the solution by employing a computer-controlled dip coater (KSV LMX2). The dipping and withdrawing speeds of the substrates were 60 mm min^{-1} with a holding time of 5 s between the dipping and withdrawing periods. Following deposition, the films were dried in air for 10 min and then that preheated films annealed at 180 °C for 15, 30, 45, and 60 min in air. The deposition and drying cycles were successively repeated 5 times. The raise of annealing time effected the adhesion properties of the chalcopyrite $Cu(In,Ga)Se_2$ structure to the soda–lime–silicate glass substrate. The thickness of the films was measured by the surface profilometer. The thickness of CIGS thin films was $\sim$150 nm.

The transmittance (% T) and the reflectance (% R) of the CIGS thin films were measured by using a double-beam spectrophotometer between the ranges of 190 and 1100 nm. Atomic force microscope (AFM) images were determined to characterize the surface morphology of thin films. Phase analysis of the CIGS thin films deposited on soda lime silicate glass was made by qualitative X-ray diffraction (XRD) using Cu-Kα radiation. X-ray diffractometer was operated at 40 kV and 30 mA with a constant scan rate of 2°/min. XRD analyses were performed to scan the samples in a θ range between 0° and 90°. The surface morphology of the films was examined using a JEOL JSM 7000F scanning electron microscope (SEM). The thickness of the thin film was measured by using a Veeco Dektak 6M surface profilometer. The thickness of CIGS thin films was in the range ~150 nm.

17.3 Results and Discussion

17.3.1 Structural Characterization

The XRD planes at (112) and (211) presented slightly two diffraction peaks, and the intensity of diffraction peaks increased slowly which annealed at 180 °C for 60 min The XRD patterns are shown in Fig. 17.1. This sample exhibits one peak corresponding to $CuInSe_2$ (112) (Park et al. 2011) and one peak corresponding to Cu(ln, Ga)Se_2 (211) phase (Javed 2007). There was not a considerable crystalline structure for the other annealing times less than 60 min.

AFM images of the CIGS thin film annealed at 150 °C for 60 min performed an optimization at optical transparency. It was analyzed for the identification of surface morphology. The surface morphology of CIGS thin films was evaluated by using Park Systems model AFM which was chromium-/gold-coated AFM device tip (HQ: NSC15/CR-AU) radius <35 nm. AFM images of the CIGS thin film derived from a colloidal solgel solution presented not so uniform surface morphology as shown in Fig. 17.2. In the general trend, the thin films exhibited spherical morphology on the surface of the CIGS grains similar to the literature (Yu et al. 2013).

Figure 17.3 shows the surface morphology of the CIGS films annealed at 180 °C for 60 min which is not so uniformly distributed with round CIGS grains with size about varied in between 0.25 and 1 m. Small cracks and pinholes can be found in this sample, owing to the significant Cu deficient feature of the film (Yu et al. 2013).

Particle distribution along the matrix is not so uniform that some parts of the matrix are free from the particles.

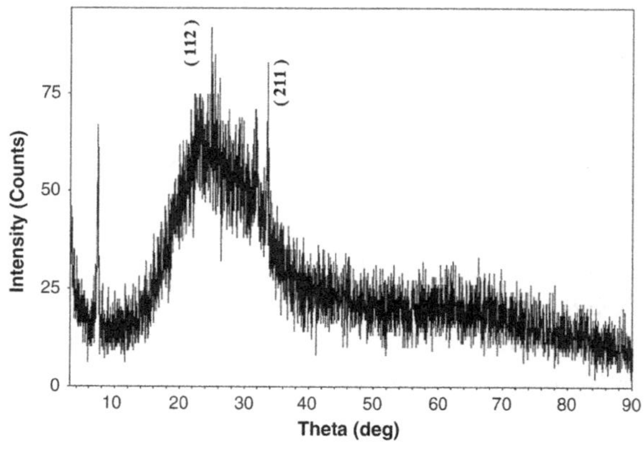

Fig. 17.1 XRD analysis of the CIGS thin film annealed for 60 min at 180 °C

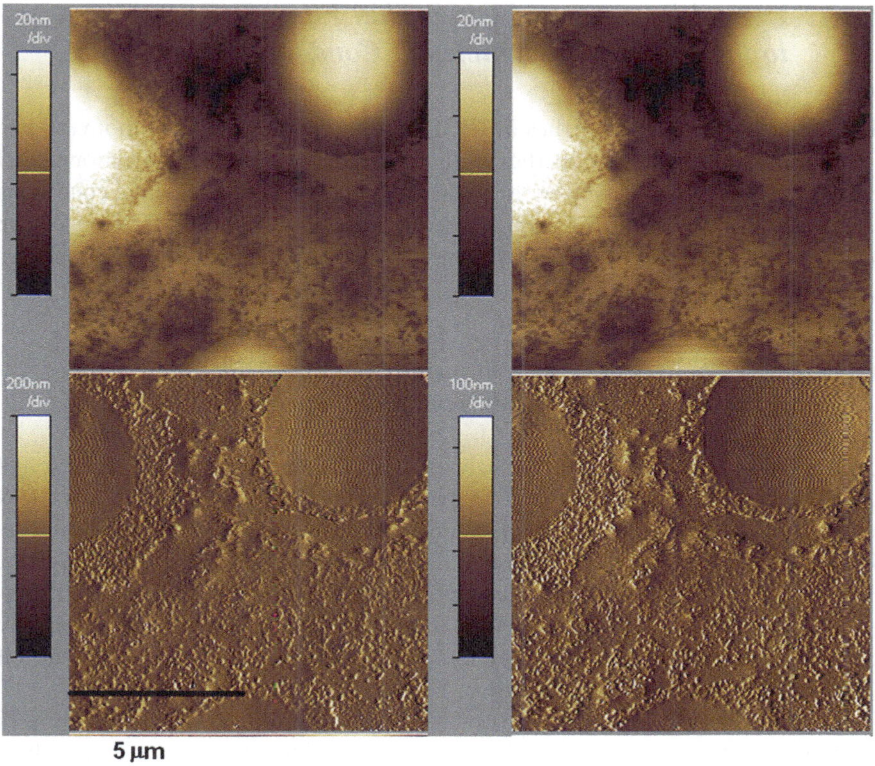

Fig. 17.2 AFM images of the CIGS thin film annealed for 60 min at 180 °C

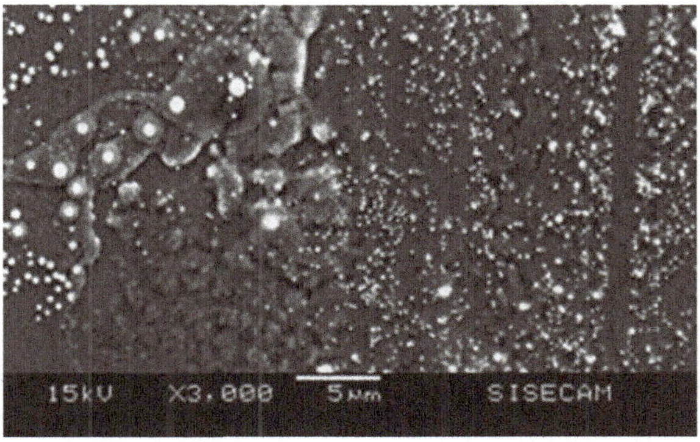

Fig. 17.3 SEM micrographs of the CIGS thin film annealed for 60 min at 180 °C

17.3.2 Optical Properties of CIGS Thin Films Exposed to the Specific Environmental Conditions

The changes of optical properties of CIGS thin films were examined to investigate the time effect on optical properties of the thin films at the specific environmental conditions. CIGS thin films were exposed to a variety of environmental conditions (such as the increase of the aging time of solgel solution and the rise of the annealing time of the thin film). Handling time has improved the optical properties of the thin film.

17.3.2.1 The Changes of Optical Properties of CIGS Thin Films with the Rise of the Aging Time of the Solution

Optical transparent electrically conductive CIGS thin films were produced by solgel dip-coating method. The films were deposited by aged colloidal solution. The CIGS solution was kept at −5 °C in a dark environment in order to extend the useful life. The CIGS solution was aged at three different times such as 18, 27, and 35 days. Besides, each of CIGS thin film samples annealed at different annealing temperatures including 150, 175, and 200 °C. All CIGS thin film samples were optically characterized by obtaining their transmission plots. The transmission plots for all the samples are shown in Fig. 17.4a–c.

From the transmittance, it can be observed that aging of the solution made losses in the optical transmittance with the enhancement of the absorbance at the films. The optical properties of the aged solution were compared with the optical properties of the un-aged solution. The optical transmittance of the film decreases by the increase of the aging time of the solution. There were the changes in transmittance and absorbance of the thin film coated on the soda–lime–silicate glass due to the increase of the aging time. The absorption plots for all the samples are shown in Fig. 17.5a–c.

Figure 17.6a–c shows the plot of $(h)^2$ versus h. The linear fit of $(h)^2$ versus h allows us to get the value of E_g. The allowed direct band gap was determined at the solgel-derived CIGS films by solgel dip-coating process in Fig. 17.6. CIGS thin films have an optical band gap to transmit the useful solar radiation for the utilization of transparent conductor.

17.3.3 The Changes of Optical Properties of CIGS Thin Films with the Rise of the Annealing Time

The optical properties exhibited the highest optical absorbance of the CIGS thin films annealed at 180 °C. Hence, the aged solution researches on the soda–lime–silicate substrates were performed at different aging time, and then, all of the films were annealed at 180 °C in air to obtain the high absorbance.

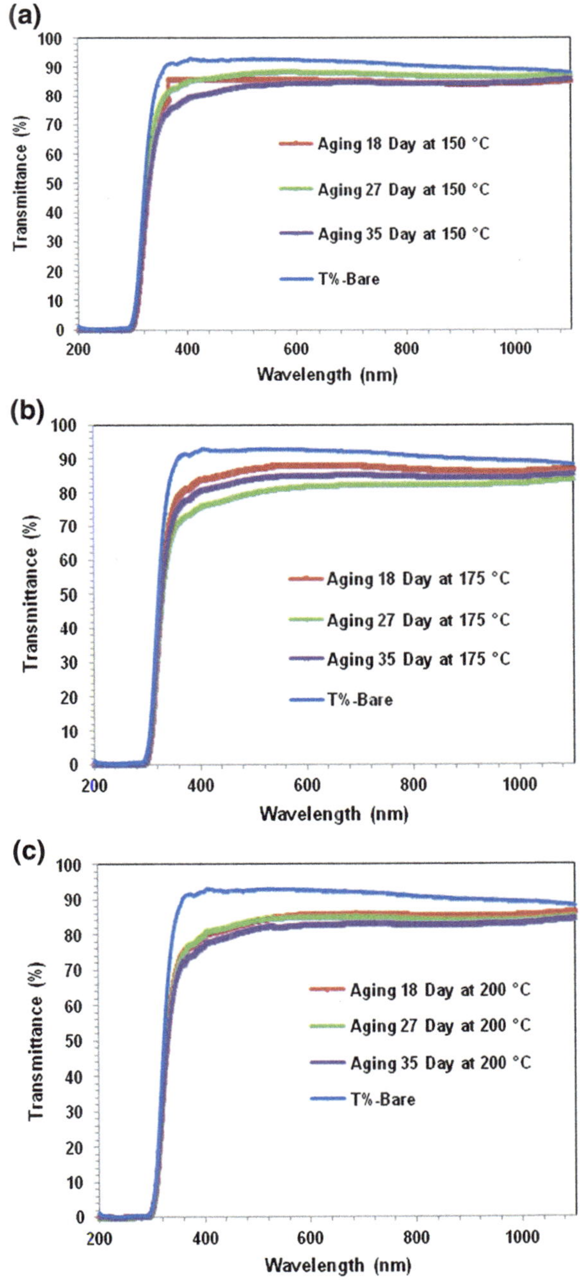

Fig. 17.4 **a** The transmittance of CIGS thin films annealed at 150 °C. **b** The transmittance of CIGS thin films annealed at 175 °C. **c** The transmittance of CIGS thin films annealed at 200 °C

CIGS thin film specimens were optically characterized by measuring their transmittance (T%) and absorbance (A%) by using a double-beam spectrophotometer. The transmittance and absorbance for all the samples are shown in

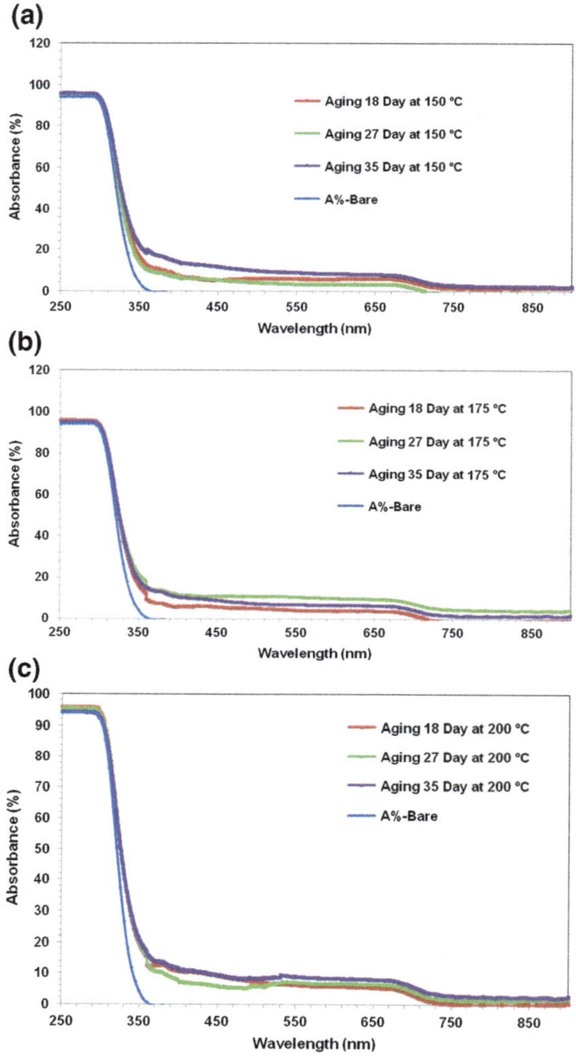

Fig. 17.5 **a** The absorbance of CIGS thin films annealed at 150 °C **b** The absorbance of CIGS thin films annealed at 175 °C **c** The absorbance of CIGS thin films annealed at 200 °C

Figs. 17.7 and 17.8. The transmittance and absorbance plots indicated that the absorption of the samples increases and transmission decreases with the rise of the annealing time at 180 °C in Figs. 17.7 and 17.8, respectively. The transmittance reaches the highest value in the range of 780–1080 nm for four different annealing times. The CIGS thin films annealed 45 min have the highest transmittance value. Besides, the films annealed for 60 min have the smallest transmittance value between 200 and 1100 nm in Table 17.1. The transmittance raises exhibiting interference patterns in the weak absorption (Colakoglu et al. 2008). The thin films were non-absorbing beyond ~295 nm for all samples which were annealed at 180 °C during four different annealing times.

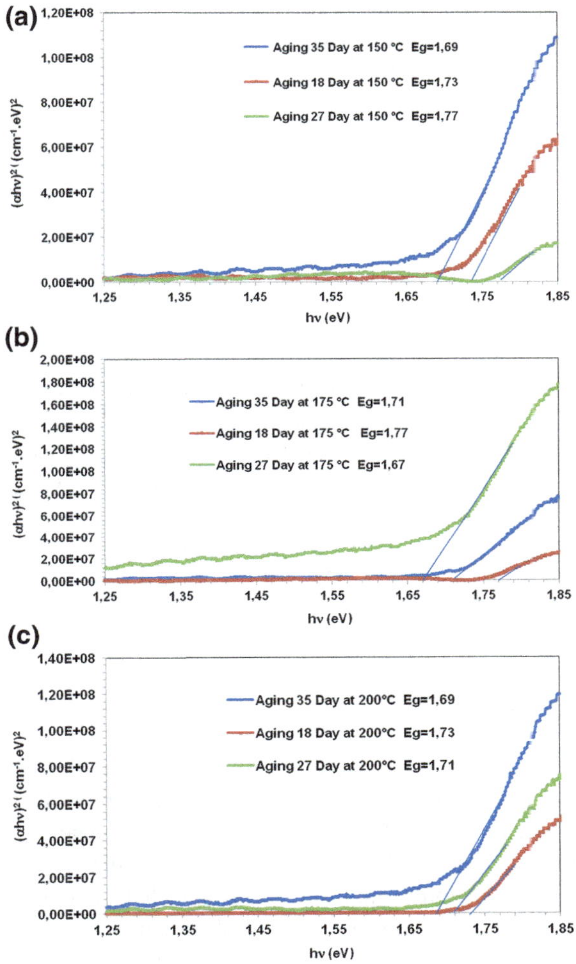

Fig. 17.6 **a** The optical band gap of CIGS thin films annealed at 150 °C. **b** The optical band gap of CIGS thin films annealed at 175 °C. **c** The optical band gap of CIGS thin films annealed at 200 °C

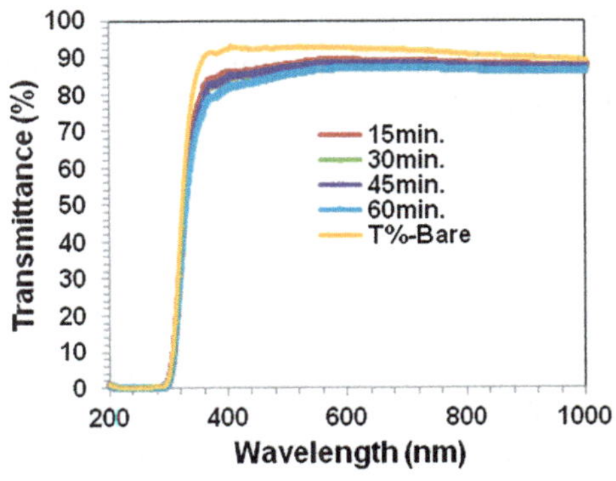

Fig. 17.7 The transmittance of the CIGS thin film at different annealed times at 180 °C

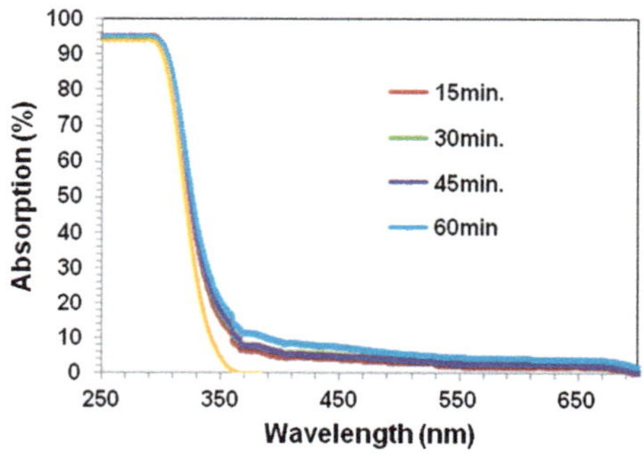

Fig. 17.8 The absorbance of the CIGS thin film at different annealed times at 180 °C

Table 17.1 The transmittance of GIGS thin films at different annealing times at 180 °C

Different annealing time (at.%)	Ultraviolet range (193–378 nm)		Visible range (380–780 nm)	Near-infrared range (780–1100 nm)
	Wavelength (nm)	Average T_{uv} (%)	Average T_{vis} (%)	Average T_{NIR} (%)
15	323–378	73.19	88.24	88.26
30	323–378	71.26	87.38	87.71
45	323–378	71.58	87.54	87.82
60	323–378	68.30	85.86	86.73
Substrate	323–378	82.02	92.44	89.85

The absorption edge of the CIGS thin films raised slightly from ~400 to ~600 nm with the increase of annealing time from 15 to 60 min. The blue shifting was significant at 15 min annealing time. The blue shifting of absorption edge indicates the CIGS thin films annealed 15 min. The films annealed 15 min exhibited lower absorption than the films annealed 60 min in the range between 200 and 1100 nm. The improvement of absorbance was defined with the rise of annealing time in Fig. 17.8. The rise of the annealing time caused to the increase of absorbance of the samples with the decrease of transmittance.

The absorption coefficient (α) used for calculating the optical band gap (E_g) of the CIGS thin film is a direct band gap semiconductor for direct transition, which is calculated from the transmission spectrum in Eqs. 17.1 and 17.2 (Abaab et al. 1999; Parlak et al. 1998).

$$\alpha d = \ln(I_0/I) \tag{17.1}$$

$$\alpha h\vartheta = A(h\vartheta - E_g)^{1/2} \tag{17.2}$$

where d is the thickness of the film, I is the transmitted light intensity, A is a constant, h is the photon energy, and E_g is the optical energy gap. Figure 17.9 shows the plot of, $(\alpha h\vartheta)^2$ versus h. The linear straight line of $(\alpha h\vartheta)^2$ versus $h\vartheta$

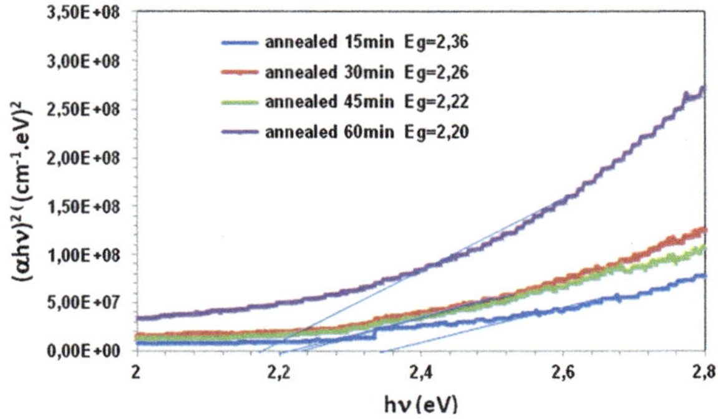

Fig. 17.9 The optical band gap of the CIGS thin film at different annealed times at 180 °C

allows us to get the value of E_g clearly. CIGS thin films have an optical band gap to transmit the useful solar radiation for the utilization of transparent conductor. The alteration in the optical band gap of the thin film depends on the increase of the annealing time of the thin film. The CIGS thin film that annealed at 180 °C for one hour has the smallest band gap which corresponds to literature.

In this study, the E_g of CIGS thin films decreased from 2.36 to 2.20 eV with the increase of annealing time from 15 to 60 min.

17.4 Conclusions

This study reports the effect of aging and annealing time on the optical properties of the CIGS thin films. Transmittance of the CIGS thin films decreased with increase in the aging time of the solution.

The decrease of the optical transmittance was distinguished with the rise of the annealing temperature after the aging process. The absorbance of the film has increased with the rise of the elapsed time after the preparation of the solution.

There is a slight decrease in the optical band gap with the rise of the elapsed time of the solution after the aging process. For this study, annealing temperature did not affect the energy of band gap so much.

CIGS thin films that derived solgel technique as an economical way are direct band gap I-III-VI_2 quaternary semiconductor which promised as an absorber layer because of the band gap match with the solar spectrum. The small amount of Ga in CIGS film effects the band gap properties of the film, and it is important for the increase of the band gap value.

This study reports the effect of annealing time on the structural and optical properties of the CIGS thin films. The XRD patterns revealed partly crystalline tetragonal chalcopyrite structure. The CIGS thin films derived by solgel dip-coating technique presented an allowed direct band gap after annealing in air at different times. The energy band gap of the thin films decreased from 2.36 to 2.20 eV with the increase of annealing time from 15 to 60 min. The rise of the annealing time of CIGS thin films at 180 °C optimizes the optical absorbance of the CIGS film. Therefore, the prepared samples can be used as an absorber layer in the fabrication of thin film solar cells.

References

Abaab, M., Kanzari, M., Rezig, B., & Brunel, M. (1999). Structural and optical properties of sulfur-annealed CuInS2 thin films. *Solar Energy Materials and Solar Cells, 59*, 299–307.

Baydogan, N., Canci, U., Akyol, S., & Cimenoglu, H. (2014). Investigation of electrical properties of cigs thin-films derived by sol-gel process. *TMS 143rd Annual Meeting supplemental Proceedings*, 143, pp. 379–381.

Benslim, N., Maasmi, A., Mahdjoubi, L., & Gouskov, L. (1992). *Proceedings of 11th European Photovoltaic Solar Energy Conference and Exhibition* (pp. 890–893), Montreux, Switzerland, October.

Brémaud, D. J. L. (2009). Investigation and development of cigs solar cells on flexible substrates and with alternative electrical back contacts (Doctor of Sciences, Zurich). DISS. ETH No. 18194.

Canci Matur, U., Akyol, S., Baydogan, N., & Cimenoglu, H. (2013). The characterization of se doped CIG thin films. In *1st International Semiconductor Science and Technology Conference, (ISSTC), Istanbul.*

Colakoglu, T., Parlak, M., & Özder, S. (2008). Investigation of optical parameters of Ag–In–Se thin films deposited by e-beam technique. *Journal of Non-Crystalline Solids, 354*, 3630–3636. doi:10.1016/j.jnoncrysol.2008.03.014.

Dimmler, B., Gross, E., Menner, R., Powwalla, M., Hariskos, D., Ruckh, M., et al. (1996). Thin film solar modules based on CIS prepared by the co-evaporation method. In *Proceedings 25th IEEE Photovoltaic Specialists Conference, Washington* (p. 757). doi:10.1109/PVSC.1996.564239.

Fredric, C., Gay, R., Tarrant, D., & Willet, D. (1993). Results of recent CuInSe2 module investigations. In *Proceedings 23rd IEEE Photovoltaic Specialists Conference, Louisville* (p. 437). doi:10.1109/PVSC.1993.347142.

Hall, R. N. (1981). Silicon solar cells. *Solid State Electronics, 24*, 595–616.

Hovel, H. J. (1975). *Solar cells. Semiconductors and semimetals* (Vol. 11, p. 274). New York: Academic Press.

Jackson, P., Hariskos, D., Lotter, E., Paeterl, S., Wuerz, R., Menner, R., et al. (2011). New world record efficiency for Cu(In, Ga)Se2 thin-film solar cells beyond 20 %. *Progress in Photovoltaics, 19*, 894. doi:10.1002/pip.1078.

Javed, A. (2007). Preparation and study of the structural, optical and electrical properties of Cu(In, Ga)Se2 thin films. *Turkish Journal of Physics, 31*(5), 287–294.

Mickelsen. R. A. & Chen, W. S. (1980). High photocurrent polycrystalline thin film CdS/CuInSe2solar cella. In *Proceedings 15th IEEE Photovoltaic Specialists Conference, Orlando* (pp. 371–373). doi:10.1063/1.91491.

Park, S. J., Lee, E., Jeon, H. S., Ahn, S. J., Oh, M. K., & Min, B. K. (2011). A comparative study of solution based CIGS thin film growth on different glass substrates. *Applied Surface Science, 258*, 120–125. doi:10.1016/j.apsusc.2011.08.016.

Parlak, M., & Ercelebi, C. (1998). The effect of substrate and post-annealing temperature on the structural and optical properties of polycrystalline InSe thin films. *Thin Solid Films, 322*(1–2), 334–339. doi:10.1016/S0040-6090(97)00929-2.

Tarrant, D. E., Gay, R. R., Hummel, J. J., Jensen, C., & Ramos, A. R. (1991). CuInSe2 module environmental reliability. *Solar Cells, 30*(1–4), 549–557. doi:10.1016/0379-6787(91)90086-5.

Todorov, T. K., Gunawan, O., Gokmen, T., & Mitzi, D. B. (2012). Solution-processed Cu(In, Ga)(S, Se)2 absorber yielding a 15.2 % efficient solar cell. *Progress in Photovoltaics: Research and Applications, 21*, 82–87. doi:10.1002/pip.

Yu, Z., Yan, Y., Li, S., Zhanga, Y., Yana, C., Liua, L., et al. (2013). Significant effect of substrate temperature on the phase structure, optical and electrical properties of RF sputtered CIGS films. *Applied Surface Science, 264*, 197–201. doi:10.1016/j.apsusc.2012.09.171.

Chapter 18
Effect of Curing Time on Poly(methacrylate) Living Polymer

Tayfun Bel, Nilgun Baydogan and Huseyin Cimenoglu

Abstract Self-healing materials increase the reliability and life span of the overall systems that they are incorporated to. This capacity of the material enhances the credibility of the in-flight health assessment in aerospace platforms. Such platforms do not require visual or acoustic inspections to recover from the damages that occur during flight. This reduces the energy requirements associated with maintenance, replacement of parts, and off-line time. Poly(methacrylate) (PMMA) living polymer mixed with nanoparticles possesses miraculous properties such as minimized gas permeability, improved heat resistance, and boosted physical performance. In this study, PMMA is synthesized by the atom transfer radical polymerization (ATRP) method and the mechanical properties of the material have been manipulated by changing the curing time. The effects of curing time on mechanical properties are examined by the stereomicroscope images and Shore D hardness tests satisfying ASTM D785 test standards.

18.1 Introduction

Modern aircraft use fly-by-wire technique that could not take off without the electronic avionic systems on board. These avionic systems handle flight control, navigation, and communications (Tooley 2007). Radiation damage on these aerospace systems has serious consequences (Iniewski 2010). Such damage has to be prevented by adequate shield. Thermoplastic polymers have good radiation shielding properties, and their performance can be enhanced by the synthesizing

T. Bel · N. Baydogan · H. Cimenoglu (✉)
Istanbul Technical University, Istanbul, Turkey
e-mail: cimenogluh@itu.edu.tr

T. Bel
e-mail: tayfunbel@gmail.com; bel@itu.edu.tr

N. Baydogan
e-mail: dogannil@itu.edu.tr

A.N. Bilge et al. (eds.), *Energy Systems and Management*,
Springer Proceedings in Energy, DOI 10.1007/978-3-319-16024-5_18

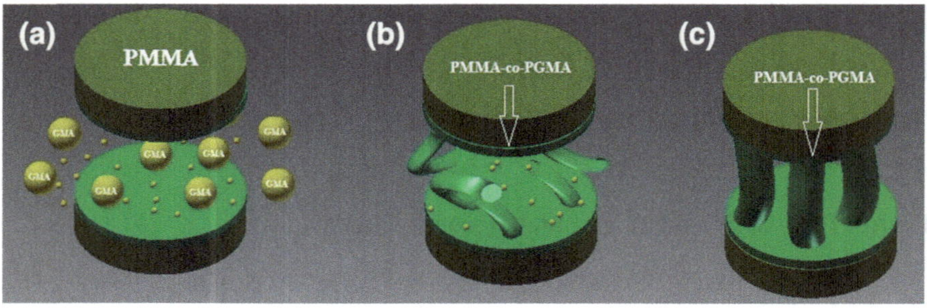

Fig. 18.1 **a** Diffusion of released GMA monomers into living PMMA matrix, **b** growth of living poly(glycidyl methacrylate) (PGMA) chains from copolymer of PGMA and PMMA (i.e., PMMA-co-PGMA), **c** crack filled with PMMA-co-PGMA

method. Among many other methods, room temperature living polymerization gives the thermoplastic self-heal without outside intervention requirement. The healing systems are based on the build-up of chemical bonds on and at the vicinity of the crack plane without a manual intervention. Previous studies show that material still exhibits self-healing capacity after one year of storage. Living polymerization possesses the capacity of recombination of new free radicals and macromolecular radicals onto the chain ends. This capacity might allow the self-healing of material from the damage induced by ionizing, ultraviolet, and electromagnetic radiation (Bel et al. 2014a, b; Zhang and Rong 2011).

Schematic drawing of crack repair by living copolymerization of GMA from the cracked surfaces of living PMMA matrix is presented in Fig. 18.1. When crack is formed within the PMMA matrix material, it follows a preprogrammed path through the microcapsules loaded with healing agent. The crack ruptures the wall of the microcapsule, and the healing agent glycidyl methacrylate (GMA) travels through the material by capillary effect. Broken polymer chains at the crack surface start to polymerize and grow until meeting the opposite crack surface. Eventually, cracked area heals and material regains its mechanical performance and physical properties (Bel et al. 2014a, b).

The capacity of self-healing materials reduces the energy requirements associated with maintenance, replacement of parts, and off-line time. New aircraft can use advanced technology at several parts of their systems such as aerodynamics and composite structures in order to obtain significant improvements in energy efficiency. Self-healing materials have attracted much attention recently because they can potentially improve the energy efficiency. The use of poly(methacrylate) (PMMA) living polymer addresses the reduction in the energy requirements related to maintenance at aerospace platforms by replacing conventional heavy and manually maintained materials. Besides, lightweight self-healing shielding materials have the potential to reduce significantly the energy consumption. Hence, synthesis of PMMA by ATRP method by using an economical process and the investigations of their mechanical properties are necessary to reduce the financial problems of energy requirements.

18.2 Experimental Procedure

18.2.1 Materials Fabrication

Materials are used as they arrived; the preparation and measurements are done inside AtmosBag four-hand, non-sterile, size L, closure-type, tape seal (Sigma-Aldrich). Tetra-n-butylammonium bromide, (98+ % Alfa Aesar), copper bromide, Puratronic®, 99.998 % (metals basis) (Alfa Aesar), methyl methacrylate, 99 %, stab (Alfa Aesar), 1,1,4,7,7-pentamethyldiethylenetriamine, 98 % (Alfa Aesar), ethyl 2-bromoisobutyrate, 98+ % (Alfa Aesar).

Extra-pure argon and regulator are obtained from local supplier and chained to a concrete wall to meet the safety regulations. Before materials measurements, AtmosBag four-hand, non-sterile, size L, closure-type, tape-seal (Sigma-Aldrich) is completely vacuumed, followed by 10-min extra-pure argon circulation into system (Fig. 18.2).

18.2.2 Synthesis of Living polymerization PMMA by ATRP

Bu_4NBr (1.031 g, 3.2 mmol) and CuBr (0.057 g, 0.4 mmol), and PMDETA (0.069 g, 0.4 mmol) were added to tube (8 × 2.5 cm) equipped inside AtmosBag. Inside the glove bag, the extra-pure argon is applied to the tube and tube is sealed with rubber septum with pressured argon and taken out of the AtmosBag. Tube merged inside an oil bath at 25 °C, and external argon appliances were connected accordingly. Previously prepared MMA (24.0 g, 0.24 mol) inside gas-tight syringe is applied to mixture under continuous argon purge under magnetic string (Fig. 18.3).

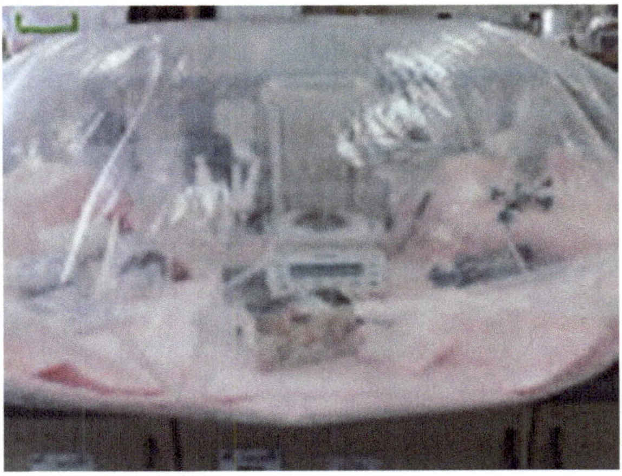

Fig. 18.2 Experimental setup forecasting of polymer in argon ambience

Fig. 18.3 Argon purging at the surface of the solution and inside the solution

Magnetically, string continued for the next 13 h, and after 13 h, EBiB (0.078 g, 0.4 mmol) was added. The mixture has been degassed completely through glass frit with argon. External argon sources prevent the mixture to get contact with atmosphere. The tube is sealed with rubber septum followed by parafilm.

After 4 h, the resultant polymer was a transparent solid exhibiting a light green color. Polymer has been moved from oil bath to AtmosBag and poured into molds. As the contact of the PMMA specimens with atmosphere was prevented the samples were achieved by ATRP (Fig. 18.4).

Fig. 18.4 Living polymer PMMA produced by ATRP method

Table 18.1 The hardness test results of the produced ATRP are presented for living polymer PMMA

Time (h)	Hardness test
13	65–70
19	81–85

18.3 Results and Discussion

The PMMA samples presented significant improvement in the hardness values by increasing the curing time. The hardness is well known indication for the overall strength of the material. Hence the hardness values of PMMA samples cured at different times were examined in this study. Further studies will be done for the impact strength values of these samples to prove the relationship between the curing time and the hardness and the impact strength. The underlying mechanism of the curing time effect on mechanical properties can be assumed as the PMMA material can form better and stronger bond structure at 19 h. However the premature ending of the polymerization will cause to weaker bonding with the decrease of the curing time to 13 h. The hardness test results of the produced PMMA are presented in Table 18.1.

18.4 Conclusions

- Curing time can affect solely the mechanical properties of PMMA produced by ATRP.
- The hardness test result of PMMA (acrylic) sheets is higher than the industrial literature, and hardness test result in industrial literature is 80 according to ASTM D785 standards.
- It is possible to develop lightweight self-healing shielding for aerospace platforms that can increase the reliability and life span of the overall systems.
- Lightweight self-healing shielding materials can reduce the energy requirements associated with maintenance at aerospace platforms by replacing conventional heavy and manually maintained materials.

References

Bel, T., Baydoğan, N., & Çimenoğlu, H. (2014a). Production of PMMA via living polymer with ATRP method. Paper presented at the 1st international semiconductor science and technology conference (ISSTC-2014), Istanbul.

Bel, T., Baydoğan, N., & Çimenoğlu, U. (2014b). Effect of curing ambient on Poly (methacrylate) living polymer. *International Journal of Mechanical And Production Engineering, 2*(3), 55–57.

Iniewski, K. (2010). Radiation effects in semiconductors. UK: Taylor & Francis.

Tooley, M. H. (2007). *Aircraft digital electronic and computer systems: Principles operation and maintenance*. Oxford: Butterworth-Heinemann.

Zhang, M. Q., & Rong, M. Z. (2011). Self-Healing polymers and polymer composites. Hoboken: Wiley.

Chapter 19
Effects of Production Parameters on Characteristic Properties of $Cu(In,Ga)Se_2$ Thin Film Derived by Solgel Process

Sengul Akyol, Utku Canci Matur, Nilgun Baydogan and Huseyin Cimenoglu

Abstract $Cu(In,Ga)Se_2$ (CIGS) thin films were obtained by solgel method on soda-lime glass substrates, economically. The optimum optical properties of CIGS thin films are obtained by varying the film layers. Besides, the solgel-derived CIGS thin films were thermally treated at different temperatures from 135 °C up to 200 °C. These results indicate that the transparent CIGS thin films derived by solgel process can be good candidates for the applications in optoelectronic devices.

19.1 Introduction

Chalcopyrite compound $Cu(In,Ga)Se_2$ (CIGS) is one of the most promising absorber materials for thin film solar cells, which exhibits the highest efficiency among all materials for the use of thin film in solar cells (Brémaud 2009). The light absorbing value and the photoelectric conversion efficiency are high enough that could be compared with silicon multicrystalline solar cells. CIGS is a direct band gap material and has a light absorption coefficient higher than 10^5 cm^{-1} (Park et al. 2012; Chuang et al. 2011). The main method that used for deposition is conventional vacuum-based processes, i.e., a multistage co-evaporation and a two-step process of sputtering and

S. Akyol · H. Cimenoglu
Department of Metallurgical and Materials Engineering, Istanbul Technical University, Istanbul, Turkey
e-mail: akyolsh@gmail.com

H. Cimenoglu
e-mail: cimenoglu@itu.edu.tr

U.C. Matur
Energy Institute, Gedik University and Istanbul Technical University, Istanbul, Turkey
e-mail: utku.canci@gedik.edu.tr

N. Baydogan (✉)
Energy Institute, Istanbul Technical University, Istanbul, Turkey
e-mail: dogannil@itu.edu.tr

A.N. Bilge et al. (eds.), *Energy Systems and Management*,
Springer Proceedings in Energy, DOI 10.1007/978-3-319-16024-5_19

selenization which has high production cost (Tolan et al. 2009; Chung et al. 2011). Solgel method is one of the best candidates for producing CIGS thin films for the solar modules economically (Park et al. 2012; Tolan et al. 2009; Chung et al. 2011).

The aim of this study was to make a contribution for the improvement of the cost-effective method by using a solgel deposition technique and to obtain polycrystalline CIGS thin films. Layers 3, 5, 7, 9, and 11 were deposited on different glass substrates. The optical properties of CIGS thin films can be controlled by varying the film layer. CIGS thin films with 30 at.% Se concentration were examined at different annealing temperatures such as 135, 150, 160, 170, 180, and 200 °C.

19.2 Experimental Procedure

CIGS solution was prepared by using Copper(II) Nitrate Hydrate ($Cu(NO_3)_2 \cdot 3H_2O$; 99.999 %), Indium(III) Nitrate Hydrate ($In(NO_3)_2 \cdot 3H_2O$; 99.99 %) and Gallium(III) Nitrate Hydrate ($Ga(NO_3)3H_2O$; 99.9 %) as the starting material with a ratio 1:0.7:0.3:2 (Park et al. 2011). The ethyl-cellulose was added as a stabilizer in solution. Hence, it was produced an economic CIGS thin film by using solgel dip-coating process (Park et al. 2012). This viscous solution was mixed with Cu-, In-, and Ga-containing ethanol solution with a magnetic stirrer for 1 h. In 20 mL ethanol, ethyl-cellulose (EC) (Aldrich, 0.75 g) and terpineol (Fluka, 15 mL) was dissolved and this viscous solution was mixed with the Cu-, In- and Ga- containing ethanol solution (Park et al. 2011). Selenium(IV) Oxide SeO_2 (Merck; 99.999 %) was mixed in 10 mL ethanol separately, and the solutions with different selenium concentrations were performed. Preparation of solgel solution produced CIGS thin film is depicted in Fig. 19.1. The preparation of solution was performed at room temperature in air. The relative humidity of the laboratory environment was held constant at 43 % by using a dehumidifier while preparing the solution and during the dip-coating process. Diethanolamine (DEA) was used as the stabilizing agent. The pH value was controlled with aid of acetic acid and hydrochloric acid (Akyol 2015).

CIGS thin films were produced by using solgel dip-coating technique. The cleaned soda-lime silicate glass substrates were withdrawn into the solution at a rate of 1 $mm.s^{-1}$, and were kept in the solution for 5 s. The films were then dried at room temperature for 10 min in an air ambient. Coating and drying process were repeated 3, 5, 7, 9, and 11 times. Finally, heat treatment was performed at 135, 150, 160, 170, 180 and 200 °C for 30 min under air enviroment.

19.3 Results and Discussion

The optical properties of the thin films changed with the increase of the annealing temperature in Fig. 19.2. The transmittance and absorbance changed at different annealing temperature. The changes of energy band gap were presented after the

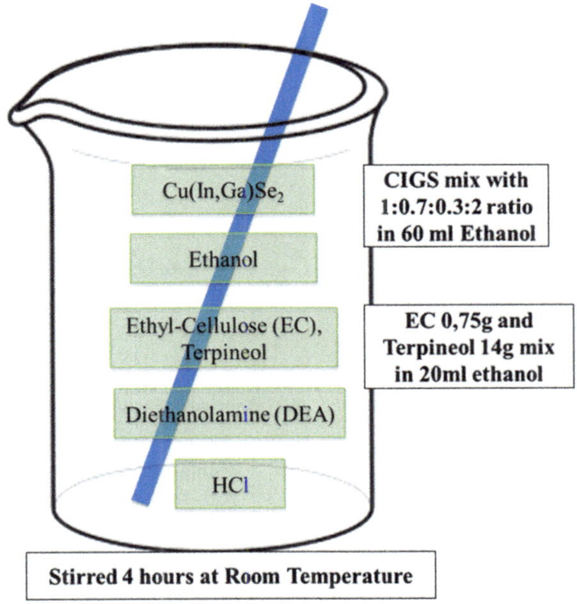

Fig. 19.1 Preparation of GIGS solution

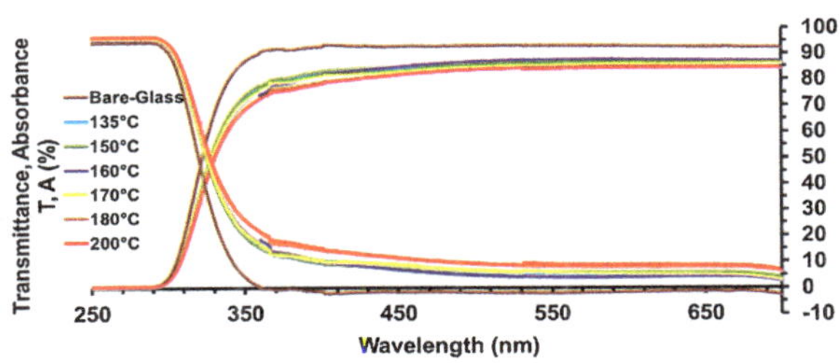

Fig. 19.2 Effect of annealing temperature on optical properties

annealing process at 150 °C and with no annealing process as shown in Fig. 19.3. The details on the changes of transmittance and absorbance of CIGS thin film annealed at different temperatures were presented in different ranges of electromagnetic spectrum in Table 19.1.

Surface resistivity of the films was measured by four-point probe technique. Hence, it was possible to determine the surface conductivity of the CIGS thin films. The changes in electrical conductivity at different annealing temperatures such as (a) no anneal, (b) 137 °C, (c) 150 °C, and (d) 200 °C are shown in Fig. 19.4.

XRD patterns of CIGS thin films with no annealing and with annealing at 150 and 200 °C for 30 min are shown in Fig. 19.5. The XRD analysis of the CIGS thin films annealed at ~150 °C, ~200 °C and no annealing are illustrated in Fig. 19.5.

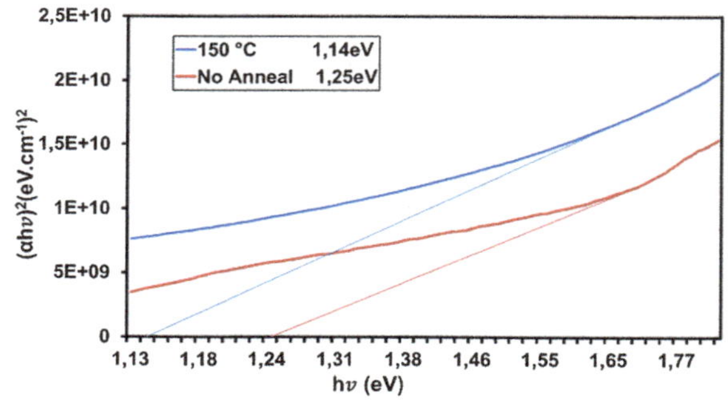

Fig. 19.3 The annealing effect on the energy band gap. The energy band gap at 150 °C annealing temperature and with no annealing process

Table 19.1 The changes of transmittance and absorbance with the rise of the annealing temperature

Anneal temperature (°C)	UV (190–380 nm)		Visible (380–780 nm)		IR (780–1100 nm)	
	T (%)	A (%)	T (%)	A (%)	T (%)	A (%)
135	22.98	71.24	85.60	5.08	86.05	0.03
150	23.41	71.05	85.48	5.46	85.51	1.00
160	22.67	71.55	85.90	4.33	85.73	−1.17
170	22.85	71.62	84.61	5.39	85.63	−0.28
180	21.73	73.33	83.23	8.20	83.96	2.97
200	21.89	73.46	82.89	8.76	82.89	3.22
Bare	27.90	64.86	92.44	−2.29	89.85	−3.18

The XRD analysis of the CIGS thin films was performed by qualitative X-ray diffraction (XRD) using Cu-Kα radiation. X-ray diffractometer was operated at 40 kV and 30 mA with a constant scan rate of 2°/min. XRD analyses were performed to scan the samples in a 2θ range between 0° and 90°. Collected XRD patterns of the CIGS thin film derived the solution at ~2-pH value presented partly crystallization when the films were annealed at 150 °C. The thin films in Fig. 19.5 exhibited $CuInSe_2$ (112) (Park et al. 2012) and Cu(In,Ga)Se_2 (211) (Javed 2007). Diffraction planes on the surface of the CIGS thin film are similar to the literature.

Surface morphology of CIGS thin films was investigated by using SEM micrographs at different annealing temperatures in Fig. 19.6a–c. SEM image of the CIGS thin film presented the cracked surface before the annealing process as shown in Fig. 19.6c. Annealing process caused to agglomeration of the CIGS grains on the surface of the film (in Fig. 19.6a, b). The cracks on the surface of the film disappeared by the annealing treatment at 150 °C in air as the annealing treatment caused the diffusion on the surface of the film (Yu et al. 2013). SEM image of the film annealed at 200 °C was presented in Fig. 19.6a. The increase of the annealing temperature from 150 to 200 °C resulted with the finer grains on the surface.

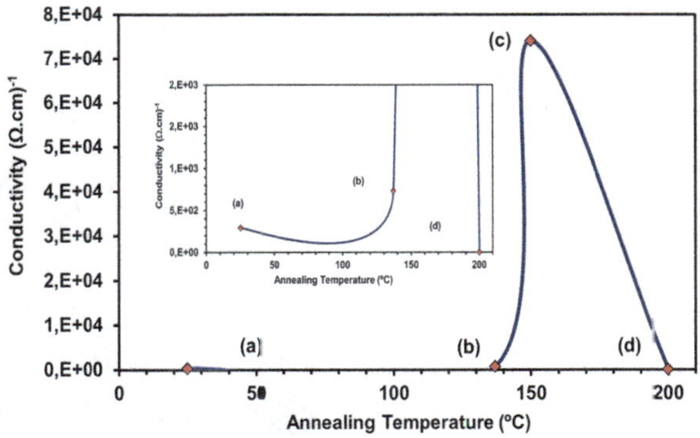

Fig. 19.4 Effect of annealing temperature on electrical properties

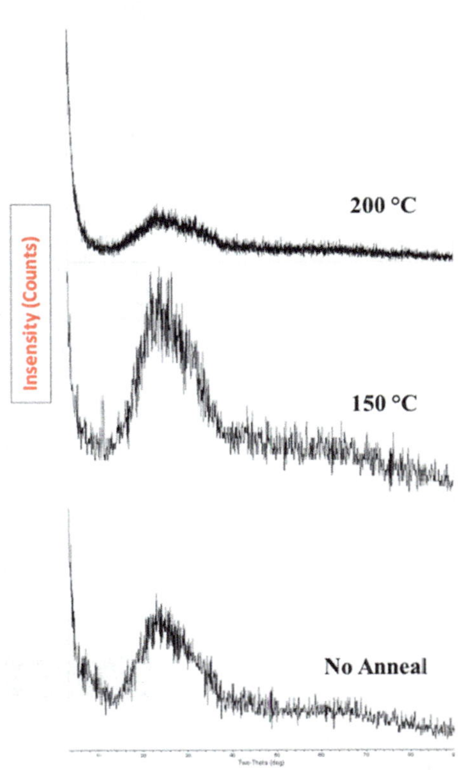

Fig. 19.5 XRD analysis of CIGS thin films

The thickness of the film was measured by using surface profilometer. The thickness of CIGS thin film with 3 layers was in the range ~80 nm. It was determined that there was a decrease in the transmittance of the CIGS thin film with the rise of the film layers in Figs. 19.7 and 19.8. However, the rise of the layers of the CIGS thin films resulted with the increase in the absorbance depending on the enhancement of the thickness in Figs. 19.7 and 19.8.

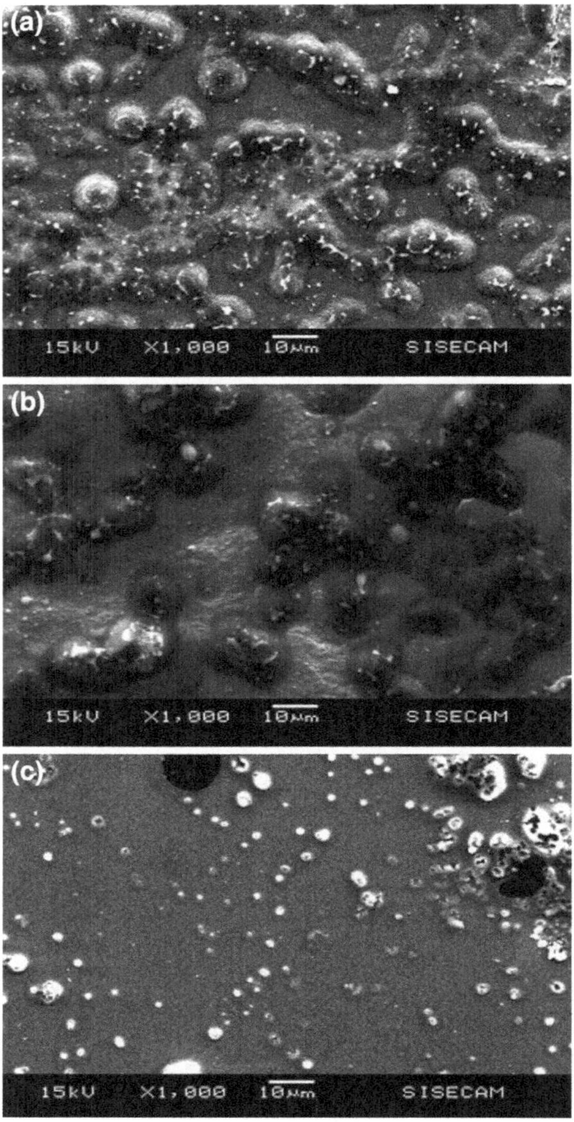

Fig. 19.6 **a** SEM images of CIGS thin film annealed at 200 °C, **b** SEM images of CIGS thin film annealed at 150 °C, and **c** SEM images of CIGS thin film no annealing

The optical transition was determined as direct transition at four different layers in CIGS thin film. There was a change in optical band gap in CIGS thin films with the increase of the film layers from 3 to 7 layers in Fig. 19.9.

The average transmittance and absorbance of CIGS films at 3, 5, 7, 9, and 11 layers were presented at ultraviolet, visible, and near infrared ranges in Table 19.2. There was a decrease in transmittance of CIGS thin film at ultraviolet range of electromagnetic spectrum.

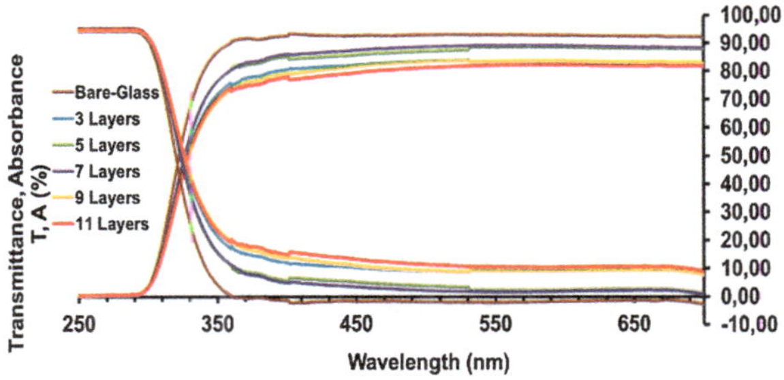

Fig. 19.7 Transmittance and absorbance changes with wavelength for 3 layers, 5 layers, 7 layers, 9 layers, and 11 layers

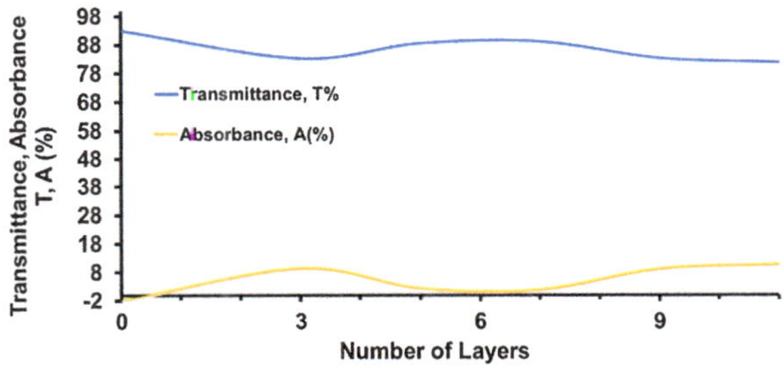

Fig. 19.8 Transmittance and absorbance values at 550-nm wavelength versus number of layers

Fig. 19.9 Optical band gap of 3, 5, and 7 layers CIGS film

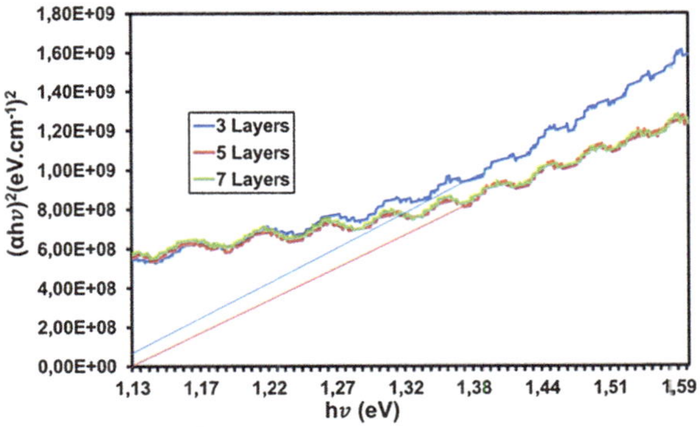

Table 19.2 The average transmittance and absorbance of CIGS film at different layers

Number of layers	UV (190–380 nm)		Visible (380–780 nm)		NIR (780–1100 nm)	
	T (%)	A (%)	T (%)	A (%)	T (%)	A (%)
3	22.93	71.22	82.49	8.99	81.7	5.27
5	24.61	69.01	87.10	2.65	86.19	0.94
7	24.55	69.07	87.99	1.82	86.57	1.18
9	22.18	71.65	82.29	9.09	81.97	5.07
11	22.44	71.61	80.71	10.68	81.26	5.67
Bare	27.90	64.86	92.44	−2.29	89.85	3.18

19.4 Conclusions

The prepared samples can be used as an absorber layer in the fabrication of thin film solar cells with respect to the production parameters. These results indicate that the solgel-derived CIGS thin films obtained the conductive film are good candidates for the applications in solar cells and other different optoelectronic devices. In the results of the study, it concluded as below:

- XRD analysis indicates that there is a continuous growth and improvement in the structural formation of quaternary CIGS compounds. The crystal-like structure achieves its highest crystallinity at 150 °C which is indicated by the SEM images as well.
- The optimum number of layers for the higher transmittance and absorbance is between 7 and 9. The optical band gap of CIGS absorber layer is determined ~ 1.14 eV by extrapolating the plot of $(\alpha h\upsilon)^2$ as a function of hυ with 7 layers.
- Seven layers of CIGS thin films were dip-coated annealed at different temperatures to determine the suitable annealing temperature for the coating. The thin films after annealing at different values A% and T% plots exhibit low values of transmittance and suitable absorption in the wavelength range of interest. The optimum temperature for the higher transmittance and absorbance is 170–180 °C. The optical band gap of CIGS absorber layer is determined as ~ 1.14 eV by extrapolating the plot of $(\alpha h\upsilon)^2$ as a function of hυ at 150 °C.
- These results of the solgel-derived CIGS thin films indicate that the CIGS thin films produced by solgel process can be good candidates for the applications in different optoelectronic devices with the improvement of the optical properties.

References

Akyol, S. (2015). Growth and characterization of CuIn1-xGaxSe2 (CIGS) thin-films for solar cell structures and the effect of beta radiation, *MSc Thesis*, Istanbul Technical University.

Brémaud, D. J. L. (2009). Investigation and development of CIGS solar cells on flexible substrates and with alternative electrical back contacts. Doctor of Sciences, Zurich. DISS. ETH No. 18194.

Chung, Y. D., Cho, D. H., et al. (2011). Effect of annealing on CdS/Cu(In,Ga)Se_2 thin-film solar cells. *Current Applied Physics, 11*, 65–67.

Chuang, C. L., et al. (2011). Method for preparing solgel solution for CIGS solar cell, United States Patent, Patent No:US7,922,804 B2.

Javed, A. (2007). Preparation and study of the structural, optical and electrical properties of Cu(In, Ga)Se_2 thin films. *Turkish Journal of Physics, 31*(5), 287–294.

Park, S. J., Lee, E., et al. (2011). A comparative study of solution based CIGS thin film growth on different glass substrates. *Applied Surface Science, 258*, 120–125.

Park, M., Ahn, S., et al. (2012). Characteristics of Cu(In,Ga)Se_2 (CIGS) thin films deposited by a direct solution coating process. *Journal of Alloys and Compounds*, *513*, 68–74.

Tolan, G. J., et al. (2009). Development of p, i and n-type CuInGa(Se_2) layers for applications in thin film solar cells. Solar Energy Group, Materials and Engineering Research Institute.

Yu, Z., Yan, Y., Li, S., Zhanga, Y., Yana, C., Liua, L., et al. (2013). Significant effect of substrate temperature on the phase structure, optical and electrical properties of RF sputtered CIGS films. *Applied Surface Science*, *264*, 197–201. doi:10.1016/j.apsusc.2012.09.171.

Chapter 20
Production of Poly(Imide Siloxane) Block Copolymers

Turkan Dogan, Nilgun Baydogan and Nesrin Koken

Abstract This work aims some challenges in the manufacturing of flexible substrates which will be used in solar cells as a substrate. Poly(imide siloxane) block copolymers were produced with the same bis(aminopropyl) polydimethylsiloxane (APPS). The polyimide hard blocks were composed by using 4,4′-oxydianiline (ODA) and benzofenon-3,3,4,4-tetrakarboksilik dianhydride (BTDA). Besides, the polysiloxane soft blocks were derived by using APPS and BTDA. APPS and BTDA formed the polysiloxane soft block in the structure. The length of polysiloxane soft block increased with increase in the length of polyimide hard block. Hence, it was possible to obtain copolymer structure and the changes in physical properties of the copolymers. These copolymers were characterized by using FT-IR analysis to evaluate the structure of flexible substrates.

20.1 Introduction

Solar energy has many advantages, and this type of renewable energy is clean, safe, and accessible. On the other hand, conventional solar energy has some disadvantages such as high cost, nonflexible, rigid, and heavy thin modules that cause some problems.

By this work, it is aimed to produce polymer material form with flexible, lightweight, low-cost, high and stable thermal expansion properties which can be used as a substrate in solar cells. By using lightweight, flexible substrates in solar cells with high thermal expansion properties, we can reduce some disadvantages which we come

T. Dogan · N. Baydogan (✉) · N. Koken
Istanbul Technical University, Istanbul, Turkey
e-mail: dogannil@itu.edu.tr

T. Dogan
e-mail: t.dogan@hotmail.com; t.dogan@itu.edu.tr

N. Koken
e-mail: nesrin@itu.edu.tr

A.N. Bilge et al. (eds.), *Energy Systems and Management*,
Springer Proceedings in Energy, DOI 10.1007/978-3-319-16024-5_20

across in solar energy technology. The usage of flexible substrates brings out better consequences in the application of solar cells, for example, in the building integration, space applications, and more compatible, portable, flexible solar energy devices.

Contemporary society requires flexible, low-cost, and lightweight energy storage systems. Flexible energy storage systems are very significant for the use in several kinds of applications ranging from portable consumer electronics to large industrial-scale power and energy management. Flexible substrates can be accepted as promising candidates due to their significant properties such as lightweight, versatile shape, and eco-friendly. The future invites the developed energy systems containing the flexible substrates. The use of flexible substrates is preferred for the main applications in solar cells because roll-to-roll processing makes them very interesting to reduce the production costs as well as the energy payback time. Poly (imide siloxane) materials are accepted as suitable candidates for use in solar cell technology (Ghosh et al. 2009, 2010; Pei et al. 2013; Cazacu et al. 2001; Arnold et al. 1989; Wohl et al. 2012 Freidrich et al. 2004).

Polyamides are available polymers that withstand high temperatures. They are insulator, lightweight, and flexible. They have smoother surfaces than metal substrates. But they suffer from reduced thermal stability and comparatively high coefficient of thermal expansion (CTE).

Therefore, for some polyimide, the semiconductor thin-film deposition temperature can be limited by thermal expansion mismatch. For that reason, the most important criteria for the selection of flexible substrates are well-matched CTE and sufficient thermal resistivity ($T \geq 400$ °C).

Poly(imide siloxane) has higher thermal stability than polyimide. In addition, this material has a chemical inertness. Poly(imide siloxane) material has high flexibility, low dielectric constants, excellent adhesion property, impact resistance, and resistance to decompose under any aggressive oxygen environment. Thus, when they are used as a substrate in solar cells, they never corrode. It has sufficient humidity barrier, and this material has high hydrophobicity. By this property, substrates will protect the active solar cell layers against the penetration of water vapor. For such properties, poly(imide siloxane) copolymer materials are accepted as the suitable candidates for use in solar cell technology (Ghosh et al. 2009, 2010; Pei et al. 2013; Yoo and Park 2010). Hence, this study was initiated to get poly(imide siloxane) copolymers as a polyimide form which have better properties than polyimide.

20.2 Experimental Procedure

Poly(imide siloxane) copolymer blocks were synthesized from the chemicals below,

- C_5H_9NO, Methyl-2-pyrrolidinone (NMP),
- $C_{17}H_6O_7$,—Benzofenon-3,3,4,4-tetracarboxylic dianhydride (BTDA),
- $C_{12}H_{12}N_2O$, 4,4′-Oxydianiline (ODA),

- $H_2N(CH_2)_3Si(CH_3)_2O[Si(CH_3)_2O]nSi(CH_3)_2(CH_2)_3NH_2$,Poly(dimethylsiloxane) (PDMS),
- $C_6H_4Cl_2$, 1-2 Dichlorobenzene (ODCB)

When BTDA anhydride was dissolved completely, ODCB was added into a different beaker on a heater. 3.52 gr bis(c-aminopropyl)polydimethylsiloxane (APPS) was added into solution. 3.26 gr BTDA + 15 mlt NMP (by a 5 mlt pipet) + 10 mlt ODCB (by an injector was pulled out) was added into the three-necked flask tube. The last mixture was added into the first mixture which was in the three necked flask tube (thus, within 45 min, first mixture was mixed completely). The heater was adjusted to 185 °C in order to mix within 10 h. The vaporization started and color change occured depending on the chemical reaction. 12 mlt water was extracted from the produced material because of the vaporization. The water was collected successfully in dean-stark. The collecting water ratio in the dean-stark slowed down, and the height of the water was nearly 0.5 mlt in the Dean Stark. It means that the vaporization started to slow down and chemical reaction was about to complete. The extract was diluted with NMP and trickled into excess ethanol with stirring to afford a precipitate. The precipitate was collected and dried under vacuum at 120 °C for 10 h to give the randomly segmented copolymer R-1. The structure of copolymer R-1 was identified by FT-IR.

20.3 Results and Discussion

Produced randomly segmented copolymer R-1 was presented in Fig. 20.1. Synthetic route of the randomly segmented poly(imide siloxane) copolymer was determined with the formulation, and the results of the produced poly(imide siloxane)

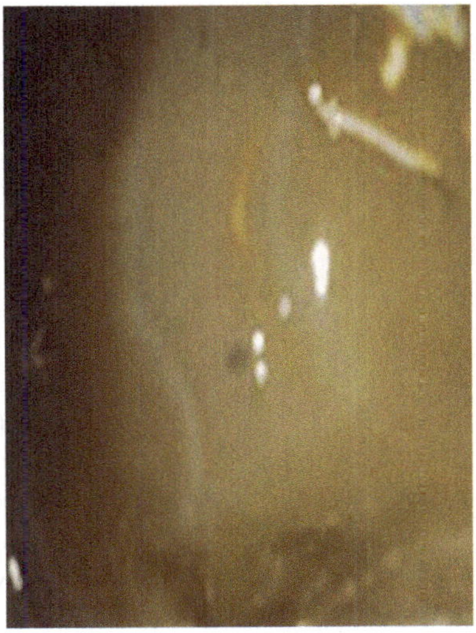

Fig. 20.1 Produced random copolymer

copolymer were presented in Fig. 20.2. Randomly segmented block copolymers were synthesized in the cosolvent system consisting of NMP and ODCB. APPS was first added to the solution of dianhydride to effectively cap the APPS, and then, nonsiloxane diamine ODA was added to the free dianhyride and anhydride-capped APPS. In the randomly segmented copolymer R-1, the APPS was randomly distributed in the polymer chain. As a result, the length of polysiloxane soft block was determined by molecular weight of polysiloxane diamines. The length of polyimide hard block was a function of composition. Thus, the obtained copolymers were called randomly segmented poly(imide siloxane) copolymers. Copolymerization is one of the most general and successful ways for preparation of new materials with specifically desired properties (Pei et al. 2013). Combination of polysiloxane and polyimide increased solubility and processability of polyimide. The enhancement of processability improved the flexibility of polysiloxane substrates.

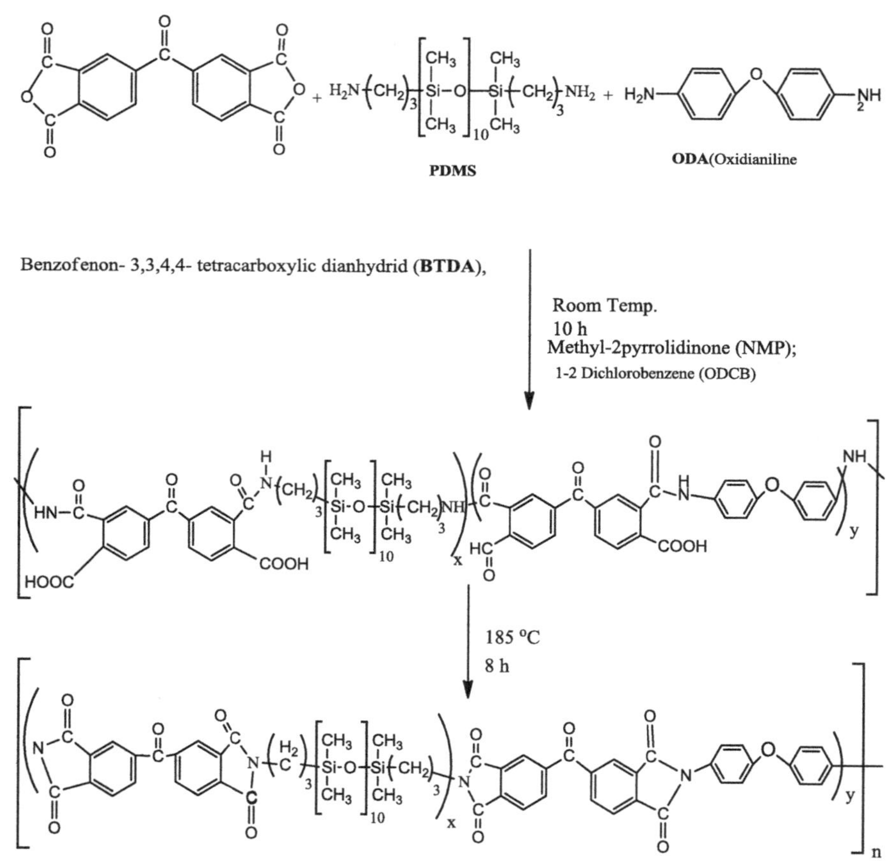

Fig. 20.2 Synthetic routes of the randomly segmented poly(urethane siloxane) copolymer R1 and poly(imide siloxane) copolymer R2

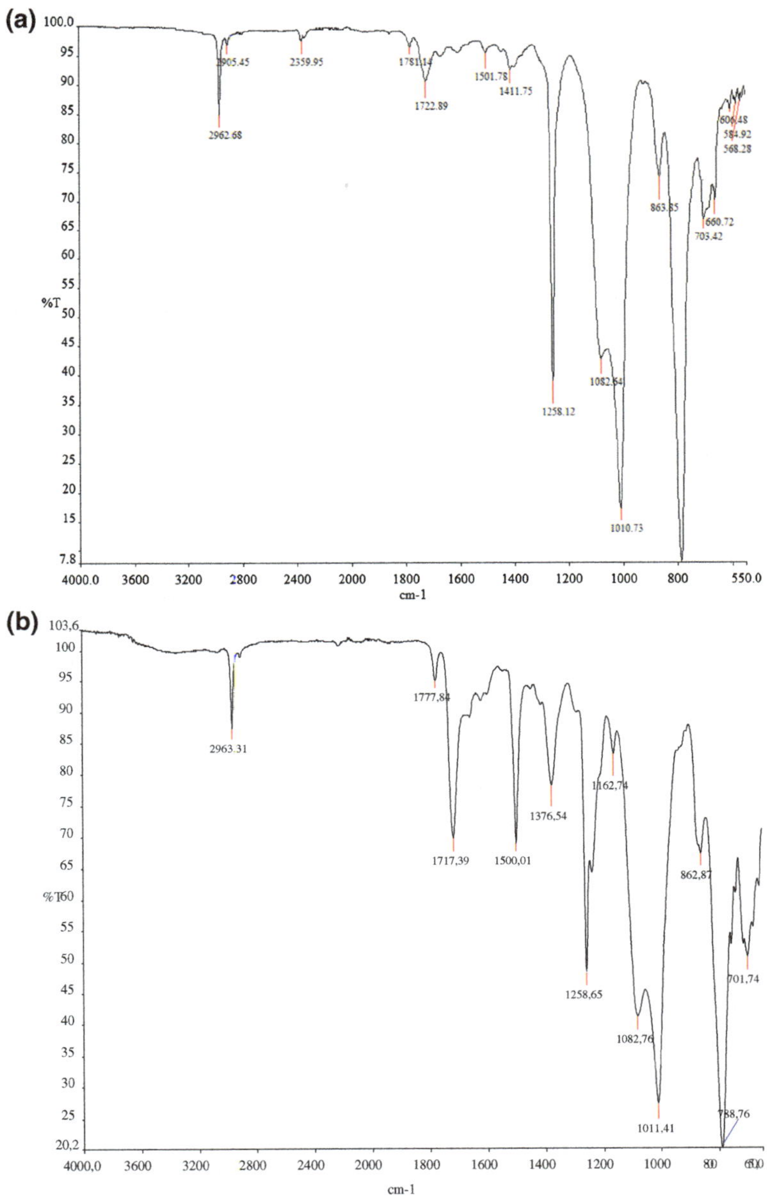

Fig. 20.3 **a** FT-IR analysis of the randomly segmented poly(urethane siloxane) copolymer R1. **b** FT-IR analysis of the randomly segmented poly(imide siloxane) copolymer R2

The results of the analysis of Fourier transform infrared (FT-IR) spectroscopy for randomly segmented poly(urethane siloxane) copolymer R1 presented random peaks, and they were evaluated as 2962 cm^{-1} (aliphatic C–H stretching), 1781 cm^{-1}

(asym C=O stretching), 1722 cm^{-1} (sym C=O stretching), 1082 cm^{-1} (asym Si–O–Si stretching), 1010 cm^{-1} (sym Si–O–Si stretching), and 790 cm^{-1} (Si–C stretching) in Fig. 20.3a. The FT-IR analysis of the poly(imide siloxane) copolymer R2 presented similar results with the FT-IR analysis of randomly segmented poly(urethane siloxane) copolymer R1 in Fig. 20.3b. In the FTIR spectra of R1, the peaks at 1781 and 1722 cm^{-1} are due to the carbonyl of urethane groups (Fig. 20.3a) and the peak at 1500 cm^{-1} is due to N–H of urethane. These urethane absorption peaks at 1781 and 1722 disappear at the end of the capping reaction with –COOH, indicating the formation of R2 copolymers (Fig. 20.3b). The peaks at 1777 and 1717 cm^{-1} in all R2 copolymers were due to imide groups. The peak at 1500 cm^{-1} in R2 copolymer is due to N–H of imide.

20.4 Conclusions

The produced flexible substrates indicated contemporary importance of this piece of work to make a progress in thin-film semiconductors at solar cell applications. The results of FT-IR analysis of the randomly segmented poly(urethane siloxane) copolymer R1 and the randomly segmented poly(imide siloxane) copolymer R2 have addressed the production of soft and hard flexible substrates. The poly (dimethylsiloxane)s are also extremely nonpolar and have very low experimental solubility parameters when compared with other organic polymers. This block copolymer leads to the thermodynamic incompatibility of PDMS with almost all other organic polymers. This type of copolymer may be used as a compatibilizer since the aromatic imide segments are compatible with many organic polymers. Therefore, these copolymers may be used as compatibilizer polymers and additives in many applications.

The formula of the randomly segmented poly(urethane siloxane) copolymer R1 was determined. Besides, the formula of the randomly segmented poly(imide siloxane) copolymer R2 was obtained to compare with the structural properties of poly(urethane siloxane) copolymer R1. In the synthesis of bock copolymers, APPS links through BTDA to form terminated anhydride. The results of FT-IR analysis indicated the diamine peaks that determine the end of polymer chain ring.

The characteristic chemical structure of block copolymers is generally composed of siloxane Si–O and carbon chain-type structure. The information on properties of randomly segmented PIS and effect of the PI hard block and soft block lengths on the PIS supports the progress in new generation materials. Some new discoveries can be found in further studies. These discoveries are very important to maintain the processability and achieve desirable physical properties of the material. This enables us to get better flexible, long-term stability substrates in order to use with semiconductor thin-film modules at solar cell applications and solar energy with the high efficient ratio.

References

Arnold, C. A., Summers, J. D., Chen, Y. P., Bott, R. H., Chen, D., & McGrath, J. E. (1989) Structure-property behaviour of soluble polyimide-polydimethylsiloxane segmented copolymers. *Polymer, 30*, 986.

Cazacu, M., Carmen, R., Vlad, A., & Marcu, M. (2001) Segmented poly(siloxane-ester-imide)s. *European Polymer Journal, 37*, 2465.

Friedrich, K., & Dominik, R. (2004) Technological aspects of flexible CIGS solar cells and modules. *Solar Energy, 77*, 685–695.

Ghosh, A., Banerjee, S., Komber, H., Schneider, K., Heaußler, L., & Voit, B. (2009) Synthesis and characterization of fluorinated poly(imide siloxane) block copolymers. *European Polymer Journal, 45*, 1561.

Ghosh, A., Banerjee, S., Heaußler, L., Voit, B. (2010) *Journal Macromolecular Science Part A: Pure and Applied Chemistry, 47*,671.

Pei, X., Chen, G., Fang, X. (2013) Synthesis and properties of Poly(imide siloxane) block copolymers with different block lengths. *Journal of Applied Polymer Science*. doi:10.1002/APP.38918, 3718–3727.

Wohl, C. J., Atkins, B. M., Belcher, M. A., & Connell, J. W. (2012) Synthesis, characterization, topographical modification, and surface properties of copoly(imide siloxane)s. *High Performance Polymers, 24*, 40.

Yoo, H., & Park, S. (2010). The fabrication of highly ordered block copolymer micellar arrays: Control of the separation distances of silicon oxide dots. *Nanotechnology, 21*(24), 245304.

Part III
Energy Management, Economics and Policy

Chapter 21
Government Incentives and Supports for Renewable Energy

Münci Çakmak and Begüm İsbir

Abstract Although energy is part of our life, we paid big amount of cost for it. As it has a great economic value, we produced energy in an unconscious way, so we polluted earth. At the end, we understood the value of environment and gave up our bad habit of extreme producing methods including pollution. Secondly, we realized that energy resources are not unlimited. These situations delivered us into clean energy consuming and producing. Governments must promote for renewable energy consumption and production. Renewable energy consumption and production cannot increase without government help or government actions so governments have to be a pioneer for renewable energy. Legal provisions are also important for civil rights (both public and private). Renewable energy actions may violate fundamental rights and freedoms such as property, health right, and also commercial rights like competition and right of initiation. These issues must be regulated by governments under the administrative law principles. Clean energy sector is not only a need but also it is a market. Although it is mostly in private sector area, it still has public interest notion.

21.1 Introduction

Renewable energy issues are very important. As the fossil energy sources are limited, both governments and investors begin to prefer renewable energy sources. Renewable energy is not only clean and sustainable but also it is cheap. In the future, we will get rid of fossil energy sources and world will be cleaner.

There are many circumstances for governments to promote renewable energy production and consumption. Governments adjust their energy policies according to

M. Çakmak (✉) · B. İsbir
Faculty of Law, Gazi University, Ankara, Turkey
e-mail: muncicakmak@outlook.com

B. İsbir
e-mail: begumisbir@hotmail.com

A.N. Bilge et al. (eds.), *Energy Systems and Management*,
Springer Proceedings in Energy, DOI 10.1007/978-3-319-16024-5_21

renewable energy because all fossil resources are limited and we have no chance to renew them; secondly, today, states are sensitive about environmental issues and finally the energy need of humankind is increasing (Kaya 2012).

Governments have a duty of developing renewable energy production and consumption as it is a public service. Energy and indirectly renewable energy markets especially functions with private law rules and marketing mechanisms, but that does not mean markets are out of control and completely deregulated. An effective renewable energy management (production, consumption, investments, research and development projects, security, etc.) needs future planning and strategies; this is a sophisticated mission, and this mission belongs to governments. At least, energy is a national issue and related to national security. There are many factors controlling future energy strategies. Technological developments, economic growth rates, population growth rates, energy prices, current energy policies, and consumer behavior are considered when predictions are being made (Kaya 2012).

The primary purposes of the renewable energy utilization are depletion of fossil resources in the future, wanting to reduce dependence on foreign resources (Uluatam 2010), and stopping environmental pollution.

Help and promotion of governments are compulsory because of the inadequate utilization of renewable energy and high costs of the initial setup of the facilities (Selek 2011).

Renewable energy is also important for industry because renewable energy means low-cost energy; low-cost energy means low production costs. This will increase the benefit of the firms. For example, in the ammoniac and cement production sector, energy cost accounts for between 50 and 60 % of the total cost (Söğüt et al. 2012). If we decrease the energy costs, it will be reflected in prices; low price means more sales and high benefit ratios.

The world begins to understand the value of the renewable energy sources and clean energy systems so countries are setting targets for the next years. For example, Germany plans to replace 50 % of its total energy generation with renewable energy systems by 2050; China is planning to invest US$740 billion in the next 10 years to increase the proportion of renewable energy by 2020; the United States will invest US$150 billion for replacing 25 % of its total energy generation with renewable energy by 2025 (Hong et al. 2014).

Since 1990s, many policies have been introduced to promote renewable energy utilization such as renewable portfolio standard (quota obligation), feed-in tariff, bidding scheme, tradable certificate production tax credit, tax incentives, net metering, green pricing, green power, green power certificates, and green power fund (Renewable Energy Promotion Policies Final Report 2012).

In particular, three kinds of promotion methods are preferred. They are (1) price- or quantity-based policies (feed-in tariff and renewable energy portfolio standards), (2) cost-reducing investment policies (subventions, discounts, tax rebates), and (3) public investments and promotions which help developing renewable energy market (Uluatam 2010).

Although these policies have helped growing of market until now, we think renewable energy market needs more promotion policies as promotion diversity attracts investments into energy market. These main policy types will not be enough in the future, and renewable energy market needs more help. Today, governments are trying to make the market more attractive by using other kinds of supports such as discounts, privileges, subventions, and other methods. Policy maker is only government, and spontaneous law in the social and commercial life can also develop its own policies in the market, but this is not the renewable energy need. Renewable energy market is challenging with profitable and rewarding fossil resources, high costs, and undeveloped technology. There are many barriers slowing down renewable energy trade, production, and consumption such as "high initial capital costs coupled with lack of fuel price risk assessment, imperfect capital markets, lack of skills or information, poor market acceptance, technology prejudice, financing risk and uncertainties, high transaction costs and some regulatory and institutional factors" (Beck and Martinot 2004). Governments must promote renewable energy as much as possible, both legally and commercially.

One important issue is promotions especially involving price must not be too much complex and mixed structure. For example, in Austria, there are about 100 different tariffs for only a tiny portion of the total electricity production (Reiche and Bechberger 2004). This kind of promotion diversity may have adverse effects on market because investors may become estranged about market and market may loose its attractiveness. Balance should be considered between diversity and complexity.

21.2 Renewable Energy and Public Service Notion

Energy, which is gained from renewable energy sources, can be utilized mainly in two ways. Either governments produce energy from renewable sources and then sell energy to people as a public service or the private sector produces energy from renewable sources and sells it to public or to governments. This production and marketing systems are selected according to the ideological structure of the state.

In liberal systems, generally second way is preferred. In these systems, energy is produced by private sector and market is executed and regulated by governments. In particular, there are governmental bodies whose function is to regulate the energy sector by administrative orders and regulations. Also one or more ministries may have duties related to the energy sector.

Private sector produces energy especially under the concession contract system, public–private partnership system, license system, or other kinds of private law contracts. License system is not completely a public service serving method, but it is used in energy sector.

At this point, we want to point out the fact that energy generators/mostly electricity generators/producers are not completely free and independent. Energy sector still has public service notion (Şanlı 2007) and so private firms must not violate individual rights. They should produce and serve energy under public law view and discipline. Individuals are "public service beneficiaries"; they are not just "customers." Although governments regulate energy sector and have a control over these firms, they should also consider themselves as a governmental or public producers. This may not be important for liberal values, but it is important for individuals and public. There is a discussion about electric services that whether it is a service or it is a good (Işıksungur 2011). In Turkish law, electricity is accepted as chattel (Işıksungur 2011), but we think that electricity services still involve public service character. Generating distribution or transmission stages can be serviced by private sector under the control of state; this is a choice of public service offering way and is under the discretional power of state. This situation does not prevent public service notion of the electric energy utilization. Renewable energy is indirectly related to electricity market because renewable sources are used for mostly electric production, but we can say renewable energy utilization has also public service character.

Why do we insist on to accept electrical energy as a public service? Because we support that governments promote electricity and renewable energy production under the public service point of view. This will protect public and individuals. This is not against the private firm rights, liberal values, and market notion. These also get benefits from public service notion. For example, uncontrolled marketing actions decrease the service quality of private firms; this will cause negative beliefs against the private firms.

At this point, governments have to give importance to law. All kinds of promotions, which help private sector for marketing actions and energy efficiency provisions, must be regulated by acts to provide legal security. Besides, orders regulating both energy and renewable energy utilization must regulate promotions, which involve duties of private firms intended to public (Sacker 2011). For renewable energy, technology is not enough for development; a healthy market is also required; otherwise, investors will not find the market attractive. No investor means no development.

21.3 Duties of Governments About Renewable Energy

Some of the governmental duties about renewable energy can be summarized as giving services about renewable energy, regulating renewable energy market, determining targets for renewable energy producing and consumption, educating public and firms about renewable energy, encouraging and supporting domestic industry for renewable energy producing and consuming, giving licenses to firms for renewable energy production, controlling firms, finding policies and strategies for renewable energy and energy productivity, monitoring renewable energy resources capacity, and promoting research and development projects.

Education about renewable energy is very significant because renewable energy in our minds has narrow place. People think that renewable energy is only about sun and wind and just for electricity, but in reality, renewable energy is also related to sustainability, clean energy, environmental issues, biomass sources, and global warming. Governments must educate people and create awareness about renewable energy. Education programs must be set up; also some programs have to begin from primary schools in order to give information about importance of renewable energy for the next generations. Private sector should also take some roles in education programs. This is one of the "public service" structures of energy sector. Firms have to join these education programs; this is not only an advertisement for them but also a charity for clean world.

Municipalities are important for social life and services so they also promote renewable energy production and consumption. They can be called key actors of renewable energy if they play an important role in renewable energy policies (Schönberger 2013). Most of our daily living activities and services are related to municipalities. Municipalities can produce renewable energy and consume it; also they promote people and investors for production and consumption. Most of the energy facilities have legal relation with municipalities. For example, in Turkish law and administration system, zone plannings are made by municipalities, which means every kind of zones in city planning has a relation with municipalities including renewable energy sources. Also public transportation is a great sector for renewable energy consumption, and municipalities have the power of transforming these vehicles into green vehicles. Municipalities have areas that have geothermal, wind, solar, and biomass energy potential.

There may be legal conflicts derivated from acts, regulations, or by-laws between ministries and municipalities (or between municipalities and metropolitan municipalities). In particular, zone planning, city planning, historical issues, and environmental issues have potential for this kind of legal conflicts. This is a serious issue because as there is no hierarchical relation between ministries and municipalities, these conflicts must be solved in special ways. Generally, they solve their conflicts between themselves, but there is no guarantee for this kind of dispute solution. As renewable energy settlements/facilities need area and construction actions, conflicts must be solved immediately to disable crisis. It is not hard to set up a commission for renewable energy actions and legal conflicts.

21.4 Renewable Energy Promotion Examples from Turkey

Turkey's energy statistics in a 30-year period:

Electricity production of Turkey (Teias 2013):

Electricity production	1984	2012
Private sector (%)	12.8	62.2
State (%)	87.2	37.8

Renewable energy production of Turkey (Teias 2013):

	1984	2012
Renewable energy production (hydrolic + renewables) (GWh)	13426.3	64625.2

By 2023, Turkey will be seeking to provide 30 % of its electricity from renewable energy sources (Global Wind Report 2013).

Turkey has a great renewable energy potential. "Turkey is one of the biggest onshore wind markets in Europe with an 11 GW pipeline of wind power projects" (Global Wind Report 2013).

Turkish energy policy depends on resource diversity, environmental sensitivity, and competitive market system (Kaya 2012). In order to achieve these targets, renewable energy law must be developed under the administrative law principles and values. In particular, competence market system cannot be thought without administrative law rules (production licenses, public private partnerships, market regulation, public procurements, etc.). Renewable energy, development, consumption, and production must not be left to market rules.

Some promotions for renewable energy in Turkish Law:

Turkey has developed its legal incentives about renewable energy for the last ten years. In Turkish legislation, the main act related to renewable energy and renewable energy sources is No: 5346 Act on Utilization of Renewable Energy Sources for the Purpose of Generating Electrical Energy (shortly Renewable Energy Act-REA/Yenilenebilir Enerji Kanunu-YEK). In this Act, Renewable Energy Act Support Mechanism (YEK destekleme mekanizması) regulates supports and promotions about renewable energy. They are as follows:

Article 4 which forbids zone plannings that will effect utilization and effectiveness of renewable energy resource areas on the public and state areas.

Article 6 guarantees feed-in tariff for a period of ten years (generation licenses that are or will be given before December 31, 2015).

Natural and legal entities have a right to benefit from tariffs for ten years; if they give their excess electricity produced from renewable energy sources to grid circuit, distribution companies are obliged to buy this electricity.

Article 6/B of the Act offers special and high feed-in tariff rate schemes for renewable energy producers if they use domestic equipments and hardwares in their facilities or projects.

Article 7 regulates an exemption for some service fees about facilities, which produce electricity from renewable sources for their own consumption (maximum 1000 kW production capacity).

Lease, easement, and usufruct right fees have a discount of 85 % for ten years, for renewable energy power transmission lines on the forestry, treasury, and state areas.

No: 6446 Electricity Market Act, No: 5627 Act of Energy Efficiency, Regulation for Administration of Geothermal Resources and Natural Mineral Waters, and Regulation for Utilization of Geothermal Resource Lands for Electrical Energy Generation also contain legal promotions for renewable energy resources and generation. Some of the acts and other regulations have also regulated renewable energy sources indirectly. For example, agricultural regulations which promote corn, sugar beet, and sugar cane production also indirectly promote renewable energy when we produce biomass source from them.

Research and development projects are also promoted by Turkey. A protocol has been signed between The Scientific and Technological Research Council of Turkey (Tübitak), Ministry of Science, Industry and Technology, and Ministry of Energy and Natural Resources on August 13, 2012, in order to develop research and development projects by the next ten years of period (Teke 2013). Also according to No: 5627 Act of Energy Efficiency, Tübitak supports renewable energy research and development projects primarily.

Turkey is willing to use energy effectively so that firstly Energy Efficiency Act, some regulations, and by-laws enacted in order to achieve this aim. This reformation process is in harmony with 25 energy efficiency policy recommendations (under 7 priority areas) of International Energy Agency (IEA) as cross-sectoral regulations, buildings, appliances and equipments, lighting, transport, industry, and energy utilities (25 Energy Efficiency Policy 2011).

21.5 Conclusion

Renewable energy needs promotion and support because it still needs high technology. Economical and technological gaps between fossil resources and renewable energy resources may only be closed by the help of governments. Otherwise, firms will be seeking for fossil resource utilization. Our energy history showed us that only few people care about environment, and it is nearly impossible to utilize fossil resources in clean methods.

As a result, we can summarize the duty of governments and incentives for renewable energy sources as follows:

1. Decreasing the ratios of taxes and other public fees, or totally making persons privileged (especially firms) from taxes in order to promote renewable energy production and consumption.
2. Governments must protect the competition in the energy market. Although renewable energy is not powerful today, in the future it will be more effective in the market. Competition regulations must be ready for the future from now on.
3. Private sector must be encouraged for renewable energy production. Liberalism and capitalism are useful for development, but that does not mean we should leave the market completely to uncontrolled mechanisms. The important subject is clean world and productive renewable energy production and consumption, not always profit.
4. Zone plannings must be set up according to renewable energy sources (especially solar and wind).
5. Promotions regulating quantity and price must not be changed too much. This will cause unstable situations, and unstable markets are not preferred by investors. This will affect market growth and the following economic benefits.
6. Lack of regulation is also a barrier in front of renewable energy utilization and market system. Investors seek legal security and stable market system controlled

by regulations. Needed acts, regulations, and ordinances must be enacted as quickly as possible.

7. Research and development is very important for renewable energy actions. Governments should also promote research and development investments, projects, and activities.
8. Governments must also promote and regulate energy storage systems. Renewable energy promotions are especially for solar, wind, and biomass sources, mostly for production of electricity, but storage is also a very important issue and must not be underestimated.
9. Government facilities and vehicles should be transformed into new technologies (buildings, cars, factories, public transportation, etc.) compatible with renewable energy.
10. Municipality facilities and vehicles should be transformed into new technologies compatible with renewable energy technologies.
11. Helping municipalities, which has lower budget for their transformation into renewable energy.
12. Educating people about clean energy, starting from primary schools.
13. Giving privileges to firms in public procurements who prefer renewable energy technologies.
14. Off-grid systems are also important, so governments should also promote these systems by providing financial incentives targeting both end users and local entrepreneurs (Iorec 2013).

References

25 Energy Efficiency Policy. (2011). *Recommendations, 2011 update*. International Energy Agency. Website: www.iea.org.

Beck, F., & Martinot, E. (2004). Renewable energy policies and barriers. *Encyclopedia of Energy, 5*, 365–383.

Global Wind Report. (2013). *Annual market update 2013*. GWEC. Website: www.gwec.net.

Hong, T., Koo, C., Kwak, T., & Park, H. S. (2014). An economic and environmental assessment for selecting the optimum new renewable energy system for educational facility. *Renewable and Sustainable Energy Reviews, 29*, 286–300.

Iorec. (2013). *International Off-Grid Renewable Energy Conference 2012, Key Findings and Recommendations*. Website: www.irena.org/Publications.

Işıksungur, Ö. D. (2011). Elektriğin Hukuki Niteliği. EÜHFD, C. XV, S. 3-4, 249–266.

Kaya, İ. S. (2012). Uluslararası Enerji Politikalarına Bir Bakış: Türkiye Örneği, TBB Dergisi, S. 102, 269–288.

Reiche, D., & Bechberger, M. (2004). Policy differences in the promotion of renewable energies in the EU member states. *Energy Policy, 32*, 843–849.

Renewable Energy Promotion Policies Final Report. (2012). *APEC peer review on low-carbon energy policies (PRLCE) phase 1*. APEC Energy Working Group, December 2012.

Sacker, F. J. (2011). Enerji sektörünün liberalizasyonunda rekabet otoriteleri ile sektör düzenleyici kuruluşların koordinasyonu, Conference, 13 April 2011, pp. 109–124. Website: www.rekabet.gov.tr.

Schönberger, P. (2013). *Municipalities as key actors of German renewable energy governance: An analysis of opportunities, obstacles, and multi-level influences*, Wuppertal papers, N. 186. Wuppertal: Wuppertal Institute.

Selek H. (2011). Yerli Katkı Payı ve Yenilenebilir Enerji Kanunu, İzmir Rüzgâr Sempozyumu ve Sergisi, 23–24 Aralık 2011. Website: www.mmo.org.tr.

Söğüt, Z., Oktay Z., & Karakoç H. (2012). Yenilenebilir enerji sistemlerinde farklı bir bakış: Enerji geri kazanımı. TMMOB MMO Mühendis ve Makina Dergisi, C. 33, S. 633, 52–59.

Şanlı, Y. (2007). Elektrik piyasası ya da elektrik kamu hizmeti, Mülkiye, C. 31, S. 254, 51–71.

Teiaş. (2013). Türkiye Elektrik Enerjisi 5 Yıllık Üretim Kapasite Projeksiyonu 2013–2017. Website: www.teias.gov.tr.

Teke, O. (2013). Dünyada ve Türkiye'de yenilenebilir enerji AR-GE stratejilerinin değerlendirilmesi, Mühendis ve Makina, C. 54, S. 640, 54–62.

Uluatam, E. (2010). Yenilenebilir enerji teşvikleri. In Ekonomik Forum, Ekim, pp. 34–41.

Chapter 22
Comparison of the Relationship Between CO_2, Energy USE, and GDP in G7 and Developing Countries: Is There Environmental Kuznets Curve for Those?

Mahdis Nabaee, G. Hamed Shakouri and Omid Tavakoli

Abstract The increasing attention to greenhouse gas (GHG) emission all around the world has led researchers to investigate its causes. Energy use and economic growth can be categorized as the most important roots for this problem. The objective of this study is to investigate the causal relationship between energy use (EU), economic growth (GDP), and CO_2 emission within two groups of countries: The G7 and the developing countries. Then, we examined the environmental Kuznets curve for these countries to see whether this hypothesis (EKC) is proved for these countries or not. To do so, seven developed and six developing countries are selected, and the annual data samples covering the period between 1993 and 2011 are gathered. To investigate the casual relationship between the variables, EU, GDP, and CO_2, we applied the techniques of panel data analysis, examining unit root and cointegration tests, based on EKC equation. The results show the casual and long-term relationship for the two groups of countries. Also, the hypothesis of Kuznets curve is proved for G7 countries and it is shown that the relationship between these variables is inverted U shaped, with a nonzero negative coefficient for the square of GDP. However, for developing countries, this hypothesis is rejected, because the coefficient of GDP^2 is found close to zero, and thus, a linear relationship is proved between these variables.

22.1 Introduction

Studying the relationship between energy consumption, greenhouse gas emissions, and economic growth has become one of the today's important issues for climate change researchers. In recent years, there has been a vast increase in awareness in

M. Nabaee (✉) · G.H. Shakouri · O. Tavakoli
University of Tehran, Tehran, Iran
e-mail: Mahdis_nabaee@ut.ac.ir

A.N. Bilge et al. (eds.), *Energy Systems and Management*,
Springer Proceedings in Energy, DOI 10.1007/978-3-319-16024-5_22

the environment and its interaction with economy. This subject becomes more important when we notice to statistics: Almost 65 % of total GHG emissions are caused by the energy production and use. Hence, investigation on the relationship between GHG emission, energy consumption, and economic growth is required, so that society can pave the road toward decreasing the GHG emissions and achieving a sustainable development.

There are big challenges between the advocates of environment and economic growth. Some of them believe that higher economic growth needs more energy consumption, and this leads to more emissions and environmental destruction. Therefore, the former believe that the government must decrease the growth for improving the environment, while the former expect more economic growth, since they believe that higher GDP is along with both decreasing the pollutant and improving of the quality of environment. The environmental Kuznets curve (EKC) tests the hypothesis of existence of such impact.

Thereafter, economic growth becomes more even along with economic growth, so that the curve starts with positive slopes but gradually converts to zero and negative slops. Indeed, the EKC is a hypothesis to survey the relationship between various indicators of environmental degradation and per capita income. In the early stages of economic growth, degradation and pollution increase, but beyond some level of per capita income (which will vary for different indicators in different countries) the trend reverses, so that at high income levels economic growth leads to environmental improvement.

This implies that the environmental impact indicator is an inverted U-shaped function of per capita income. The EKC is named for Kuznets (1955), who hypothesized income inequality first rises and then falls as economic development proceeds. If this hypothesis is confirmed for our understudy relationship, then economic growth in its higher levels leads to a tool for environment improvement instead of a threat to it. This paper is to investigate the relationships weather it establishes for developed and developing countries or not.

22.2 A Short Review on Previous Studies

The relationships between the economic growth and energy consumption are considered on various studies with a vast diversity of methods. One can refer to Zhang and chow, 2009, who studied the casual relationship between energy consumption and economic growth. They used panel data in 1995–2008 period of time and showed that in some of China's regions there is a positive relation (Zhang and Xu 2012).

There are also many other researches showing a direct relation between the two variables. Marrero did it by applying a dynamic panel data for 24 countries in 1990–2006, (Marrero 2010). Coers and Sanders considered the relationship on duration of 40 years for 30 of OECD countries and showed that the direction of causality is from economic activity to energy consumption (Coers and Sanders

2013). Similar work was done by Dagher and Yacoubian for Lebanon (2012). The results found by Shaari, Hussain and Ismail in 2012 showed a significant long-term relationship between these variables and pollution in Malaysia (2012), (Azlina and Mustapha 2012). Apergis and Payne proved a long-term relationship between GDP, renewable, and non-renewable electricity consumption, real gross investment, and labor for developing countries during 1990–2007 (Apergis and Payne 2011). Tugcu, Ozturk, and Aslan did a similar survey for G7 countries in 2012. They also identified which kinds of energy resources, the renewable or non-renewable, are necessary for economic growth, (Tugcu et al. 2012). Many other researches can be found in this area (Menyah and Wolde-Rufael 2010; Ozturk and Acaravci 2010).

Studies on the EKC hypothesis have led to four different results. Some results showed that there are not any linear or nonlinear relationships between economic growth and pollutants emission (Luzzati and Orsini 2009). Moreover, Halkos and Tzeremes (2009), who studied EKC between environmental efficiency and national income for 17 countries of OECD in 1980–2002 by generalized moment method (GMM), concluded there is not any evidence of EKC for them (Halkos and Tzeremes 2009).

The second category includes the studies proving a relationship between the variables. However, the proven relationships are divided into three categories: The first are so-called univocal relationships between economic growth and pollutants (Halkos and Tsionas 2001). The second group concluded that there is an inverted U relationship between income and pollutants (Huang et al. 2008; Vollebergh et al. 2009).

Finally, in the last group, the studies show an N-shaped rare relationship (Brajer et al. 2008; Egli and Steger 2007). The N-shaped curve means that after passing EKC phase (the inverted U), more economic growth leads to destruction of environment again, same as the first phase.

22.3 Methodology and Data Preparation

Panel data refer to data sets consisting of multiple observations on each sampling unit. The estimated panel data model is as follows:

$$y_{it} = \beta_{i0} + \beta_1 x_{1it} + \beta_2 x_{2it} + \cdots + \beta_k x_{kit} + \varepsilon_{it} \quad i = 1, \ldots, N; \quad t = 1, \ldots, i$$

where i is an index for the understudy cross-sectional variable, i.e., the countries, t is the time dimension, y_{it} is the dependent variable, and x_{kit} refers to the k independent variable of the i country at time t. Finally, β_i are the coefficients that should be estimated. In this study, first, the unit root tests for surveying the stationary in the data panel variables are applied. Then, causality and long-term relationship between the variables is tested, and the coefficients of the equation are estimated, leading to the EKC hypothesis test.

22.3.1 The Panel Unit Root and Cointegration Tests

Panel unit root tests are used for surveying whether a time series is stationary or not. Levin and Lin, Im and Pesaran, and augmented Dicky Fuller (LLC, IPS, and ADF) tests are used among all. To follow up the test methods, three different stationary models should be built:

*M*1: Model with common time effect; *M*2: model with a trend and common time effect; *M*3: model with no trend and no time effect. Also, to investigate causality and long-run relationship between variables, Kao residual cointegration test and Johansen Fisher panel cointegration test are applied.

Also, random effects and fixed effects are two different approaches for estimating the panel data. The model is estimated based on one of these effects by implementing Housman test. In fixed effect, ε_{it} is constant and is calculated for each country. The null hypothesis is that the individuals and time effects are not correlated with x_{it}.

22.3.2 Testing the Hypothesis of EKC

For investigating the environmental Kuznets curve (EKC) hypothesis, the long-run relationship between the above-mentioned variables can be examined by the following panel data model:

$$\mathrm{GHG}_{it} = \alpha_{it} + \beta_{1i}\mathrm{EU}_{it} + \beta_{2i}\mathrm{GDP}_{it} + \beta_{3i}\mathrm{GDP}^2 + \varepsilon_{it} \quad t = 1, \ldots, T$$

The subscript $i = 1, \ldots, N$ denotes the counties, while $t = 1, \ldots, T$ denotes the time period. The output variable, GHG is greenhouse gas emissions; EU stands for energy consumption; GDP is the real GDP, and GDP^2 is the square of real GDP The parameters β_1, β_2, and β_3 are the coefficients that should be estimated. Based on the EKC hypothesis, β_1 is expected to be positive, and β_2 is expected to be positive too, whereas a negative sign is expected for β_3.

22.3.3 Data Description

Data used in the analysis are annual time series during the period of 1993–2011 on GDP energy use and CO_2 emissions (all are per capita) for the seven “G7” countries and six developing countries. The data were obtained from World Development Indicators,[1] published by the World Bank. The choice of the starting period was

[1] WDI.

constrained by the availability of data on some countries and variables. The variable's notations and definitions are as follows:

CO_2	CO_2 emission in ton per capita.
GDP	Gross domestic product per capita (based on $2000 constant US).
GDP^2	Gross domestic product per capita square GDP.
EU	Energy use per capita in kilograms of oil equivalent per capita.

22.4 Results and Discussion

In this section, the relation between variables for two groups of countries has been investigated based on previous sections and methodologies.

22.4.1 Results of G7 Countries

The G7 countries consist of the following: Canada, France, Germany, Italy, Japan, the United Kingdom, and the United States. The results of these countries are illustrated in Tables 22.1, 22.2 and 22.3, and the results of unit root tests are illustrated in Table 22.1.

As can be seen in the tables, the unit root hypothesis is not rejected when the variables are taken in levels. But when we use these tests for variables in first differences, the large values for the statistics indicate rejection of the null of non-stationary at 1 % level for all models. It may, therefore, be concluded that the four variables CO_2, GDP, GDP^2 and EU are unit root variables of order one, or $I(1)$ or short.

Based on the results of E-views software, the null hypothesis (lack of cointegration) is rejected and the long-term relationship is proved for variables. Considering all three rejected hypotheses, we conclude that none of the variables contain stochastic trends after all. The results are illustrated in Table 22.2. Also, the null of random effect is accepted and the fixed effect model is rejected because of the amount of 0.19; it is higher than the amount of possibility 0.1. Then, the model of G7 countries is based on random effect.

Also, the results of this test for each country are presented in Table 22.3 separately:

The long-term relationship is rejected for all cross sections except France (the second country). These tests are done based on ($M3$). Also, with the use of Kao[2] test and under individual intercept and no trend condition, the value of statics is

[2] Kao Residual Cointegration Test.

Table 22.1 The unit root test results for G7 countries

First differences			
Condition	*M*1	*M*2	*M*3
CO_2 emissions			
LLC	−3.34962[a]	−3.68131[a]	−5.70894[a]
IPS	−3.32353[a]	−3.48999[a]	–
ADF chi-square	35.5881[a]	36.4512[a]	54.8530[a]
Energy use			
LLC	−0.21986	−1.00663	−6.14051
IPS	−2.52136	−5.05274[a]	–
ADF chi-square	28.0901	49.1919[a]	55.42882[a]
GDP			
LLC	−4.80793[a]	−5.86667[a]	−4.90506[a]
IPS	−2.95438[a]	−3.52925[a]	–
ADF chi-square	32.1586[a]	35.9331[a]	44.5478[a]
GDP^2			
LLC	−4.78722[a]	−5.80162[a]	−4.89100[a]
IPS	−3.02527[a]	−3.61547[a]	–
ADF chi-square	32.8802[a]	36.6609[a]	45.1715[a]

[a] The null hypothesis is rejected (the rejection of unit root for variable)

Table 22.2 The results of panel cointegration test for G7 countries (Unrestricted cointegration rank test)

Hypothesized No. of CE	Fisher stat (from trace test)[a]	Probability
None	131.3	0.0000
At most 1	69.37	0.0000
At most 2	39.28	0.0003
At most 3	29.60	0.0087

[a] Probabilities are computed using asymptotic chi-square distribution

Table 22.3 The results of panel cointegration test for each of G7

Cross section	Trace test statics	Probability
Hypothesis of no cointegration		
1	57.6178	0.0004
2	33.5122	0.1991
3	72.0185	0.0000
4	75.5764	0.0000
5	67.0252	0.0000
6	77.1424	0.0000
7	49.8112	0.0041

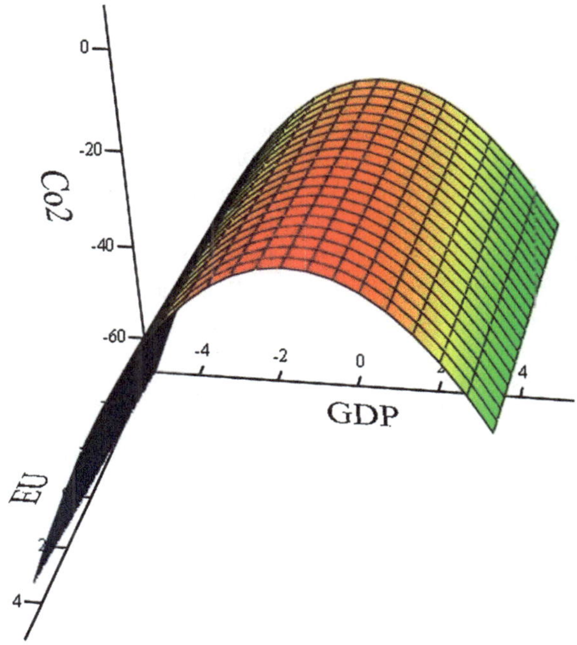

Fig. 22.1 The relationship between CO_2, GDP, and EU variables for G7 countries: the EKC hypothesis is proved for G7 countries. It means economic growth in its higher levels leads to a tool for environment improvement instead of a threat to it. This implies that the environmental impact indicator is an inverted U-shaped function of per capita income (based on the results of this paper)

−1.885583 and the long-term relationship is proved. Finally, the estimated long-term equation is as follows for G7 countries,

$$CO_2 = 0.013 + 0.84\text{EU} + 3.65\text{GDP} - 1.72\text{GDP}^2 \tag{22.1}$$

EU and GDP coefficients are positive, and GDP^2 coefficient is negative. Therefore, the inverted U curve (KEC hypothesis) is proved and accepted for G7 countries. The graph of this equation is illustrated in Fig. 22.1.

22.4.2 Results of Developing Countries

Developing countries consist of China, Egypt, India, Indonesia, Iran, and Thailand. These countries are illustrated in the following tables as 1, 2, … 6, respectively. We apply all the described methodologies used for G7 countries, and the following results are gotten:

The results of units root tests are illustrated in Tables 22.4, 22.5 and 22.6:

As can be understood from these tables, the unit root hypothesis is not rejected when the variables are taken in levels. But when we use these tests for variables in the first and second differences, the large values for the statistics indicate rejection of the null of non-stationary at 1 % level for all models.

Table 22.4 The unit root test results for developing countries

Condition	*M1*	*M2*	*M3*
CO_2 emissions, first differences			
LLC	−2.70943[a]	−2.01190[a]	−3.30250[a]
IPS	−3.09298[a]	−1.91381[a]	–
ADF chi-square	31.6514[a]	22.3215[a]	33.8780[a]
Energy use, second differences			
LLC	−1.33676	−0.36318	−9.44076[a]
IPS	−5.44235[a]	−4.09790[a]	–
ADF chi-square	50.3784[a]	37.3503[a]	84.4027[a]
GDP, second differences			
LLC	−5.60149[a]	−4.69452[a]	−8.62944[a]
IPS	−5.09964[a]	−3.48121[a]	–
ADF chi-square	48.6428[a]	34.4578[a]	76.7302[a]
GDP2, second differences			
LLC	−4.25668[a]	−4.88789[a]	−7.66221[a]
IPS	−4.71761[a]	−4.27292[a]	–
ADF chi-square	45.8760[a]	41.6553[a]	69.6979[a]

[a] The null hypothesis is rejected (the rejection of unit root for variable)

Table 22.5 The results of panel cointegration test for developing countries (unrestricted cointegration rank test)

Hypothesized no. of CE	Fisher stat (from trace test)[a]	Probability
None	75.37	0.0000
At most 1	41.77	0.0000
At most 2	27.53	0.0065
At most 3	17.69	0.1256

[a] Probabilities are computed using asymptotic chi-square distribution

Table 22.6 The results of panel cointegration test for each of the developing countries

Cross section	Trace test statics	Probability
Hypothesis of no cointegration		
1	59.1178	0.0002
2	67.4426	0.0000
3	56.2304	0.0006
4	46.2047	0.0110
5	39.7421	0.0553
6	42.5420	0.0283

Based on the results of E-views software (Table 22.5), the null hypothesis (lack of cointegration) is rejected and the long-term relationship is proved for variables. So, we proceed assuming that there are two cointegration relationships.

The long-term relationship is rejected for all cross sections except Iran (cross section 5). The null hypothesis is accepted, and there is not long-term relationship between those variables for the fifth country (Iran). These tests are done based on the (*M*3).

Also, with the use of Kao test and under individual intercept and no trend condition, the value of statics is −2.811575 and the long-term relationship is proved. And finally, the estimated equation for developing countries is as follows, Eq. (22.2)

$$CO_2 = -1301 + 2.98EU + 1.5GDP - 0.0004GDP^2 \quad (22.2)$$

EU and GDP coefficients are positive, and GDP^2 coefficient is negative. Therefore, the inverted U is proved for G7 countries and the KEC hypothesis (inverted U) is accepted for these countries. But the value of GDP^2 coefficient is too insignificant and close to zero. The concavity is negligible and generally is concluded that the relationship is linear with a positive slope. It means the higher amount of GDP and EU lead to more CO_2 emission. The graph of this equation for developing countries is illustrated in Fig. 22.2.

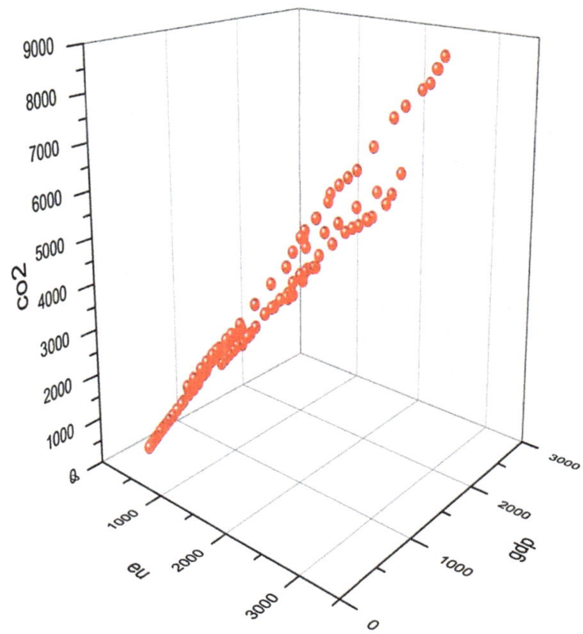

Fig. 22.2 The relationship between CO_2, GDP, and EU variables for developing countries: The EKC hypothesis is rejected for these countries. The linear proved relationship is proved (based on the results of this paper)

22.5 Conclusion

The purpose of this study was to investigate the environmental Kuznets curve hypothesis, the casual and long-term relationship among the EU (energy use), GDP, and CO_2 emission of two different groups, G7 (developed) and developing countries. The applied methodology is panel data method (the panel data unit roots and cointegration tests). First, we began with unit root tests and then by estimating the equations (based on EKC hypothesis) among those variables for both groups of countries.

The EKC hypothesis is proved for G7 countries and is rejected for developing countries (a positive linear relationship). The coefficient of EU for developing countries is higher than the developed countries; it means energy use in these countries leads to more and more CO_2 emission. Here, we can refer to the inefficiency of energy technologies, regardless of environmental impacts, lack of management politics for controlling the amounts of pollutants, investing on efficient technologies, etc. The management politics and decisions are important for decreasing emission. The government can also control emission by including tax on CO_2 and other controlling policies. Also, GDP coefficient is higher than EU coefficient in developing countries and vice versa. A large portion of GDP consisted of consuming the energy resources, and it is not due to production.

When we look at each country separately, the long-term relationship is rejected for Iran and France. France is the biggest country in generating and consuming the nuclear electricity. The share of nuclear power on electricity generation is 77.7 % of total in 2011. Also, Iran as a developing country is confronted with some issues like the following:

- Transportation sector produces a lot of CO_2 because of old vehicles, etc.
- The price of energy is less than national prices and causes more consumption and more pollutants.
- Reliance on oil and gas (fossil fuel) for generating electricity releases a lot of CO_2 into environment.

For future studies, more important variables can be taken into account for making more effective decisions such as the energy price, the amount of renewable and non-renewable, the share of energy use in production and demand, the amount of efficiency, etc. These relationships and equations can also vary in different groups of countries.

References

Apergis, N., & Payne, J. E. (2011). Renewable and non-renewable electricity consumption–growth nexus: Evidence from emerging market economies. *Applied Energy, 88*(12), 5226–5230.

Azlina, A. A., & Mustapha, N. H. (2012). Energy, economic growth and pollutant emissions nexus: The case of Malaysia. *Procedia-Social and Behavioral Sciences, 65*, 1–7.

Brajer, V., Mead, R. W., & Xiao, F. (2008). Health benefits of tunneling through the Chinese environmental Kuznets curve (EKC). *Ecological Economics, 66*(4), 674–686.

Coers, R., & Sanders, M. (2013). The energy–GDP nexus; addressing an old question with new methods. *Energy Economics, 36*, 708–715.

Dagher, L., & Yacoubian, T. (2012). The causal relationship between energy consumption and economic growth in Lebanon. *Energy policy, 50*, 795–801.

Egli, H., & Steger, T. M. (2007). A dynamic model of the environmental Kuznets curve: Turning point and public policy. *Environmental and Resource Economics, 36*(1), 15–34.

Halkos, G. E., & Tsionas, E. G. (2001). Environmental Kuznets curves: Bayesian evidence from switching regime models. *Energy Economics, 23*(2), 191–210.

Halkos, G. E., & Tzeremes, N. G. (2009). Exploring the existence of Kuznets curve in countries' environmental efficiency using DEA window analysis. *Ecological Economics, 68*(7), 2168–2176.

Huang, W. M., Lee, G. W., & Wu, C. C. (2008). GHG emissions, GDP growth and the Kyoto Protocol: A revisit of Environmental Kuznets Curve hypothesis. *Energy Policy, 36*(1), 239–247.

Luzzati, T., & Orsini, M. (2009). Investigating the energy-environmental Kuznets curve. *Energy, 34*(3), 291–300.

Marrero, G. A. (2010). Greenhouse gases emissions, growth and the energy mix in Europe. *Energy Economics, 32*(6), 1356–1363.

Menyah, K., & Wolde-Rufael, Y. (2010). CO_2 emissions, nuclear energy, renewable energy and economic growth in the US. *Energy Policy, 38*(6), 2911–2915.

Ozturk, I., & Acaravci, A. (2010). CO_2 emissions, energy consumption and economic growth in Turkey. *Renewable and Sustainable Energy Reviews, 14*(9), 3220–3225.

Shaari, M. S., Hussain, N. E., & Ismail, M. S. (2012). Relationship between energy consumption and economic growth: Empirical evidence for Malaysia. *Business Systems Review, 2*(1), 17–28.

Tugcu, C. T., Ozturk, I., & Aslan, A. (2012). Renewable and non-renewable energy consumption and economic growth relationship revisited: Evidence from G7 countries. *Energy economics, 34*(6), 1942–1950.

Vollebergh, H. R., Melenberg, B., & Dijkgraaf, E. (2009). Identifying reduced-form relations with panel data: The case of pollution and income. *Journal of Environmental Economics and Management, 58*(1), 27–42.

Zhang, C., & Xu, J. (2012). Retesting the causality between energy consumption and GDP in China: Evidence from sectoral and regional analyses using dynamic panel data. *Energy Economics, 34*(6), 1782–1789.

Chapter 23
Identification and Analysis of Risks Associated with Gas Supply Security of Turkey

Umit Kilic and A. Beril Tugrul

Abstract In this study, a detailed risk assessment study was done for the purpose of identifying risks associated with the security of gas supply in Turkey. Moreover, analysing the risks of the state according to the regulation 994/2010 was issued by European Parliament to safeguard security of gas supply that is accepted after suffering gas supply interruptions. One of the means considered in the regulation to achieve this target is performing a full risk assessment. Recent disruptions took place in Turkish gas network call for performing an extensive risk assessment for Turkey as well. With this paper, gas supply security of Turkey is discussed and its impacts on the security in the state is evaluated in the context of regulation 994/2010 issued by European Union (EU).

23.1 Introduction

Historically, energy security was primarily associated with oil supply. While oil supply remains a key issue, increasing share of natural gas among total primary energy supply of countries brings to mind supply security of this commodity as well. At present, reliable gas supply is one of the most crucial issues in Europe. This subject became even more important after gas supply interruptions and limitations took place in January of 2009 in some countries of European Union (EU).

In this study, a detailed risk assessment study was done for the purpose of identifying risks associated with the security of gas supply in Turkey. Moreover, analysing the risks of the state according to the regulation 994/2010 was issued by European Parliament to safeguard security of gas supply that is accepted after suffering gas supply interruptions. One of the means considered in the regulation to

U. Kilic (✉) · A.B. Tugrul
Istanbul Technical University, Istanbul, Turkey
e-mail: utkilic@gmail.com

A.B. Tugrul
e-mail: beril@itu.edu.tr

A.N. Bilge et al. (eds.), *Energy Systems and Management*,
Springer Proceedings in Energy, DOI 10.1007/978-3-319-16024-5_23

achieve this target is performing a full risk assessment. Recent disruptions took place in Turkish gas network call for performing an extensive risk assessment for Turkey as well.

With this paper, gas supply security of Turkey is discussed and its impacts on the security of the state is evaluated in the context of regulation 994/2010 issued by EU.

23.2 Natural Gas

Environmentally friendly feature of natural gas is an important factor that helps rise of its use. Natural gas, thanks to its gas form of state, makes a better mixture with air and easily burns up. Its gas form of state makes it more accurately controllable. Once the natural gas burned, it does not produce substantial amounts of solid waste (i.e. ash) (Genceli 1989).

Share of natural gas among available energy sources is increasing day by day (Tugrul and Cimen 2013; Tugrul 2011). Use of energy sources in electricity generation in 1973 is given in Fig. 23.1, whereas share of energy sources in power generation in 2011 is given in Fig. 23.2. Share of natural gas is almost doubled from 12 to 22 %.

23.3 Natural Gas in Turkey

The principals of natural gas market in Turkey are laid down by the Natural Gas Law No. 4646 issued on 18 April, 2001. Based on these principals, Turkish natural gas market is targeted to be liberalized. The aim is to reduce the market share of

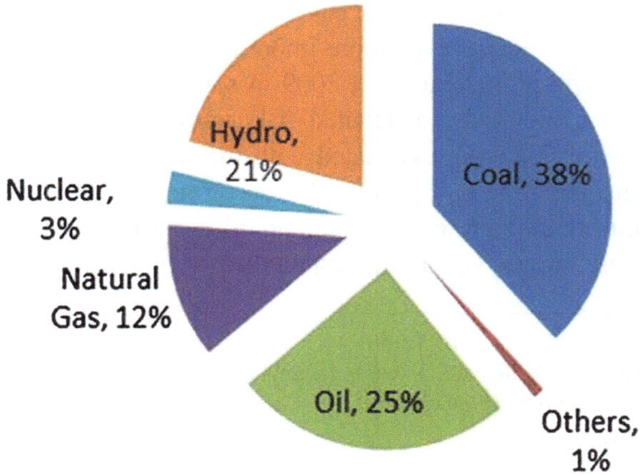

Fig. 23.1 Worldwide electricity generation by energy source in 1973 (IEA 2013)

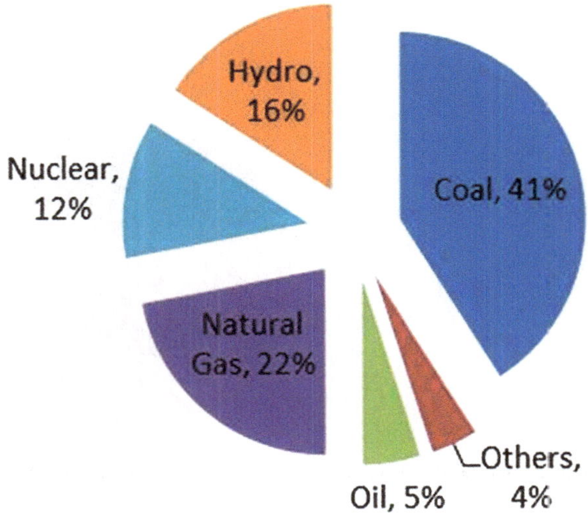

Fig. 23.2 Worldwide electricity generation by energy source in 2011 (IEA 2013)

state-owned companies in natural gas market and to form a competitive market through enhancing market access for private players (EMRA 2013).

Turkey consumed 45.2 bcm of natural gas in 2012. 0.6 bcm of total consumption supplied by local production, whereas rest of domestic demand supplied by Russia, Iran and Azerbaijan through pipelines and by Nigeria, Algeria and spot markets through LNG. As of 2012, gas storage capacity of Turkey stands at 3.1 bcm (EMRA 2013).

Natural gas exploration and production activities, as laid down by Petroleum Law No. 6323, are done via exploration and operation licences issued by General Directorate of Petroleum Affairs, a subordinate of Ministry of Energy and Natural Resources. As of 2012, Turkey's natural gas production stands at 0.6 bcm (Blue Book 2012). In Turkey, there are 14 active natural gas production sites. In 2012, Turkey's natural gas consumption stood at 45.2 bcm, 99 % of total demand covered by imports (Fig. 23.3).

Natural gas imports in Turkey are done through pipelines and LNG terminals. There are 4 International Natural Gas Pipelines and 2 LNG terminals active in Turkey as of 2012 (Fig. 23.4).

23.4 Risk Analysis

International ISO 31000 Risk Management System Standards defines the essentials of risk management and standardizes risk management measures to be implemented. To ensure that risk management is done in an effective manner, principles and generic guidelines are defined in ISO 31000 Risk Management System Standards.

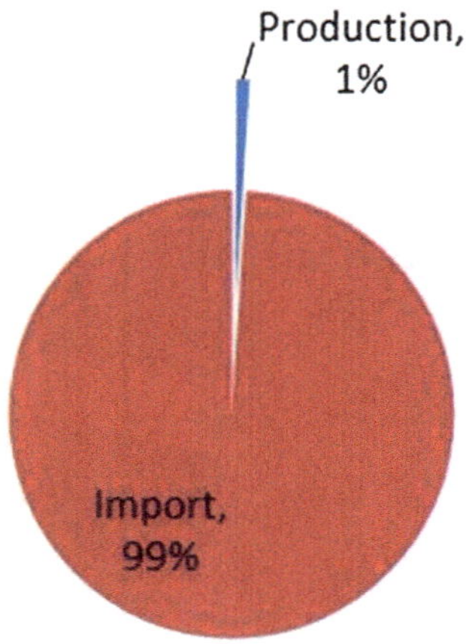

Fig. 23.3 Source of natural gas supply in 2012 (EMRA 2013)

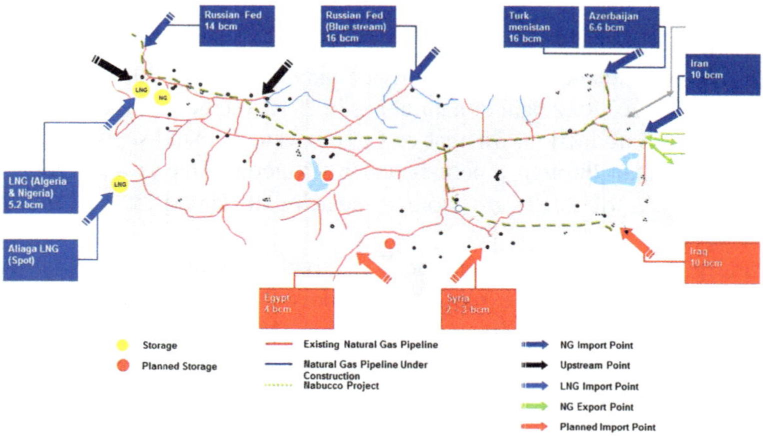

Fig. 23.4 Natural gas entry points of Turkey (BOTAŞ 2012)

The intention of the Regulation 994/2010 of the European Parliament and of the Council Concerning Measures to safeguard security of gas supply is to prevent the kind of gas crisis situations EU-27 experienced in January 2009. One of the means considered in the regulation to achieve this target is performing a full risk assessment.

Each Member State shall make a full risk assessment of the risks affecting its security of gas supply by:

- using the infrastructure and supply standards,
- taking into account all relevant national and regional circumstances,
- running various scenarios of exceptionally high gas demand and supply disruption,

such as failure of the main transmission infrastructures, storages or LNG terminals and disruption of supplies from third country suppliers, taking into account the history, probability, season, frequency and duration of their occurrence as well as, where appropriate, geopolitical risks and assessing the likely consequences of these scenarios.

Based on the regulation No. 994/2010 issued by European Parliament, in the event of disruption of the single largest gas infrastructure during a day of exceptionally high gas demand occurring with a statistical probability of once in 20 years, the rest of the system will cover the demand or should not be calculated with $N-1$ formula.

The $N-1$ formula describes the ability of the technical capacity of the gas infrastructure to satisfy total gas demand in the calculated area in the event of disruption of the single largest gas infrastructure during a day of exceptionally high gas demand occurring with a statistical probability of once in 20 years. $N-1$ criterion is calculated as follows (EU Parliament 2010):

$$N-1[\%] = \frac{\mathrm{EP_m} + P_\mathrm{m} + S_\mathrm{m} + \mathrm{LNG_m} - I_\mathrm{m}}{D_\mathrm{max}} \times 100, \quad N-1 \geq \%100 \qquad (23.1)$$

where

$\mathrm{EP_m}$ technical capacity of entry points (in mcm/d), other than production, LNG and storage facilities covered by P_m, S_m and $\mathrm{LNG_m}$

P_m maximal technical production capability (in mcm/d) means the sum of the maximal technical daily production capability of all gas production facilities

S_m maximal technical storage deliverability (in mcm/d) means the sum of the maximal technical daily withdrawal capacity of all storage facilities

$\mathrm{LNG_m}$ maximal technical LNG facility capacity (in mcm/d) means the sum of the maximal technical daily send-out capacities at all LNG facilities in the calculated area

I_m technical capacity of the single largest gas infrastructure (in mcm/d) with the highest capacity to supply the calculated area

D_max total daily gas demand (in mcm/d) of the calculated area during a day of exceptionally high gas demand occurring with a statistical probability of once in 20 years

In the event that gas disruptions in supply can be covered with market-based demand-side measures, $N-1$ formula is given as follows (EU Parliament 2010):

$$N-1[\%] = \frac{\mathrm{EP_m} + P_\mathrm{m} + S_\mathrm{m} + \mathrm{LNG_m} - I_\mathrm{m}}{D_\mathrm{max} - D_\mathrm{eff}} \times 100, \quad N-1 \geq \%100 \qquad (23.2)$$

where

D_eff: the part (in mcm/d) of D_max that in case of a supply disruption can be sufficiently and timely covered with market-based demand-side measures

Risk assessment is the determination of quantitative or qualitative value of risk related to a concrete situation and a recognized threat. Quantitative risk assessment requires calculations of two components of risk (R): the magnitude of the potential loss (L) and the probability (p) that the loss will occur. In this context, risk is mathematically showed as follows (Lester 2014):

$$R_i = I_i \cdot p(S_i) \tag{23.3}$$

where

R_i the magnitude of risk of scenario or issue i
I_i the magnitude of impact of scenario or issue i
$p(S_i)$ the probability of loss in case scenarios S_i takes place

All risks related to a situation or a threat mathematically expressed as (Lester 2014):

$$R_{\text{total}} = \sum_i I_i \cdot p(S_i) \tag{23.4}$$

Risk assessment consists of an objective evaluation of risk in which assumptions and uncertainties are clearly considered and presented. Part of the difficulty in risk management is that measurement of both of the quantities in which risk assessment is concerned—potential loss and probability of occurrence—can be very difficult to measure (Kilic 2014). In defining the probabilities, common approach is classification of existing data according to the rating table given as Table 23.1

Similarly, impact rating table is shaped based on available data, assumptions and status of the issue under analysis. Impact rating table used in this study is given as Table 23.2

Table 23.1 General probability rating table for a scenario under risk assessment

Description	Rating
Extremely improbable	1
Rather improbable	2
Less probable	3
Probable	4
Very probable	5

Table 23.2 General impact rating table for a scenario under risk assessment

Probability	Rating
Extremely improbable	1
Rather improbable	2
Less probable	3
Probable	4
Very probable	5

Table 23.3 Technical capacities of Turkey's natural gas network entry points (BOTAŞ 2014)

Entry point	Technical capacity (mcm/day)
Malkoçlar	51.4
Durusu	47.3
Türkgözü	19
Gürbulak	28.6
Marmara Ereğlisi LNG terminal	22
Aliağa LNG terminal	16.1
TPAO silivri gas storage	20
TEMI edirne production	0.96
TPAO Akçakoca-Çayağzı production	2.1
Total capacity	207.5

Natural gas network infrastructure of Turkey is analysed in this section taking into consideration technical capacities of all entry points specified in Table 23.3. Turkish natural gas network consists of 4 pipeline import points, 2 production points, 2 LNG terminals and 1 storage points, all of which makes 9 nine entry points in total. Among those entry points, the one with maximum technical capacity is Malkoclar entry point which enables the entrance of natural gas transported from Russia, through Ukraine, Romania and Bulgaria. In terms of $N - 1$ criterion, Malkoclar is the entry point with maximum technical capacity, thus becoming the most critical entry point.

In order to define maximum daily natural gas consumption which is another parameter used in $N - 1$ supply security criterion calculations, maximum daily natural gas consumption values between 2008 and 2014 are given in the Table 23.4.

Once the daily gas consumption data analysed, it is seen that maximum natural gas consumption realized as 197 mcm/day on 10 December 2013. In $N - 1$ criterion calculations, this value is taken into consideration as exceptionally high gas demand occurring with a statistical probability of once in 20 years, other parameters used in the formula are given in the Table 23.5.

Having the parameters given in Table 23.5 factored in the formula, Turkey's $N - 1$ value is calculated as 79 %. Since AB criterion indicates that $N - 1$ value

Table 23.4 Maximum daily natural gas consumption in Turkey between 2008 and 2014 (BOTAŞ 2014)

Year	Max daily consumption (mcm/day)
2008	146
2009	137.7
2010	161.6
2011	176.5
2012	186.1
2013	197.0
2014	183.7

Table 23.5 Parameters used in the $N - 1$ security criterion calculation

Parameter	Technical capacity (mcm/day)
EP_m	146.3
P_m	3.06
S_m	20
LNG_m	38.1
I_m	51.4
D_{max}	197

should be equal or higher than 100 %, Turkey cannot meet with the supply security requirements laid down by EU Regulation.

While the crisis scenarios simulated, supply and demand were taken as two main inputs and scenarios were generated accordingly historical data shows that maximum average daily gas consumption in a week between 2008 and 2014 stood at 173.8 mcm/day in December 2013, whereas maximum average daily consumption during winter months took place in 2012 standing at 158.4 mcm/day. In the light of this information, two separate demand scenarios are generated.

In demand scenario 1, it is assumed that daily demand will be 158.4 mcm and this demand will last 1 month, whereas in demand scenario 2, consumption will stand at a level of 173.8 and the same consumption level will be maintained throughout 1 week.

Ratings used to define the probabilities in crisis scenarios are given in Table 23.6.

With the simulation of supply scenarios given in Table 23.7 and two demand scenarios pointed out before, 48 scenarios in total generated and impact rates shown in Table 23.8. Taking into consideration the impact and probability matrices, risk matrix is generated and shown in Fig. 23.5.

Risk scenarios given in the risk matrix are symbolized as $S_i(a_j)$ where S represents scenario, i is demand scenario numbered with i, a represents supply scenario numbered with j. There are two demand and twenty-four supply scenarios taken into consideration in this study. Ratings of all scenarios, in other words the outcome of the risk assessment study, are simply summarized with the generation of risk matrix given in Fig. 23.6. Although most of the crisis scenarios generated fall under the left-down side of the matrix which reflects the less risky zone, there are some scenarios that requires more attention. Scenario $S_2(a_1)$ has the highest probability

Table 23.6 Probability ratings table for the crisis scenarios taken into consideration

Rating	Probability
1: Extremely improbable	Once every 50 years or less
2: Rather improbable	Once every 20 years or less
3: Less probable	Once every 10 years or less
4: Probable	Once every 3 years or less
5: Very probable	Annual or more frequent occurrence

Table 23.7 Impact matrix for supply crisis scenarios taken into consideration

Supply and demand	Duration	Impact
Supply matches or higher than demand	–	1
Demand higher than supply by less than 10 mcm/day	Up to 1 week	2
Demand higher than supply by less than 10 mcm/day	More than 1 week	3
Demand higher than supply by more than 10 mcm/day	Up to 1 week	4
Demand higher than supply by more than 10 mcm/day	More than 1 week	5

Table 23.8 Probabilities for supply crisis scenarios taken into consideration

Scenario no	Entry point	Reason	Cut (%)	Duration	Probability
1	Malkoçlar	Political	100	1 month	2
2	Malkoçlar	Political	100	1 week	3
3	Malkoçlar	Political	50	1 month	3
4	Malkoçlar	Political	50	1 week	4
5	Durusu	Political	100	1 month	1
6	Durusu	Political	100	1 week	2
7	Durusu	Technical	100	1 month	1
8	Durusu	Technical	100	1 week	2
9	Durusu	Technical	50	1 month	2
10	Durusu	Technical	50	1 week	3
11	Türkgözü	Technical	100	1 month	1
12	Türkgözü	Technical	100	1 week	2
13	Türkgözü	Technical	50	1 month	2
14	Türkgözü	Technical	50	1 week	3
15	Gürbulak	Terrorism	100	1 month	2
16	Gürbulak	Terrorism	100	1 week	3
17	Gürbulak	Terrorism	50	1 month	3
18	Gürbulak	Terrorism	50	1 week	4
19	Marmara Er.	Technical	100	1 week	5
20	Marmara Er.	Technical	100	20 days	3
21	Marmara Er.	Technical	100	1 month	2
22	Silivri Storage	Technical	100	1 week	3
23	Silivri Storage	Technical	100	20 days	2
24	Silivri Storage	Technical	100	1 month	1

and impact rating combination, meaning that it has highest risk rating among 48 scenarios. $S_2(a_5)$ and $S_2(a_7)$ have low risk ratings compared to $S_2(a_1)$; however, the impact of those scenarios is critical. The same applies for $S_1(a_{19})$ and $S_2(a_{19})$, although the risk rating of these two scenarios is low, the probabilities of them are considerably high.

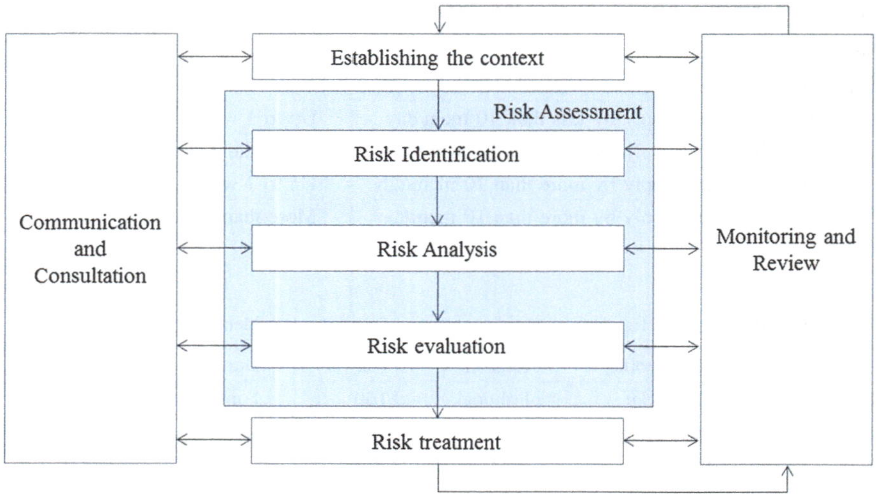

Fig. 23.5 Risk assessment methodology based on ISO 31000 risk management standard

Probability	1	2	3	4	5
5	S_1(a19), S_2(a19)				
4	$S_1(a_4)$, $S_1(a_{18})$, $S_2(a_4)$, $S_2(a_{18})$				
3	$S_1(a_3)$, $S_1(a_{10})$, $S_1(a_{14})$, $S_1(a_{16})$, $S_1(a_{17})$, $S_1(a_{20})$, $S_1(a_{22})$, $S_2(a_3)$, $S_2(a_{10})$, $S_2(a_{14})$, $S_2(a_{17})$, $S_2(a_{20})$, $S_2(a_{22})$	$S_1(a_2)$, $S_2(a_{16})$			
2	$S_1(a_6)$, $S_1(a_8)$, $S_1(a_9)$, $S_1(a_{12})$, $S_1(a_{13})$, $S_1(a_{15})$, $S_1(a_{21})$, $S_1(a_{23})$, $S_2(a_9)$, $S_2(a_{12})$, $S_2(a_{13})$, $S_2(a_{21})$, $S_2(a_{23})$		$S_1(a_1)$, $S_2(a_{15})$	$S_2(a_2)$, $S_2(a_6)$, $S_2(a_8)$	$S_2(a_1)$
1	$S_1(a_5)$, $S_1(a_7)$, $S_1(a_{11})$, $S_1(a_{24})$, $S_2(a_{24})$, $S_2(a_{11})$				$S_2(a_5)$, $S_2(a_7)$
	Impact				

Fig. 23.6 Generated risk matrix

23.5 Conclusion

In this study, gas supply security of Turkey, as a country still having negotiations on EU accession, is discussed and its impacts on the security of the state is evaluated in the context of regulation 994/2010 issued by EU and ISO 31000 Risk Standard. Within this context, 48 different crisis scenarios are simulated and the risks in regard to gas supply security of Turkey is analysed in a quantitative way.

As for the $N - 1$ criterion calculations outlined in the regulation no 994/2010 issued by EU in regard to natural gas supply security, the period between 2008 and 2014 has chosen. In other words, a seven-year data analysis has been conducted. Once the data analysed and factored in $N - 1$ formula, with its 79 % value Turkey seems to be way below the 100 % criterion pointed out by EU and cannot meet with the supply security criterion.

The risk ratings of simulated scenarios show that there are no scenarios above 15 risk ratings that need urgent attention. However, scenario $S_2(a_1)$, that has a 10 risk rating, indicates importance of an internal or external problem that will cause a disruption at Malkoclar entry point during cold winter days with high natural gas demand.

Other important scenarios that require attention are $S_2(a_1)$, $S_2(a_5)$ and $S_2(a_7)$. Although those scenarios have relatively low risk ratings, their impacts are quiet high. Those scenarios draw attention to fragile structure of Turkey's natural gas supply security in case of possible disruptions that might take place at the entry points that have high technical capacity such as Durusu and Malkoclar.

In brief, the conducted risk assessment study shows that in case of occurrence of a political, economic, technical, or geopolitical disruption at Malkoclar entry point, which has the highest technical capacity among all entry points, Turkey will not be able to match the demand during a day of exceptionally high gas demand occurring with a statistical probability of once in 20 years. In other words, Malkoclar entry point has a significance and potential of causing a breaking point in gas supply security of Turkey.

References

Blue Book. (2012).

BOTAŞ. (2012). Annual Report, Ankara.

BOTAŞ. (2014). Data acquired form BOTAS within the context of right to information Act No 4982.

EMRA. (2013). Natural gas market report, Ankara.

EU Parliament. (2010). Regulation no 994/2010 of the European Parliament and of the council. *Official Journal of the European Union*.

Genceli, O. F. (1989). *Properties of natural gas*. İstanbul: Bulletin of ITU Mechanical Engineering Faculty.

IEA. (2013). *Key world energy statistics*. Paris, France: International Energy Agency.

Kilic, U. (2014). Identification and analysis of risks associated with gas supply security of Turkey. *MSc Thesis*, Istanbul Technical University, Energy Institute.

Lester, A. (2014). *Project management, planning and control* (6th ed.), London, pp. 71–81.

Tuğrul, A. B. (2011). Nuclear energy in the energy expansion of Turkey. *Journal of Energy and Power Engineering, 5*(10), 905–910.

Tuğrul, A.B., & Çimen, S. (2013). Energy initiatives for Turkey. In: *Proceedings of International Conference on Economics and Econometrics, ICEE-2013, Dubai-BAE* (pp. 40–44).

Chapter 24
The Social Cost of Energy: External Cost Assessment for Turkey

Aylin Çiğdem Köne

Abstract The social or full costs of energy sources, which include the external cost plus the private cost, are the most important criteria for energy and environmental policy making. Energy policy making is concerned with both the supply side and the demand side of energy provision. On the energy supply side, deciding on alternative investment options requires the knowledge of the full cost of each energy option under scrutiny. On the demand side, social welfare maximisation should lead to the formulation of energy policies that steer consumers' behaviour in a way that will result in the minimisation of costs imposed to society as a whole. Demand-side policies can benefit significantly from the incorporation of full energy costs in the corresponding policy formulation process. The geographical dimension is also important since environmental damage from energy production crosses national borders. Hence, a consistent set of energy costs allows a better understanding of the international dimensions of policy decisions in these areas. This paper, focusing on classical pollutants, tries to assess external costs from human health damages, damages to buildings, crop losses and from biodiversity impacts. To this aim, first, emissions data have been drawn from European Monitoring and Evaluation Programme (EMEP) database. Then, these emissions data have been transformed into monetary terms using the results of cost assessment for sustainable energy systems (CASES) project for the years 2000 and 2010. The results have been discussed in the context of energy and sustainability.

24.1 Introduction

From resource extraction to the provision of energy services, pollutants are produced, emitted or disposed of, often with severe health and environmental impacts. Much of the current energy supply and use is chiefly responsible for urban air

A.Ç. Köne (✉)
Muğla Sıtkı Koçman University, Muğla, Turkey
e-mail: ckone@mu.edu.tr

A.N. Bilge et al. (eds.), *Energy Systems and Management*,
Springer Proceedings in Energy, DOI 10.1007/978-3-319-16024-5_24

pollution, regional acidification and global climate change (IAEA 2005). Therefore, the influence on the environment and human health due to air pollution should be adequately taken into account.

An external cost arises when the social or economic activities of one group of persons have an impact on another group and when that impact is not fully accounted, or compensated for, by the first group. Thus, an energy system that generates emissions of SO_2, NO_x, particulates, etc., causing damage to building materials, biodiversity or human health, imposes an external cost. This is because the impact on the owners of the buildings, crops or on those who suffer damage to their health is not taken into account by the supplier of the energy when deciding on the activities causing the damage. Therefore, the environmental costs are "external" because, although they are real costs to these members of society, the producer is not taking them into account when making decisions (ExternE 1999). During the EU Framework 6 programme under priority, sustainable energy systems cost assessment for sustainable energy systems (CASES) project was financed (CASES 2013). The purpose of the project was to evaluate external and private cost of electricity generation in EU-27 and other countries seeking to provide reliable data for electricity system development scenarios in case then external costs are integrated through all the chain of electricity supply system (Streimikiene et al. 2009; Streimikiene and Alisauskaite-Seskiene 2014).

This paper, focusing on classical pollutants, which are NH_3, NMVOC, NO_x, PPM_{co}, $PPM_{2.5}$ and SO_2, tries to assess external costs from human health damages, damages to buildings, crop losses and from biodiversity impacts in Turkey. To this aim, first, emissions data have been drawn from European Monitoring and Evaluation Programme (EMEP) database (CEIP-EMEP 2013). Then, these emissions data have been transformed into monetary terms using the results of CASES project for the period of 2000–2010. The results have been discussed in the context of energy and sustainability.

24.2 External Costs Calculation Methodology

ExternE methodology considers seven major types of damages. The main categories are human health (fatal and non-fatal effects), effects on crops and materials. The impact pathway approach—and coming along with this approach, the Eco-Sense model, an integrated software tool for environmental impact pathway assessment—was developed within the ExternE project series and represents its core (EC 2004).

Impact pathway assessment is a bottom-up approach in which environmental benefits and costs are estimated by following the pathway from source emissions via quality changes of air, soil and water to physical impacts, before being expressed in monetary benefits and costs. The use of such a detailed bottom-up methodology is necessary, as external costs are highly site dependent. Two emission scenarios are needed for each calculation, one reference scenario and one case scenario. The background concentration of pollutants in the reference scenario is a significant factor for pollutants with nonlinear chemistry or nonlinear dose–response functions. The estimated difference in the simulated air quality situation between the case and the reference situation is combined with exposure response functions to derive differences in physical impacts on public health, crops and building material (Schlesinner and Nielsen 1997).

It is important to note that not only local damages have to be considered—air pollutants are transformed and transported and cause considerable damage hundreds of kilometres away from the source. So local and European-wide modelling was performed during ExternE and its extensions (EC 2004). As a next step within the pathway approach, exposure response models are used to derive physical impacts on the basis of these receptor data and concentration levels of air pollutants. In the last step of the pathway approach, the physical impacts are evaluated in monetary terms. According to welfare theory, damages represent welfare losses for individuals. For some of the impacts (crops and materials), market prices can be used to evaluate the damages. However, for non-market goods (especially damages to human health), evaluation is only possible on the basis of the willingness-to-pay or willingness-to-accept approach that is based on individual preferences. To complete the external cost accounting framework for environmental themes (acidification and eutrophication), a complementary approach for the valuation of such impacts based on the standard-price approach is developed and improved. This procedure deviates from the pure welfare economic paradigm followed in ExternE, but it allows estimating damage figures for ecological impacts complementary to the existing data on impacts from the same pollutants on public health, materials and crops (based on damage function approach and welfare-based valuation studies). The integration of this methodology and data into the existing external costs framework is an important extension as it also covers impact categories that could otherwise not be addressed properly in ExternE.

24.3 External Costs of Pollutant Emissions Due to Energy Sectors

Within framework of CASES project based on EcoSense model run, the external cost for classical air pollutants corresponding to an average height of release was obtained for 39 European and non-European countries and five sea regions, and for the EU-27 as an average. The values are based on parameterised results of a complex dispersion model. Results are available for emission of NH_3, NMVOC, NO_x, PPM_{co}, PPM_{25} and SO_2. The receptor domain covers the whole of Europe regarding impacts on human health and crops, damage to materials, and loss of biodiversity caused by acidification and eutrophication, implemented into the assessment framework of EcoSense.

Table 24.1 Pollutant emissions for the SNAP sectors (tonnes)

Pollutant	S1 sector		S2 sector	
	2000	2010	2000	2010
NH_3	0	201	618	621
NMVOC	4610	3426	202,450	285,546
NO_x	201,389	456,580	191,640	79,904
PPM_{co}	29,069	37,164	17,654	2707
$PPM_{2.5}$	33,694	39,617	111,017	71,522
SO_2	1,292,484	894,110	73,695	317,140

Based on information provided in Table 24.1 and the emissions data provided in Table 24.2, the external cost of pollutant emissions due to selected nomenclature for reporting of air pollutants (SNAP) Sectors namely "combustion in energy and transformation industries (S1)" and "non-industrial combustion plants (S2)" in Turkey for the years 2000 and 2010 is calculated. S1 and S2 SNAP Sectors consist of public power, district heating plants, petroleum refining plants, solid fuel transformation plants, coal mining, oil/gas extraction, pipeline compressors and commercial and institutional plants, residential plants, plants in agriculture, forestry and aquaculture, respectively. By using the data given in Tables 24.1 and 24.2, the calculated external costs of pollutant emissions due to S1 and S2 SNAP sectors are presented in Tables 24.3 and 24.4, respectively.

Table 24.2 External costs of the classical pollutants released in Turkey

Cost (Euro 2000/tonnes)	
Human health impact	
NH_3	5194
NMVOC	−33
NO_x	2780
PPM_{co}	862
$PPM_{2.5}$	19,725
SO_2	5337
Loss of biodiversity	
NH_3	607
NMVOC	−6
NO_x	102
SO_2	2
Impact on crops	
NH_3	−34
NMVOC	9
NO_x	44
Impact on materials	
NO_x	71
SO_2	259

Table 24.3 External costs of pollutant emissions due to energy sector (Million Euro 2000)

External costs	S1 sector		S2 sector	
	2000	2010	2000	2010
Human health impact				
NH_3	0.00	1.04	3.21	3.23
NMVOC	−0.15	−0.11	−6.68	−9.42
NO_x	559.86	1269.29	532.76	222.13
PPM_{co}	25.06	32.04	15.22	2.33
$PPM_{2.5}$	664.61	781.44	2189.81	1410.77
SO_2	6897.99	4771.87	393.31	1692.58
Loss of biodiversity				
NH_3	0.00	0.12	0.38	0.38
NMVOC	−0.03	−0.02	−1.21	−1.71
NO_x	20.54	46.57	19.55	8.15
SO_2	2.58	1.79	0.15	0.63
Impact on crops				
NH_3	0.00	−0.01	−0.02	−0.02
NMVOC	0.04	0.03	1.82	2.57
NO_x	8.86	20.09	8.43	3.52
Impact on materials				
NO_x	14.30	32.42	13.61	5.67
SO_2	334.75	231.57	19.09	82.14

Table 24.4 External costs of pollutant emissions by impact types (Million Euro 2000)

External costs	2000	2010
Human health impact	11,275	10,177
Loss of biodiversity	42	56
Impact on crops	19	26
Impact on materials	382	352
Total	11,718	10,611

As can be seen from Table 24.4, the highest external cost in energy sector (S1 + S2) is calculated for human health impact, while the lowest external cost belongs to impacts on crops. Comparing 2000 and 2010 results for the total external costs, it can be seen that there is a slight decrease in external costs of the Turkish energy sector. There is a decrease also in terms of the share of the external cost to gross domestic product (GDP). The total GDP of Turkey for the years 2000 and 2010 is 278,500 and 335,600 Million Euro 2000 (WBI 2013). The total external costs of pollutant emissions are approximately 4 and 3 % of Turkey's total GDP for the years 2000 and 2010, respectively. Most of this improvement can be attributed to fuel substitution (from high-sulphur content domestic coal to natural gas) in power generation.

Turkish energy system is dominated by fossil fuel use at present. The share of the utilisation of fossil fuels in total primary energy supply increased from 87 % in 2000 to 89 % in 2010. The rapid growth of fossil fuel use is a result of specially accelerated natural gas utilisation (RTMENR 2013). The emission levels of natural gas particularly for SO_2 and PM are negligible in comparison with coal and oil (EEA 2006; Chatzimouratidis et al. 2008). Therefore, natural gas can be used in many ways to help reduce the emissions of pollutants into the atmosphere, and Turkey has done that. However, since the environmental benefits of natural gas have been consumed, Turkey should focus on other ways to reduce external costs in energy sector.

Although the air pollutant emission levels of natural gas are negligible in comparison with coal and oil, the air pollutant emission levels of natural gas are higher in comparison with nuclear and renewable sources (EEA 2006; Chatzimouratidis et al. 2008). An important policy implication of this study is that the share of non-fossil fuels such as nuclear and renewable energy of Turkey in total primary energy supply must be increased to decrease the external cost of air pollutant emissions.

24.4 Policy Conclusions

Based on the above analysis, several policies can be suggested for Turkey to decrease the pollutant emissions:

- The national emissions standards for large power plants are significantly less stringent than the EU standards (EPCD 2010). The emissions standards should be strengthened to align them with the EU legislation.
- The efficiencies of air pollutant prevention equipments, such as flue gas desulphurisation systems and electrostatic precipitators, should be checked periodically by an authority independent of the power plant management to ensure the objectivity of the audit and to put pressure on the plant management to operate at maximum equipment efficiencies. In this context, the results should be made public so that local people have access to this information.
- In all energy investments and especially in power generation, the principle should be to minimise the negative effects on the environment. Future thermal power plants should be required to adopt new technologies to minimise the pollutant emissions. Research and development in the area of minimising the pollutant emissions from thermal power plants should be supported and funded by the government.
- The national air quality limit values for the air pollutants are greater than the EU standards (RTOG 2008). The national air quality limit values for the air pollutants should be strengthened to align them with EU legislation.
- The legislation that would support power generation based on renewable sources must be renewed in the light of a macrostrategy that aims to base Turkey's energy policies solidly on local and renewable sources.

References

CASES, Cost Assessment for Sustainable Energy Systems. Available from http://www.feem-project.net/cases/. Accessed December 28, 2013.

CEIP-EMEP, Centre on Emission Inventories and Projections-European Monitoring and Evaluation Programme. *Emissions as used in EMEP models*. Available from http://www.ceip.at/webdab-emission-database/officially-reported-emission-data/ Accessed December 28, 2013.

Chatzimouratidis, A. I., Petros, A., & Pilavachi, P. A. (2008). Multicriteria evaluation of power plants impact on the living standard using the analytic hierarchy process. *Energy Policy, 3*, 1074–1089.

EC, European Commission (2004). Work Package 6: Revision of external cost estimates. In *New elements for the assessment of external costs from energy technologies* (*NewExt*). Final Report. Contract No. ENGI-CT2000-00129.

EEA, European Environment Agency (2006). *EMEP/CORINAIR Atmospheric Emission Inventory Guidebook.*

EPCD, The European Parliament and of the Council Directives (2010). Directive on industrial emissions (integrated pollution prevention and control), Directive Number: 2010/75/EU.

ExternE, Externalities of Energy (1999). Vol. 7. Methodology 1998 update. Luxemburg: Office for Official Publications of the European Communities, 9–15.

IAEA (2005). (International Atomic Energy Agency), UNDESA (United Nations Department of Economic and Social Affairs), IEA (International Energy Agency), Eurostat, EEA (European Environment Agency). *Energy indicators for sustainable development: Methodologies and guidelines*, Vienna: IAEA.

RTMENR, Republic of Turkey Ministry of Energy and Natural Resources (2013). *The Overall Energy Balance 1970–2010*. Available from http://www.enerji.gov.tr/index.php. Accessed December 28, 2013.

RTOG (2008). Republic of Turkey Official Gazette, Air Quality Assessment and Management Regulations, Date: 06.06.2008, Number: 26898.

Schlesinner, L., & Nielsen, P. S. (1997). External costs related to power production technologies ExternE national implementation for Denmark Risø National Laboratory, Copenhagen 2–13.

Streimikiene, D., & Alisauskaite-Seskiene, I. (2014). External costs of electricity generation options in Lithuania. *Renewable Energy, 64*, 215–224.

Streimikiene, D., Roos, I., & Rekis, J. (2009). External cost of electricity generation in Baltic States. *Renewable and Sustainable Energy Reviews, 13*, 863–870.

WBI, World Bank Indicators, Available from http://data.worldbank.org/indicator. Accessed December 28, 2013.

Chapter 25
Energy Infrastructure Projects of Common Interest in the SEE, Turkey, and Eastern Mediterranean and Their Investment Challenges

Panagiotis Kontakos and Virginia Zhelyazkova

Abstract European Union's energy strategy for the period up to 2020 builds on eight priority corridors for electricity, gas, and oil. Accordingly, on October 2013, 248 energy infrastructure projects were selected and assessed from a European perspective as the most critical to implement. Particularly, for a project to be included in the list, it has to bear significant benefits for at least two member states, contribute to market integration and further competition, enhance security of supply, and reduce CO_2 emissions. Furthermore, by the characterization of "projects of common interest" (PCI), they are indented to benefit from more rapid and efficient permit granting procedures and improved regulatory treatment (European Union, 2014a). The aims of this working paper are, firstly, to introduce and briefly discuss the priority corridors and thematic areas that must be implemented in the coming decade to assist the EU meet its short- and long- term energy and climate objectives; secondly, to outline the key projects of common interest specifically in the South East European, Turkey, and eastern Mediterranean regions; finally, to present the major investment and financial challenges associated with the undertaking and implementation of these projects.

Acronyms

Bcma	Billion cubic meters
CBA	Cost-benefit Analysis
CEE	Central East Europe
CEF	Connecting Europe Facility
CO_2	Carbon dioxide

P. Kontakos (✉)
University of Macedonia, Thessaloniki, Greece
e-mail: pkontakos@uom.gr

V. Zhelyazkova
VUZF University, Sofia, Bulgaria
e-mail: virginia.zhelyazkova@gmail.com

A.N. Bilge et al. (eds.), *Energy Systems and Management*,
Springer Proceedings in Energy, DOI 10.1007/978-3-319-16024-5_25

DG ENER	Directorate-General for Energy
ENTSO	European Networks of Transmission System Operators
EC	European Commission
EU	European Union
LNG	Liquefied natural gas
PCI	Projects of common interest
SEE	South East Europe
Tcm	Trillion cubic meters
TEN-E	Trans-European Networks for Energy
TYNDP	Ten-Year Network Development Plans

25.1 Introduction

EU's energy strategy for the period up to 2020 builds on eight priority corridors for electricity, gas, and oil (European Union, 2012). Accordingly, on October 2013, a number of 248 energy infrastructure projects of common interest were selected and assessed as the most critical to implement from a European perspective.

The aims of this work are to introduce and briefly discuss the priority corridors and thematic areas that must be implemented in the coming decade by the energy infrastructure package to assist the EU meet its short- and long-term energy and climate objectives along with an analysis of their investment needs and financing gaps and to outline the key projects of common interest specifically in the SEE, Turkey, and eastern Mediterranean regions. Particular reference to the southern gas corridor is made.

25.2 The Energy Infrastructure Package

In October 2011, the EC listed a complete package to boost trans-European infrastructure development in the areas of transport, energy, and information society. Respectively, the energy infrastructure package is comprised of legislative guidelines, proposals for setting up infrastructure policies, and the Connecting Europe Facility (CEF) regulation, in charge of providing financial aid.

In this context, the term energy infrastructure covers electricity transmission lines, gas, CO_2 and oil pipelines, compressed natural gas, liquefied natural gas reception facilities, and electricity and gas storage.

The trans-European energy infrastructure guidelines identify eight priority corridors and four thematic areas that must be implemented in the coming decade. Further, to materialize these corridors/areas into tangible projects, the guidelines consider a new approach of identifying energy infrastructure projects that can obtain the label of projects of common interest (PCIs). Thus, to qualify for financial support under the CEF, projects must firstly be recognized as PCIs.

Based on the recommendations made by the European Networks of Transmission System Operators (ENTSO), draft regional lists of potential PCIs were established in the end of 2012 and incorporated within the Ten-Year Network Development Plans (TYNDP). The final Union-wide list was approved in October 2013 and includes 248 CPIs, based on several criteria such as cost-benefit analysis (CBA), demonstration of added value toward the achievement of EU's overall energy goals, and cross-border benefits which can further act as a mechanism facilitating cost allocation among EU members.

From 2014 onwards, PCIs will be eligible to submit a request for financial support for studies under the annual call for proposals in the CEF. Furthermore, they may also be eligible to submit a request for financial support for works under the CEF, provided they meet the following two conditions:

- the project should demonstrate through a CBA and business plan that it provides societal non-commercial benefits;
- the affected national regulatory authorities have provided a positive opinion on the CBA and on the scope of the non-commercial benefits.

25.2.1 The Eight Priority Corridors

In Fig. 25.1, the eight priority corridors included in the energy infrastructure package are depicted. These included the following:

- *Northern seas offshore grid* to facilitate transport of electricity from renewable offshore wind energy sources;
- *North–South electricity interconnections in Western Europe with the Mediterranean*, also to combine electricity from renewable sources;
- *North–South electricity interconnections in CEE/SEE* to develop the internal market and integrate generation from renewable sources;
- *Baltic energy market interconnection plan* in electricity/gas to reinforce an integrated regional gas energy market in the Baltic Sea Region;
- *North–South gas* interconnections in Western Europe to diversify routes and to increase short-term deliverability;
- *North–South gas interconnections* between the Baltic Sea region, the Adriatic and the Aegean Seas and the Black Sea *and Oil supply connections in CEE* to raise security of supply and condense environmental risks;
- *Southern gas corridor* for the transmission of gas from the Caspian Basin, Central Asia, the Middle East, and eastern Mediterranean basin to the Union to enhance diversification of gas suppliers.

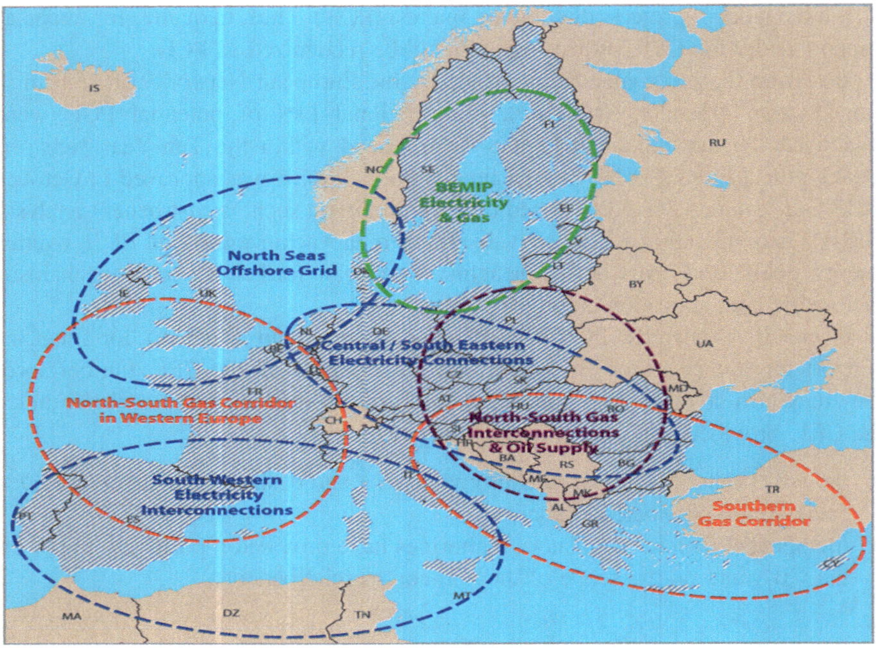

Fig. 25.1 European energy infrastructure priorities for electricity, gas, and oil (Adopted from © European Union—Directorate-General for Energy)

In Fig. 25.2, specific projects in the SEE region are depicted. Furthermore, in parallel to the southern gas corridor, there is particular interest in the development of a gas corridor for the transmission of gas produced offshore in the eastern Mediterranean (mainly Cyprus and Israel). In addition, the EuroAsia Interconnector

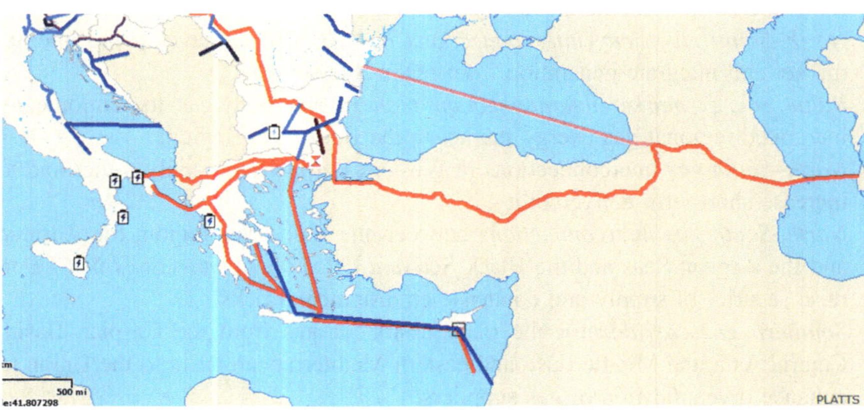

Fig. 25.2 Projects of common interest in the SEE, Turkey & Eastern Mediterranean (Adopted from © European Union—Directorate-General for Energy)

is already under construction linking Greek, Cypriot, and Israeli power grids via the world's longest submarine power cable with a length of 287 km and a power rating of 2000 MW. This interconnector is expected to be completed by 2016.

25.3 Investment Needs and Financing Gap

Based on the grounds that all permitting issues are resolved, an investment gap which is estimated in approximately €66 bn is likely to exist by 2020. The reasons for the gap are mainly due to difficult access to finance and lack of adequate risk mitigating instruments, especially for projects with positive externalities and wider European benefits, but no adequate commercial justification (EU, DG ENER calculations based on PRIMES, 2012).

To cover this gap, the European Commission has suggested working on two directions:

- Encouraging the involvement of private sources through improved cost allocation based on rules commonly agreed by the systems' regulators. The current tariff-based business model will continue to be the basic mechanism.
- Enforcing EU's partnerships with international financial institutions and introduction of innovative market-based financing solutions beyond the conventional support forms (grants, interest rate subsidies).

25.4 Southern Gas Corridor

The major strategic perspective of the southern gas corridor is twofold: to bring new gas sources to the EU and diversify particularly concentrated markets in SEE. The rationale is to directly link the EU gas market to large deposits of gas in the Caspian Region, Middle East (estimated at 90.6 tcm; for comparison, Russian proven reserves amount to 44.2 tcm), and the eastern Mediterranean basin.

Moreover, these gas fields are geographically closer than the main Russian deposits (Fig. 25.3). The planned transit routes for these gas volumes are through Turkey, the Black Sea, and the eastern Mediterranean.

Pipeline components will also be improved for transporting substantial supplementary quantities of LNG to Europe, mainly from the Middle East. Initially, this will be implemented with the construction of LNG reception points in Europe and via interconnections to the wider network.

The strategic aim of the corridor is to accomplish a supply route to the EU of approximately 10–20 % of EU gas demand by 2020, equivalent nearly to 45–90 bcma.

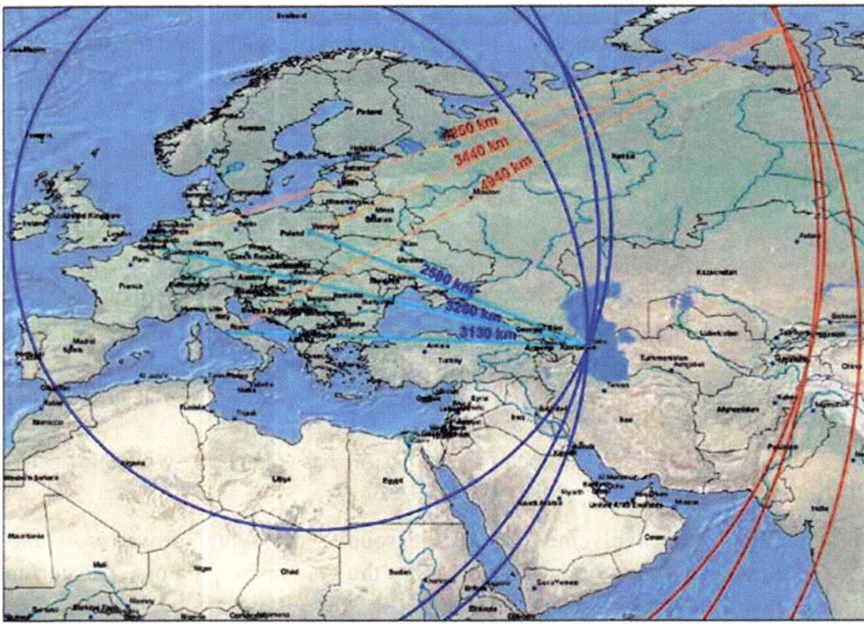

Fig. 25.3 Comparison of distances of main eastern gas supplies to main EU consumption hubs (Adopted from © European Union—Directorate-General for Energy)

Finally, the southern gas corridor will essentially be an additional supply corridor to the existing ones from Russia, Norway, and North Africa.

25.5 Conclusions

Europe is at an unprecedented junction for its energy future. To achieve its energy and climate goals of "20-20-20 by 2020" (i.e., renewable sources to contribute 20 % of total energy consumption, reduce greenhouse gas emissions by 20 %, and deliver 20 % savings in energy consumption) and ensure the transition to a low-carbon economy by 2050 while fostering growth and jobs, Europe needs to invest in the upgrading of its energy infrastructure in the next 10 years (European Union, 2014b).

The energy infrastructure package along with the Guidelines and the CEF provide an enormous prospect toward this direction. In addition to traditional grants, CEF will be the foundation for innovative financial instruments and can provide financial support from the Union's budget for the period 2014–2020, either to bridge so-called energy islands or to warrant security of supply.

References

European Union, Directorate-General for Energy. (2012). Connecting Europe: The energy infrastructure for tomorrow.

European Union, Directorate-General for Energy. (2014a). Overview of projects by country. http://ec.europa.eu/energy/infrastructure/pci/doc/2013_pci_projects_country.pdf.

European Union, Directorate-General for Energy. (2014b). Energy infrastructure priori-ties for 2020 and beyond. http://ec.europa.eu/energy/infrastructure/strategy/2020_en.htm.

Chapter 26
Incorporating the Effect of Time-of-Use Tariffs in the Extended Conservation Supply Curve

Aakash Jhaveri and Santanu Bandyopadhyay

Abstract The conservation supply curve (CSC)—a plot of the cost of conserved energy (CCE) versus cumulative energy conserved—allows for an economic comparison of multiple Energy conservation measures (ECMs) leading to the identification of economically feasible ones. Its major advantages are separation of the cost of implementing such measures from their benefit and its independence from the fuel price. Different ECMs save different amounts of energy during different periods of the day. Hence, TOU tariffs, implemented to incentivise electrical load management and energy conservation, affect their cost-saving potentials. The CSC, in its present form, does not incorporate the effect of such tariffs. The primary objective of this work was to propose a methodology to extend and to modify the CSC to incorporate the effects of these tariffs, while being able to draw the same inferences as can be drawn from the original CSC. In doing so, energy profiles of the measures (a set of saving potentials over a particular time period, e.g. 24 values for 24 h of the day) and TOU charges (rates over and above the base energy charge, calculated over the same time period) are used to arrive at a value indicating the impact of the TOU rates on said measures. These values, along with the CCEs of the measures, are used to generate the modified CSC. Applicability of the proposed methodology is demonstrated with an illustrative example. Current research is focused on accounting for multiple types of fuels saved by conservation measures.

26.1 Introduction—The Conservation Supply Curve

Energy conservation measures (ECMs) require additional investments of money and lead to energy savings for a definite period of time. For each such measure, the total additional investments when divided by the total energy-saving potential give

A. Jhaveri (✉) · S. Bandyopadhyay
Indian Institute of Technology, Bombay, India
e-mail: jhaveri.aakash@gmail.com

S. Bandyopadhyay
e-mail: santanub@iitb.ac.in

A.N. Bilge et al. (eds.), *Energy Systems and Management*,
Springer Proceedings in Energy, DOI 10.1007/978-3-319-16024-5_26

a cost per unit energy value defined as the cost of conserved energy (CCE). A supply curve of conserved energy, also called the conservation supply curve (CSC) for energy conservation, is a means of expressing energy conservation potentials of different ECMs on one plot in increasing order of the aforementioned CCE. Energy is a broad term, and for the sake of simplicity and due to its relation with time-of-use tariffs, this paper focuses only on electrical energy.

Figure 26.1 shows a typical CSC plot for electricity conservation. The plot consists of two lines—the CSC (shown in blue in Fig. 26.1) and the energy price line (shown in green in Fig. 26.1). Each horizontal segment (step) of the CSC is used to represent a unique ECM. The length of the segment along the abscissa is the amount of energy that the ECM is expected to save in a particular time period—in this case, a year. The y-ordinate of the segment is its CCE.

The CSC in Fig. 26.1 shows eight different ECMs with successively increasing CCEs.

The area under the CSC is the yearly cost of implementing all the measures, and the area under a particular section of the curve is the yearly cost of incorporating the ECM represented in that section of the curve. The energy price line depicts the cost of purchasing energy. The area under this line is the yearly benefit (avoided cost) that can be attributed to conservation.

The CSC has been used by many to depict ECMs for industrial, commercial and residential sectors alike. Koomey et al. (1991) analysed energy conservation options for the residential sector in the USA using the CSC. Worrell et al. (1999) evaluated ECMs for the iron and steel industry in the USA using the CSC, and recently, Hasanbeigi et al. (2010) analysed the energy efficiency opportunities for the cement industry in the Shandong province in China.

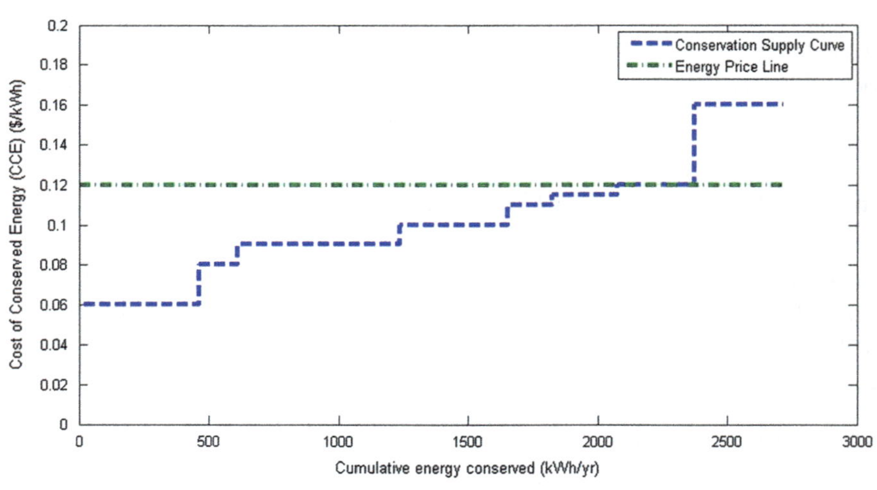

Fig. 26.1 A typical conservation supply curve

26.2 The Conservation Supply Curve Compared to Other Methods of Evaluating Energy Conservation Measures (ECMs)

Many statistics have been developed to quantify the quality of an investment. In the context of ECMs, such statistics transform information regarding the cost of the measure, its expected duration of functioning, the energy savings that one can attribute to it and the prices of energy during its lifetime into a single statistic. Their final aim is to help determine which ECMs are feasible, and out of them which one is the most beneficial. It would be prudent to note that different statistics use different assumptions for the same base case and hence may provide different 'feasible' and 'most beneficial' measures from among the same set of options. Some commonly used statistics are the net present value (NPV), the internal rate of return (IRR), the benefit-to-cost ratio (BCR), return on investment (ROI) and the simple payback period (SPP) (Ross et al. 2007).

These objective statistics are purely financial in nature. In the case of energy conservation, convenience and robustness are at times more important than profit estimates. Furthermore, as energy conservation is generally a long-term commitment, assessment of conservation measures in the light of different possible future scenarios is useful and important. While doing so, it is also important to track energy-related parameters of conservation measures. These statistics provide little scope for such analyses and comparisons. Lastly, single-value statistics seldom give insight into the cost and benefit sides of particular ECMs.

On the other hand, CSCs are a set of curves that depict all the conservation measures under consideration on the same plot. Multiple sets of curves can also be drawn for different scenarios, on the same plot. Additionally, these curves separate the cost and benefit (avoided cost) numbers for these ECMs, effectively adding another dimension of comparison. Lastly, and importantly, it allows a correction for the impact of one conservation measure on another during its construction, giving a more realistic estimate of the actual cost of conserving energy based on the order in which the conservation measures are implemented.

Although the CSC has been in use since its proposal, very little work has focused on developing the curves themselves. In the meanwhile, the energy sector has undergone considerable changes, one of the most important ones being the imposition of time-of-use tariffs. Time-of-use tariffs are used to make electricity relatively expensive during the peak periods to incentivise reduction in consumption, and cheaper during off-peak periods to incentivise consumption. Such tariff structures do not affect the cost of conservation directly, but make it more attractive during peak hours. Since the CSC only considers the CCE for deciding the order of implementation, the effect of these time-of-use tariffs is not captured. It would be prudent to note that time-of-use tariffs can be accounted for while using one of the traditional single-statistic measures like the NPV or the IRR. As a result, the incorporation of time-of-use tariffs in the CSC is important in order to maintain and enhance its value proposition. This work aims to provide a method to incorporate these effects of time-of-use tariffs on the CSC while allowing users to draw the same inferences as can be drawn from the CSC at present.

26.3 Time-of-Use Tariffs and Energy Conservation Measures

Utilities today have a time-of-use-based tariff structure. In the state of Maharashtra, India, the electricity tariff ($/kWh) is composed of two parts. The first is the base electricity rate, which is fixed for all electricity drawn from the utilities. In May 2014, it was 0.117$/kWh (Maharashtra State Electricity Distribution Company Limited 2014). The second is the time-of-use charge, which varies with the time of the day, from +0.0183$/kWh to −0.0166$/kWh. The variation of the total tariff during the day can be seen in Table 26.1 and Fig. 26.2. In Fig. 26.2, the base electricity rate is shown as the green dot dash line.

ECMs generally save different amounts of energy/electricity during different periods of the day or year. Some measures may provide majority of their energy savings during the sunlit period of the day, while others may provide energy savings at night. Some measures may also provide different savings during different months of the year, owing to climate extremes. The province of Ontario in Canada, for example, has extremely cold and harsh winters with the sun available for only a small fraction of the day. This could affect any ECMs dependant on the sun. Further, the need for heating also alters the electricity demand profile considerably, leading to an inversion of the tariff structure in Ontario (Ontario Energy Board 2014).

Table 26.1 Time-of-use tariffs for the state of Maharashtra, India (MSEDCL 2014)

Hour	Base tariff ($/kWh)	Time-of-use charge ($/kWh)	Total time-of-use tariff ($/kWh)	Hour	Base tariff ($/kWh)	Time-of-use charge ($/kWh)	Total time-of-use tariff ($/kWh)
1	0.117	−0.017	0.100	13	0.117	–	0.117
2	0.117	−0.017	0.100	14	0.117	–	0.117
3	0.117	−0.017	0.100	15	0.117	–	0.117
4	0.117	−0.017	0.100	16	0.117	–	0.117
5	0.117	−0.017	0.100	17	0.117	–	0.117
6	0.117	−0.017	0.100	18	0.117	–	0.117
7	0.117	–	0.117	19	0.117	0.018	0.135
8	0.117	–	0.117	20	0.117	0.018	0.135
9	0.117	–	0.117	21	0.117	0.018	0.135
10	0.117	0.013	0.130	22	0.117	0.018	0.135
11	0.117	0.013	0.130	23	0.117	−0.017	0.100
12	0.117	0.013	0.130	24	0.117	−0.017	0.100

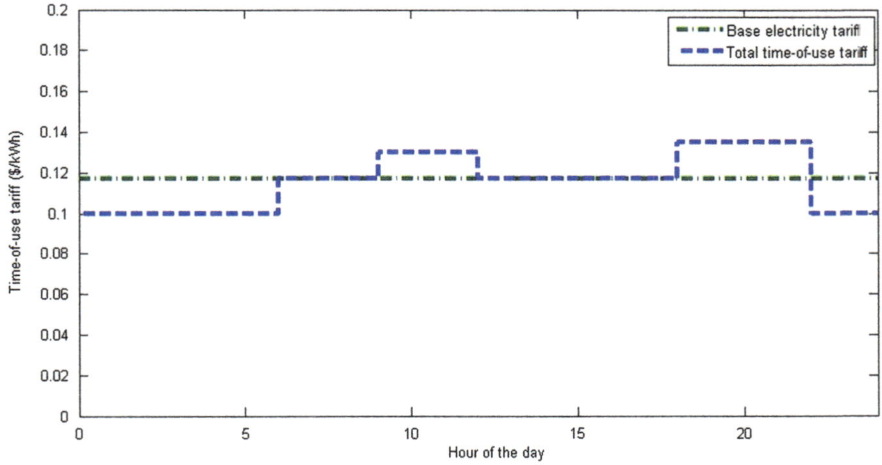

Fig. 26.2 Time-of-use tariffs for the state of Maharashtra, India (MSEDCL 2014)

Since the difference between the CCE and the cost of the alternative source (in this case, electrical energy) is different during different periods of the day or year, each conservation measure would save a different amount of electricity cost even if all of them saved the same total amount of electricity during the year. Hence, it is the difference between these costs that should ideally determine which conservation measure to implement first, and not just the implementation cost of the measure. Further, it is more accurate to compare an ECM's CCE with the average cost per unit electricity that it helps avoid, and not the overall average cost per unit of electricity.

In the next section, we propose and explain a method to incorporate this concept in the CSCs using an example in the Indian context.

26.4 Proposed Modifications to the Conservation Supply Curve and the Method of Its Construction

Apart from the data used for the construction of the original construction supply curve, the time-of-use tariffs and the time-varying energy conservation potentials of different conservation measures need to be gathered or calculated in order to modify the CSC to incorporate the effect of the time-of-use tariffs. Underlying data for such calculations are now readily available due to the advances in data collection and storage and the resulting measurement of important environmental parameters over the past one or two decades. In addition, advances in simulation and forecasting have rendered forecasts more accurate and have made detailed forecasting possible.

Once the data have been collected, we propose the following steps for the modification:

Step 1: Once we have the required time-varying energy conservation potentials (either through simulations in case of certain retrofits, or predictions based on predicted climate variables for renewable energy-based devices, or others) and time-varying tariffs (from utilities or governments), we define an average tariff corresponding to each measure. This tariff is the average cost of buying electricity from the grid to make up for the electricity not saved if the particular measure were not in place. It is calculated as follows:

For each conservation option (*M* in total) denoted by variable *a*

- E_{ai} = Electricity savings attributable to the *a*th *conservation option* for the *i*th *time period*;
- R_i = Time-of-use tariff for the *i*th *time period*;

The average tariff is

$$\sum_{i=1}^{T}(R_i * E_{ai}) \Big/ \sum_{i=1}^{T}(E_{ai}) \text{ for all } a$$

Further, if we subtract the base electricity rate from this average tariff, we get an average time-of-use charge for the conservation option. This average charge is essentially the average charge over all the electricity saved by the conservation measure. It is positive if the measure saves more electricity during the peak tariff periods and negative if it saves more electricity during the off-peak periods.

Step 2: At this point, we can also calculate, as per the original procedure described by Meier (1982), the CCE for each measure by dividing the annualised cost of implementing the measure by the annual energy conservation potential of the measure.

Step 3: Now, with the average tariff calculated for each measure, instead of adding the measure with the lowest CCE first onto the CSC, we add the one with the highest difference between the average tariff and the CCE first onto the CSC.

Step 4: Further, we update the CCEs of the remaining measures after the first measure has been added onto the CSC to account for the effect of its implementation on other measures, as per the original procedure described by Meier (1982). Then, the second measure to be added onto the curve is also selected from among the remaining measures based on the difference between the average tariff and the CCE.

Step 4 is continued until all measures have been added onto the curve.

26.5 Illustrative Example

As an example, we have a study based on hypothetical time-varying electricity conservation profiles for 8 ECMs, assumed to be implemented in the state of Maharashtra, India. For the sake of simplicity, we have also assumed that all the 8 measures are mutually independent; that is, the implementation of one measure

Table 26.2 Electricity conservation profiles for all the 8 conservation measures for the first 12 h of the day

Hour of the day		1	2	3	4	5	6	7	8	9	10	11	12
Electricity conserved in the particular hour during the year (kWh/year)	ECM 1	20	20	20	20	20	20	20	20	10	10	10	10
	ECM 2	20	20	20	20	20	20	20	10	10	10	10	10
	ECM 3	0	0	0	0	0	10	10	20	20	20	10	10
	ECM 4	5	0	0	0	0	10	20	30	30	30	30	30
	ECM 5	0	0	0	0	0	0	0	0	10	10	20	40
	ECM 6	0	0	0	0	0	0	0	5	10	20	40	50
	ECM 7	0	0	0	0	0	0	10	10	30	30	10	10
	ECM 8	3	5	66	8	17	1	9	4	17	3	11	2

Table 26.3 Electricity conservation profiles for all the 8 conservation measures for the last 12 h of the day

Hour of the day		13	14	15	16	17	18	19	20	21	22	23	24
Electricity conserved in the particular hour during the year (kWh/year)	ECM 1	10	20	30	30	30	20	20	20	20	20	20	20
	ECM 2	20	30	30	30	30	20	10	10	5	5	20	20
	ECM 3	0	0	0	0	0	20	10	10	10	10	10	0
	ECM 4	30	30	30	30	30	30	40	50	60	60	40	10
	ECM 5	50	50	40	40	30	10	0	0	0	0	0	0
	ECM 6	50	50	50	30	20	10	5	0	0	0	0	0
	ECM 7	10	0	0	0	10	20	10	0	0	0	0	0
	ECM 8	2	1	5	27	13	0	18	1	3	0	29	5

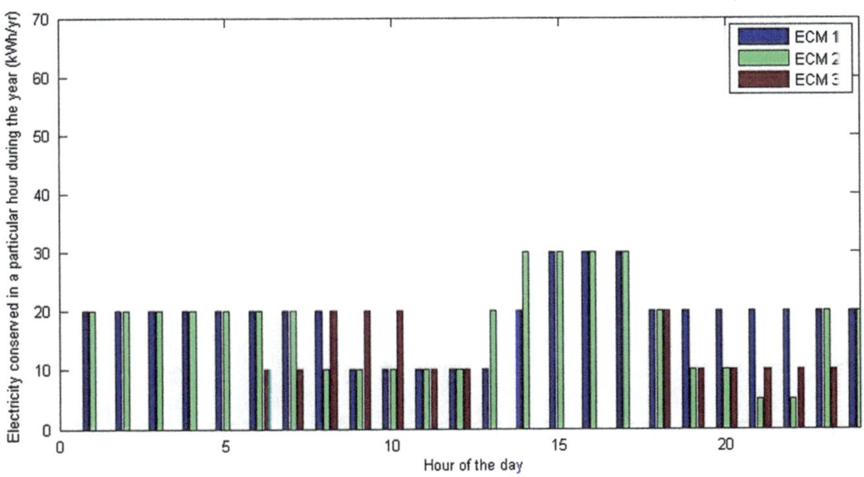

Fig. 26.3 Electricity conservation profiles for ECMs 1, 2, and 3

does not affect the energy conservation potential of another. The time-of-use tariffs for the state of Maharashtra, India, can be seen in Fig. 26.2 and Table 26.1. The electricity conservation profiles of the 8 measures can be seen in Tables 26.2, 26.3 and Figs. 26.3, 26.4, 26.5. These profiles show the amount of electrical energy a measure saves during a particular hour of the day, during the whole year. For example, if a measure saves 1 kWh of energy every day from 1 to 2 pm (the 13th hour of the day), then the number in the energy profile corresponding to the 13th hour would be 1 * 365 = 365 kWh.

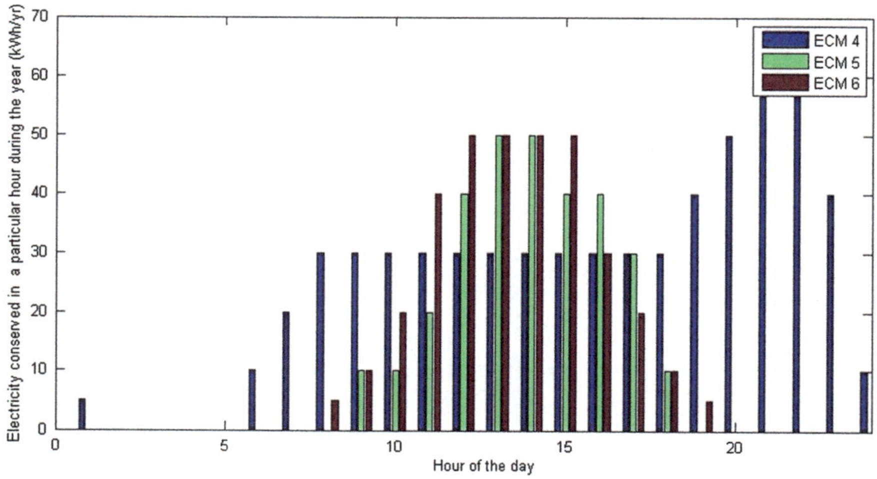

Fig. 26.4 Electricity conservation profiles for ECMs 4, 5, and 6

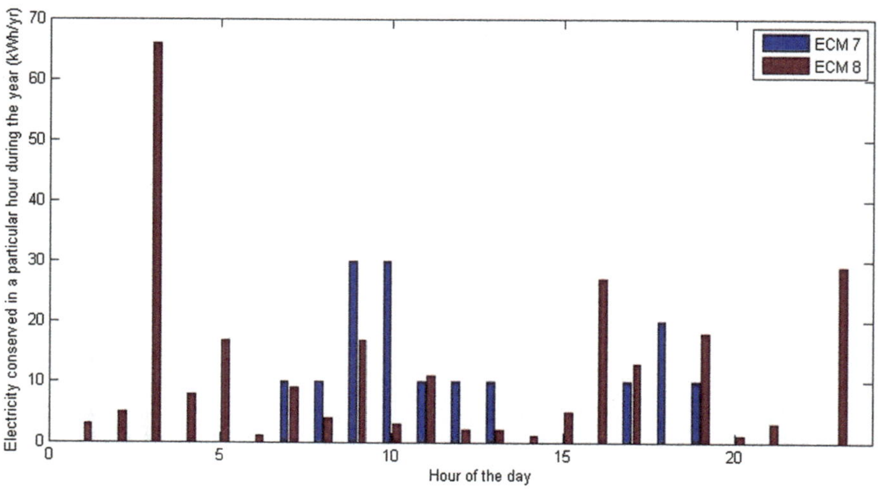

Fig. 26.5 Electricity conservation profiles for ECMs 7 and 8

26.6 Results and Discussion

Following the modified procedure and using the numbers from the sample example, we obtain results as shown in Table 26.4 and Fig. 26.6. The measures have been arranged in decreasing order of the difference between the average tariff and the CCE.

As can be seen by comparing the values in column 3 (CCE) and column 6 (average tariff—CCE) of Table 26.4, lowest CCE ordering is not the same as the highest (average tariff—CCE) ordering. This is important in the light of the fact that conservation measures are often implemented on a lowest CCE basis, but seldom on a net benefit basis. Further, there are many ECMs that are not deemed attractive as

Table 26.4 Results table—application of the proposed methodology to the example based on values in Tables 26.2 and 26.3

	Total yearly electricity conservation potential (kWh/year)	Cost of conserved energy ($/kWh)	Average tariff (with base 0.117 $/kWh)	Average time-of-use charge ($/kWh)	(Average tariff) (CCE) ($/kWh)
ECM 1	460	0.058	0.115	−0.002	0.057
ECM 2	150	0.082	0.121	0.005	0.040
ECM 3	625	0.087	0.123	0.006	0.037
ECM 4	420	0.077	0.113	−0.004	0.036
ECM 5	170	0.095	0.122	0.005	0.027
ECM 6	248	0.107	0.110	−0.006	0.004
ECM 7	300	0.118	0.120	0.003	0.002
ECM 8	340	0.158	0.123	0.006	−0.036

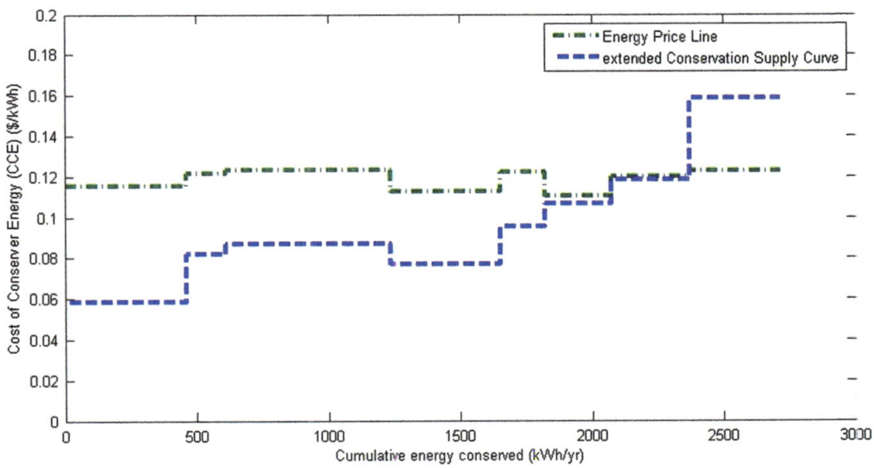

Fig. 26.6 The extended conservation supply curve for the illustrative example

their CCE is higher than the base energy or electricity price, but may actually be attractive if they generate most of their savings during periods of peak demand and higher tariffs.

26.7 Conclusion

Time-of-use tariffs are now an important part of the energy pricing structure in most countries. The time-varying component of these tariffs indirectly affects ECMs by making non-conservation either more or less expensive at different times of the day. Such tariffs can be easily incorporated while calculating a financial statistic like the NPV or the IRR for ECMs, but these differences in benefits could lead to reordering in the CSC and hence a change in their optimal order of implementation. Since the CSC in itself is a cost curve, and because cost of implementing measures is not affected by the time-of-use tariffs, the CSC as is cannot incorporate these effects. The proposed method, as demonstrated, allows one to incorporate these effects without compromising the ability to draw all the other inferences that can be drawn from a simple CSC. This can be attributed to the fact that the ordering of the measures is still independent of the base electricity price and only depends on the average time-of-use charge.

References

Hasanbeigi, A., Price, L., Lu, H., & Lan, W. (2010). Analysis of energy-efficiency opportunities for the cement industry in Shandong Province, China: A case study of 16 cement plants. *Energy Policy, 38*, 3461–3473.

Koomey, J. G., Atkinson, C., Meier, A. K., McMahon, J. E., Baghosian, S., Atkinson, B., Turiel, I., Levine, M. D., Nordman, B. & Chan P. (1991) The potential for electricity efficiency improvements in the US residential sector. Lawrence Berkeley Laboratory, University of California, LBL-30477, UC-350.

Maharashtra State Electricity Distribution Company Limited. (10th June 2014). http://www.mahadiscom.com/tariff/Tariff-Booklet-aug-2012.pdf

Meier, A. K. (1982) Supply curves of conserved energy. Ph.D thesis, Energy and Environment Division, Lawrence Berkeley Laboratory, University of California.

Ontario Energy Board. (10th June 2014). http://www.ontarioenergyboard.ca/oeb/_Documents/For%20Consumers/TOU_prices_Summer.pdf

Ross, S. A., Westerfield, R. W. & Jordan, B. D. (2007). Net present value and other investment criteria. In S. A. Ross (Ed.), *Fundamentals of corporate finance* (pp. 264–302). New York: McGraw-Hill Irwin.

Worrell, E., Martin, N. & Price, L. (1999). Energy efficiency and carbon dioxide emissions reduction opportunities in the U.S. iron and steel sector. Ernest Orlando Lawrence Berkeley National Laboratory Report 41724

Chapter 27
Management of Distribution System Protection with High Penetration of DGs

Abdelsalam Elhaffar, Naser El-Naily and Khalil El-Arroudi

Abstract As a result of deregulation and emerging distributed generation (DG) in distribution networks, some protection problems and challenges emerge which need immediate solutions by design engineering. The present generation of numerical protection relays allows the implementation of adaptive settings for distribution system protection especially with systems of a high penetration of DG. This chapter presents an overview of the use of communication infrastructure to improve some aspects of distribution system protection, especially adaptive protection. One of the key issues has been to verify that traditional network protection schemes and settings are simply not adequate when there are DG systems connected to the network. The impact of the DG unit increases with the size of the generator and with the length of the line section between the DG unit and the fault. Results have provided a clear indication of the potential protection problems that need to be solved by careful protection design. Some solutions have been proposed in this chapter through communication and intelligent electronic devices (IEDs) based on power system simulation studies using ETAP software. The results will be used to focus further research and development in protection systems and concepts.

27.1 Introduction

In the last decades, electrical power systems (EPS) have enlarged to meet the growing demand for electrical power. After deregulation, EPS also become more interconnected and difficult to operate, which increases competition among agents. Thus, the

A. Elhaffar (✉) · K. El-Arroudi
University of Benghazi, Benghazi, Libya
e-mail: absalam.elhaffar@uob.edu.ly

K. El-Arroudi
e-mail: khalil.elarroudi@mail.mcgill.ca

N. El-Naily
H.V. Networks, General Electricity Co. of Libya, Benghazi, Libya
e-mail: naseralnaile@yahoo.com

A.N. Bilge et al. (eds.), *Energy Systems and Management*,
Springer Proceedings in Energy, DOI 10.1007/978-3-319-16024-5_27

tasks of operating, planning, and protecting these systems can no longer be held in the traditional manner. The EPS protection should be taken into account the changing topology and loading of these networks hence, requiring adaptive techniques to improve its effectiveness. Otherwise, the protection of the system may become unreliable for certain fault situations. The concept of adaptive protection is presented as a philosophy that adjusts the protection relays to suit the varying conditions of the electrical system. The use of adaptive protection is feasible when numerical relays are applied, considering their ability to change their setting online and processing speed. However, some aspects may be improved with the use of artificial intelligence (AI) techniques in conjunction with such devices. Modern substation relays and reclosers can have their settings changed by remote control. Having a telecommunication link with the controller allows the user to use these capabilities.

General Electricity Company of Libya (GECOL) is a Libyan utility company active in generation, transmission, and distribution of electricity. Its grid is composed of the following voltage levels: transmission grid (400/220 kV), sub-transmission grid (132/66/30 kV), and regional distribution grid (11 and 0.4 kV). Its production is mainly composed of gas and steam turbine generators.

During the last year, the shortage of generation has been encountered which requires extending its production with embedded distributed generation (DG). They are very often small or medium capacity power production units which are located where the resources and fuel are available. Since the need for power is high, some medium-sized gas turbine units have been installed at important substation where the power demand is high. Most of them are connected to the 11 kV bus of the regional distribution grid. This 11 kV grid has a radial topology point.

Many literature papers have been published on this subject; in the research work (El-Arroudi et al. 1999), the authors propose an expert system (ES) that designs protection systems. This ES deals with a vast database that is divided into data about relays (specifications, names, and settings) and the equipment being protected (transmission lines, transformers, and busbars). The inference mechanism examines and evaluates some functions using the rules and search strategies during the reasoning process. It is able to work in the presence of uncertainties and to explain its conclusions. A graphical interface allows the interaction with the user. Through the interface, the user can enter the necessary information or run the software. The outputs/responses from the ES are reports and specifications of the devices necessary to achieve a comprehensive protection of the system in question. In (El-Arroudi et al. 2007), a technique for islanding detection of DG is proposed based on recognizing the patterns of the sensitivities of some indices (or features) at a target location to prescribed credible events. The proposed technique uses the data-mining technology to extract information from the large data sets of these indices after they are screened off-line via massive event analyses using network simulations.

Distribution system protection should not be drastically impacted by a low (10 % or less) DG penetration; however, it can become a concern with higher DG penetration especially if the DG type is a generator-based system. These types of systems can continue to feed a fault due to their inertia if they are not disconnected and can affect the relay/fuse coordination of the distribution network. However, power electronic inverter-based DG provides the ability to act extremely fast to limit their current or shut down to avoid this problem for nearby faults. If the fault is not nearby, it may be preferable to keep the DG connected especially if it is capable of producing both active and reactive power. With the non-active power capability, the DG provides increased voltage support to the distribution system when it is most needed.

System protection is an important factor for distribution systems with DG, since their capacity can be a significant portion of the feeder capacity. However, DG and system protection should be coordinated even in the case of low DG penetration. The current standards take the position that DG should be disconnected when there is a system disturbance that drops voltage or frequency to certain levels or for a given period of time. However, DGs can extend the margin-to-voltage collapse by supporting the distribution system during such a system disturbance (Rizy et al. 2010). During the past decade, DGs were obliged to immediately disconnect from the grid in case of network disturbances or faults. However, the disconnection of all these on-site power generators due to each temporary fault in the transmission level may lead to severe stability problems and is not acceptable any longer by steady increase in DG penetration. In order to supply adequate short-circuit current and efficiently support grid stability, several guidelines in different countries require fault-ride-through (FRT) capabilities for DG connected at high (HV) and medium voltage (MV) levels. If the protection system of DG units is able to detect the fault and rapidly disconnect from the network, DG will not interfere with the normal operation of the protection system. Most interconnection standards and guidelines, therefore, require disconnection of DG if a fault occurs (IEEE PES 2004). Under fault conditions, DG remains connected to the grid and contributes to grid stability by injecting reactive current and support grid voltage for a definite time. If the fault is not cleared after a certain time, instead of DG disconnection from the grid a suitable control system may also initiate intended islanded operation of a grid section to uninterruptedly supply power to as many loads as possible (Degner et al. 2010).

Overcurrent protection is the predominant protection method used for distribution feeders. The standard time–current curves, pickup values, and time dial settings are selected to time-coordinate the operation of multiple protection relays on radial feeders. The objective is to operate as fast as possible for faults in the primary zone, while delaying operation for faults in the backup zone. The engineer derives the protection settings from knowledge of the available short-circuit current and the desired coordination time interval between relays. DG penetration results in system configurations that may render the protection settings ineffectively. Power flow direction is reversed in some areas when the loads are fed from DG sources. Additionally, the available fault current levels may be significantly different in the new configuration (Greer et al. 2011).

One of the principal applications of distribution automation is the remote controlling of feeder relays and reclosers. Modern reclosers and substation relays can have their settings changed by remote control. Having a telecommunication link with the controller allows the user to use these capabilities. Most of the controllers have the capability to register several alternate settings allowing the protection device to be efficient in various conditions, for example, when the system configuration is changed. This flexibility has to be managed to avoid protection malfunction (Antonova et al. 2012).

This paper investigates some aspects of distribution system protection with DGs. The traditional network protection schemes and settings are simulated when there are DG systems connected to a typical Libyan distribution network. The potential protection problems that need to be solved by careful protection design have been emphasized. Some solutions have been proposed in this paper through communication and intelligent electronic devices (IEDs) based on power system simulation studies using ETAP software.

27.2 The Case Study System

The study case system is a portion of the Libyan regional distribution network at Benghazi, a single-line diagram is shown in Fig. 27.1. The characteristics of this system are suitable to study the impact of DG on the overcurrent protection of the distribution system. This study system consists of five buses with radial connection network. The main injection infeed is an equivalent 220 kV system with short-circuit power of 6528MVA at bus 1. A 25MVA synchronous generator-based DG under study is connected at bus 4 (namely: Hotel Bus) through a 30/11 kV, 25MVA power transformer. Short-circuit studies and coordination analyses are simulated on different buses, and the results will be presented on Sect. 27.6.

27.3 Impact of DG Units on Protection

A typical design of a graded overcurrent protection system to a typical distribution system without DG, applying chronometric selectivity for the instantaneous units (IEC 50), and time–current selectivity for temporized units (IEC 51), for coordinate

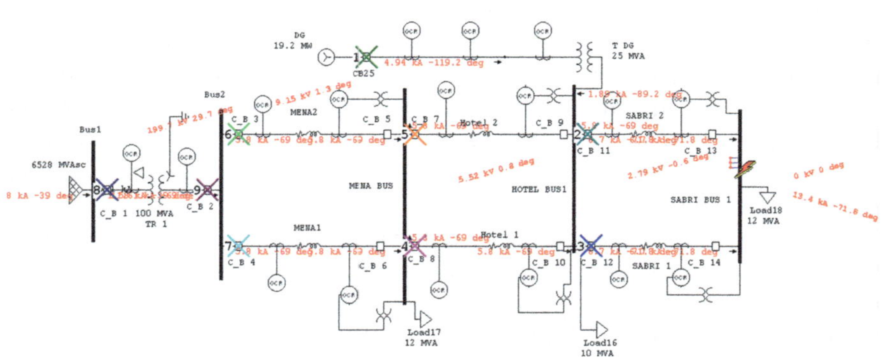

Fig. 27.1 The study case distribution system with DG

relays. Using ETAP software, the protection coordination is checked, viewing time–current characteristic (TCC) curves and sequence of relay operation for faults in different buses. Then, the DG unit is connected with different loading conditions to identify its effect on protection coordination. The presence of the DG causes a decrease in the short-circuit current flowing through some branches, which leads to the loss of sensitivity of the protective devices.

The connection of DG to distribution feeders changes the fault currents in faulted feeders. The rate of change of the fault currents strongly depends on the ability of the DG to contribute to the fault current. DG based on an asynchronous generator does not provide a sustainable fault current during a grid disturbance. The same holds mostly for inverter-connected DG such as micro-turbines, fuel cells and PV systems, from which the fault current contribution can be neglected. A generator type that contributes a sustainable fault current is the synchronous generator.

In 2013, GECOL installed few small gas turbine generators (with light fuel supply option) on several substations, as the demand is high with a shortage in the utility generation. This was one of the drivers behind this research. The DG contribution obviously changes the fault current distribution in upstream lines. Hence, the relay operation time will vary accordingly.

27.3.1 Protection Problems

The grid contribution to the total fault current will be reduced because of the contribution of DG. Due to this reduction, it is possible that the short circuit stays undetected because the grid contribution to the short-circuit current never reaches the pickup current of the feeder relay. Overcurrent relays as well as directional relays and reclosers rely their operation on detecting an abnormal current. Hence, all protective systems based on these protection devices can suffer malfunctioning because of the reduced grid contribution (see Table 27.1). This mechanism is called blinding of protection and belongs to the first category of protection problems. In this case, distance protection can be used with adaptive setting to overcome variable infeed from the DG. It can be concluded that DG with a relevant contribution to the fault current directly affects the sensitivity of a protective system and therefore the reliability of the protective system.

Table 27.1 The impact of the DG penetration level on the short-circuit currents from Mena bus to Hotel bus

DG penetration level in [%] of rated power	Two lines Mena-Hotel fault current (kA)
0	12.886
36	12.812
70	12.758
96	12.726

27.4 Prevention of Detection and Selectivity Problems

Fault detection problems do have a relation with the amount of generation connected to the distribution grid and the local short-circuit power. To prevent fault detection problems, a first attempt is to modify the relay settings of the relays and reclosers. The generator contribution leads to a reduction of the grid contribution to the fault current. Hence, the pickup current of the relays has to be reduced. However, fault detection problems might be solved by reducing the pickup current. The sensitivity, and the security of the protective system are decreased and might lead to false tripping in case of a fault in an adjacent feeder. By reducing the pickup current, blinding of protection is solved but at the same time it introduces false tripping for faults in a certain area (Gaonkar 2010). A proposed solution is to install protection devices with an additional time delay to give the feeder including the DG a longer fault clearing time (Deuse et al. 2007). Alternatively, adaptive overcurrent relay setting can be carried out which decreases the pickup current as the output of the local generation increases.

If the relays have multiple setting groups as the case is with numerical protective relays, then different setting groups should be used. The groups should be initiated by inputs associated with the condition corresponding to the new DG level of penetration and/or network topology. For instance, if the DG contribution is about 50 % of its rating, the DG relay sends a signal corresponding to its output power to the affected relays to modify its setting to group 1 that is appropriate to this case. In our case, three setting were selected at 0, 50, and 90 % of the DG rating. The proposed adaptive protection system modifies automatically the relay settings based on the system topology and DG loading to maintaining a coordinated overcurrent protection with the optimum selectivity and sensitivity of the protective relays. The internal logic functions (called continuous function chart CFC) inside the numerical relay are programmed to detect the event and send the signal through either a binary output or global system for mobile communications (GSM) message to the target relays. In the case of multiple DG sources in the network or in adjacent feeders, more than one setting group is needed to properly adjust all pickup levels and directional settings, as presence or absence of DG sources can significantly impact the short-circuit current that can flow in the direction of a local fault (Antonova et al. 2012).

27.5 Impact of Communication Media on Protection

For data transmission, already existing media and infrastructure shall be used (e.g., fiber optics, Ethernet, DLC, or wireless communication). By the utilization of (for instance) fiber optics, there shall be no negative impact on the transmission of energy and standard protocols shall be used. The data speed must be sufficient for the application. In this research, the wireless communication infrastructure was utilized in addition to the possibility of adding the fiber-optic network of the substation automation when available.

27.6 Experimental Setup

The test setup consists of two feeder manager numerical relays and two GSM modems. Relay A is the DG feeder manager relay, and relay B is the Mena-Hotel line relay. The power injected by the DG is simulated via Omicron test set with analog current and voltage inputs as shown in Fig. 27.2. In this part, the configuration design for new parameters of the feeder manager relays to their masking binary input and output (I/O configuration matrix) is presented.

27.6.1 Configuration Procedure of the Relays

The configuration setting of the two relays is carried out using DIGSI software. A new parameter group is inserted first in relay A. Under this level, three internal single-point indications (power+, power−, setting scheme) are generated with an internal double-point indication (P = arrow). The internal single-point indications (power+, power−) are assigned to a source CFC and destination CFC, while (setting scheme) signal is assigned to binary output (NO 1) and to source CFC to change the relay setting for feeder A, it is also assigned to LED and also in the event recorder.

A new parameter group is inserted first in relay B. Under this group, two single-point indication (P > received, received signal) and two internal single-point indications (send setting, P > sent) are generated. The signal (P > received) means that the active power at relay A exceeded a prescribed limit (>80 % for instance), and it is assigned to a binary input and to the destination CFC. The signal (receive

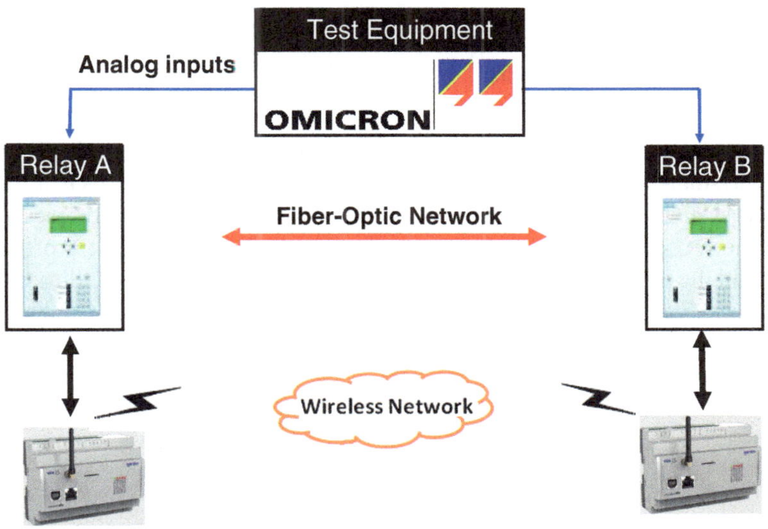

Fig. 27.2 Test setup of the adaptive relay coordination setting

setting) means that the group setting change is activated at relay B, and this signal is assigned to a binary input NO 1 and to a destination CFC. The signal (P > sent) means that the active power at relay A exceeded the limit and sent to relay B. (DG feeder A is producing power p more than a prescribed limit.). The signal (send setting), which is responsible for sending a signal to change relay B group settings.

The testing units used are as follows:

1. Omicron CMC 156 secondary test set for injecting currents and voltages to the relay to simulate DG injected power rise.
2. Two numerical protection relays Type SIEMENS 7SJ63 (Siemens 2007).
3. Two GSM modem-type ENTES GEM15.
4. Personal computer equipped with SIPROTEC software DIGSI 4.

Relay A: The following conditions have been simulated:

- If the exported power has exceeded the set value, which is 80 %, it is represented as a binary output 1 at relay A.
- If the exported power has exceeded the set value, which is 50 %, it is represented as a binary output 2 at relay A.
- If the exported power i has the set value, which is 0 %, it is represented as a binary output 3 at relay A.

Relay B: The following conditions have been simulated:

- The exported power of relay at location A must exceed the set value, which is 80 %, which is represented as a binary input to relay B.
- The relay setting is changed from group 1 to group 2, while the power is more than a pre-calculated setting for the new configuration.

27.6.2 Simulation Results

The system under study of Fig. 27.1 has been simulated using the Electrical Transient Analysis Program (ETAP) with and without DG. With DG in the distribution system, there are several configurations that need special protection settings. The DG penetration levels are simulated at 0, 36, 70, and 96 % of the DG rating, and the impact of these variations on the relay coordination are illustrated on Figs. 27.3, 27.4, 27.5, and 27.6 for a fault simulated on the bus 5 (namely: Sabri substation).

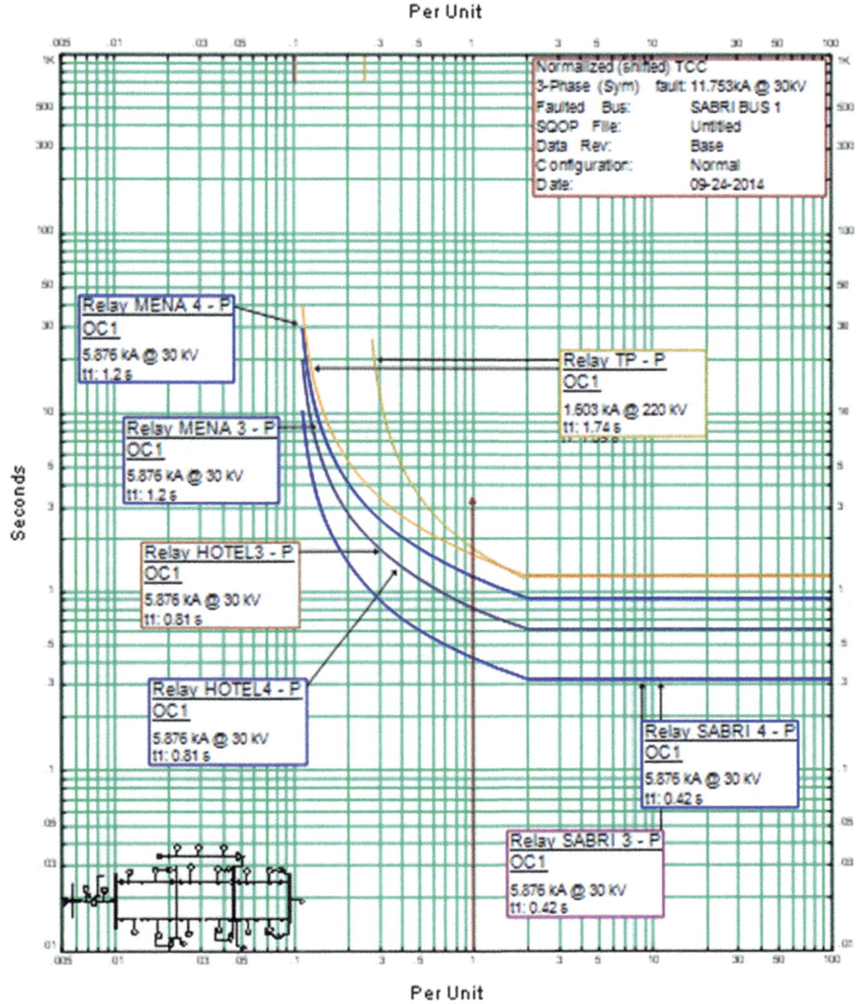

Fig. 27.3 No DG penetration level (0 %) at bus 4 with fault at bus 5

27.7 Protection Management for Distribution Networks with DG

The protection settings are adapted to the changes in the network topology using the available communication media. In this paper, the GSM wireless network operated by (Elmadar network), which has a very good coverage in the area under study, has been adopted as the main link among the protection relays. The setting remains at

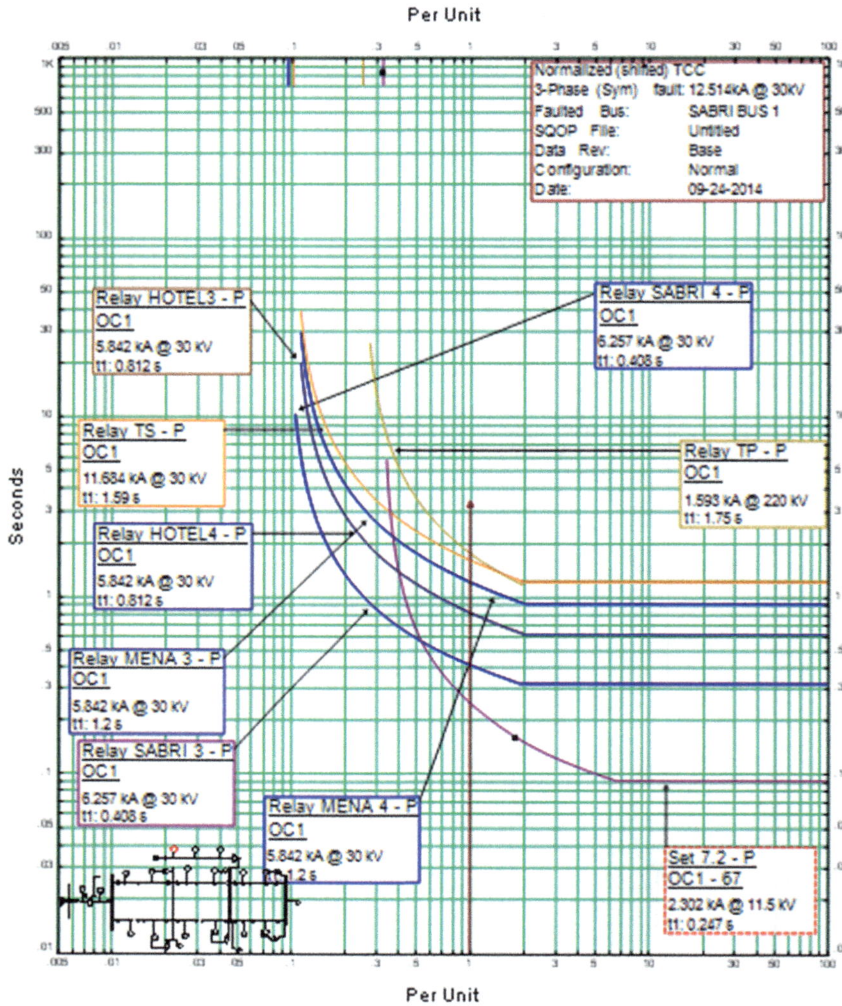

Fig. 27.4 36 % DG penetration level at bus 4 with fault at bus 5

the studied configuration as long as the DG does not reach a pre-defined value. For any new configuration, the overcurrent relay settings are calculated using time–current coordination method. The setting of the network is calculated at four configurations and saved in the relay group settings 1–4. The setting is held until the DG power is stable at the setting using CFC function inside the relay logic. The results show that the methodology and the proposed method for protection system management are applicable and work effectively for existing Libyan distribution

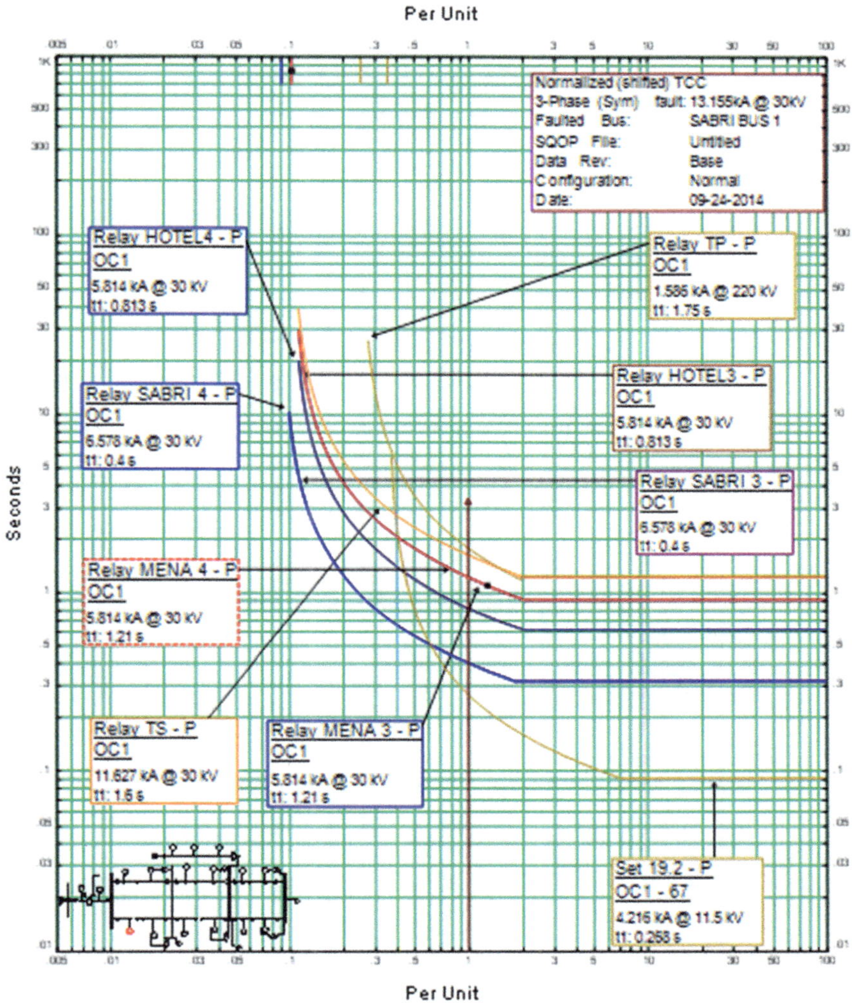

Fig. 27.5 70 % DG penetration level at bus 4 with fault at bus 5

systems with the recently installed DGs. The dynamic protection setting concept opens a new range of possibilities, but is essentially a significant departure from the traditional coordination approach based on a firm belief that a technician should always be at the device while making setting changes to the relay.

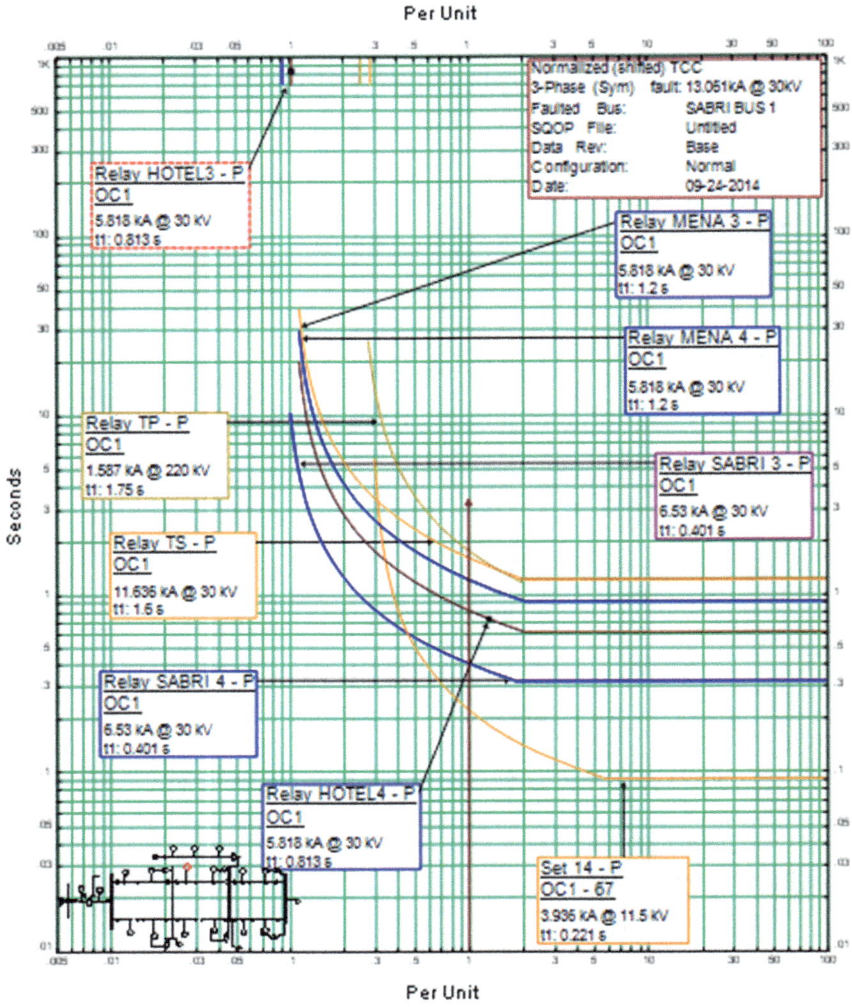

Fig. 27.6 96 % DG penetration level at bus 4 with fault at bus 5

27.8 Conclusions

This paper identifies the impact of DG on distribution system protection at different penetration levels. The relay group setting is interchanged according to the system topology and DG penetration level. The communication media were used to transmit the changes from the DG or feeder relays. For future work, the reliability and time delay of the communication media should be investigated to make the protection system more robust and selective.

References

Antonova, G., et al. (2012). Distributed generation and its impact on power grids and microgrids protection. In *65th Annual Conference for Protective Relay Engineers, April 2–5, 2012*, College Station, TX, USA.

Degner, T., Schäfer, N., Jäger, J., Keil T., & Shustov, A. (2010). Adaptive protection system for distribution networks with distributed energy resources. In *10th International Conference on Developments in Power System Protection,* 29 March–1 April 2010.

Deuse, J., Grenard, S., Bollen, M., Häger, M. & Sollerkvist, F. (2007). Effective impact of DER on distribution system protection, CIRED 19th International conference on Electricity Distribution, May 21–24, Vienna.

El-Arroudi, K., McGillis, D., & Joos, G. (1999). A methodology for power system protection design based on an intelligent system approach. In *1999 IEEE Canadian Conference on Electrical and Computer Engineering*, vol. 2, no., pp. 1164–1169.

El-Arroudi, K., Joós, G., Kamwa, I., & McGillis, D. T., (2007). Intelligent-based approach to islanding detection in distributed generation. *IEEE transactions on power delivery*, *22*(2), 828–835.

Gaonkar, D. N., (2010, chap.5). *Distributed Generation*. InTech.

Greer, R., Allen, W., Schnegg, J., & Dulmage, A. (2011). Distribution automation systems with advanced features. In *Rural Electric Power Conference (REPC), 2011*. IEEE.

IEEE PES, 2004 WG7, Power System Relay Committee Report. (2004). *Impact of distributed resources on distribution relay protection*. August 2004.

Rizy, D. T., et al. (2010). *Properly understanding the impacts of distributed resources on distribution systems*. Power and energy society general meeting, IEEE, July 25–29, 2010.

Siemens, (2007). SIPROTEC relay Manual, Multi-Functional Protective Relay with Local Control, 7SJ62/63/64 V4.6

Chapter 28
Assessment of Total Operating Costs for a Geothermal District Heating System

Harun Gökgedik, Veysel İncili, Halit Arat and Ali Keçebaş

Abstract District heating system (DHS), especially geothermal, is an important class of heating, ventilating, and air conditioning systems. This is due to the fact that in many countries and regions of the world, they have been successfully installed and operated, resulting in great economic savings. In recent years, such systems have received much attention with regard to improving their energy efficiency, equipment operation, and investment cost. Improvement in performance of a geothermal district heating system (GDHS) is a very effective mean to decrease energy consumption and to provide energy saving. To perform the potential energy savings in a GDHS, the advanced exergoeconomic analysis is applied to a real GDHS in the city of Afyon/Turkey. Then, it is evaluated based on the concepts of exergy destruction cost and investment cost. The results show that the advanced exergoeconomic analysis makes the information more accurate and useful and supplies additional information that cannot be provided by the conversional analysis. Furthermore, the Afyon GDHS can be made more cost effectiveness, removing the system components' irreversibilities, technical-economic limitations, and poorly chosen manufacturing methods.

28.1 Introduction

Geothermal district heating system (GDHS) has recently been given increasing attention in many countries. These systems are simple, safe, and adaptable systems, minimum negative environmental impact, low operating cost, decentralized production advantages, and simplicity of their technologies. Numerous successful

H. Gökgedik · V. İncili · A. Keçebaş (✉)
Muğla Sıtkı Koçman University, Muğla, Turkey
e-mail: alikecebas@mu.edu.tr

H. Arat
Afyon Kocatepe University, Afyonkarahisar, Turkey
e-mail: halitarat@gmail.com

A.N. Bilge et al. (eds.), *Energy Systems and Management*,
Springer Proceedings in Energy, DOI 10.1007/978-3-319-16024-5_28

GDHS projects have been reported. Experience by researchers and engineers still plays an important role in the system analysis, design, and control (Lund and Freeston 2001; Hepbasli 2010). In particular, the heat economic losses in GDHSs cause the fast energy consumption, eventually environmental problems. Improvement in performance of GDHSs is a very effective mean to decrease energy consumption. The importance of energy efficiency is also linked to environmental problems, such as global warming and air pollution. Energy efficiency is a rather general term and in practice various energy performance indicators are used, usually grounded in thermodynamics or economics. With geothermal direct utilization in thermal systems, controlling the thermodynamic efficiency, the energy consumption and the product costs are an unavoidable topic. To achieve sustainable development, the focus on thermal system efficiency is moving from thermal analysis to economic analysis studies that assess both thermodynamic inefficiencies and economic benefits. The effectiveness of an energy conversion system can be evaluated by conventional exergy-based analyses (thermal, economic, and environmental). However, these analyses do not provide enough information about the relations between the components and they are inadequate in determining the real improvement potentials. Briefly, thermodynamic, economic, and environmental analysis methods, which are called the advanced exergy-based analyses, were developed to resolve the deficiencies in the conventional exergy-based analyses (Açıkkalp et al. 2014).

For example, the exergy destruction, the exergy costs, the investment, and the environmental effect for any component can be considered to be a result of the component itself or other components. The advanced exergy-based analysis simultaneously provides everyone in the formation about the improvement limits of the considered component or the system, which resulted from technical, economic, and ecological constraints. In this study, it is focused on the advanced exergy-based analysis methods especially for economic constraint. Advanced exergoeconomic analysis is a new method and it uses the results of the corresponding conventional exergy-based analyses, but the advance examination process by introducing new calculation steps to reveal component interactions and potential for improvement (Tsatsaronis and Park 2002; Cziesla et al. 2006; Kelly et al. 2009). In the literature, its applications to various energy conversion systems are relatively low in numbers (Cziesla et al. 2006; Kelly et al. 2009; Tsatsaronis 2008; Wei et al. 2012; Manesh et al. 2013; Keçebaş and Hepbasli 2014). Thus, the main purposes of this study are to (i) analyze the advanced exergoeconomic aspects of a GDHS, (ii) apply the advanced exergoeconomic analysis to the Afyon GDHS in Turkey, and (iii) assess its total operating costs.

This paper is organized as follows: Sect. 28.2 briefly describes the conducted system. In Sect. 28.3, the study methodology for evaluating total operating costs of a GDHS according to the advanced exergoeconomic analysis is given. The results of the study are discussed in Sect. 28.3. Finally, conclusions of the study are presented in Sect. 28.4.

28.2 System Description

To provide residential heating for buildings through geothermal water, the Afyon GDHS was installed in the city of Afyonkarahisar, 1994. While the Afyon GDHS was initially designed for 10,000 residences with a potential of 48.3 MW, there are only 4613 residences nowadays that have been heated. Its heat source originates the geothermal fluid with 225 kg/s and 105 °C from the Ömer-Gecek geothermal field. In this study, the Afyon GDHS is investigated, and its schematics, which mainly consists of three cycles, namely (i) the energy production circuit (EPC), (ii) the energy distribution circuit (EDC), and (iii) the energy consumption circuit (ECC), is illustrated in Fig. 28.1. In the EPC, the geothermal fluid at an average flow rate, temperature, and pressure of 630 ton/h, 95 °C, and 8 bar (for 14,650 m length) is pumped to the Afyon GDHS. Next, because the maximum discharge mass flow rate of the residential heating (630 ton/h) is beyond the total re-injection mass flow rate (440 ton/h) the

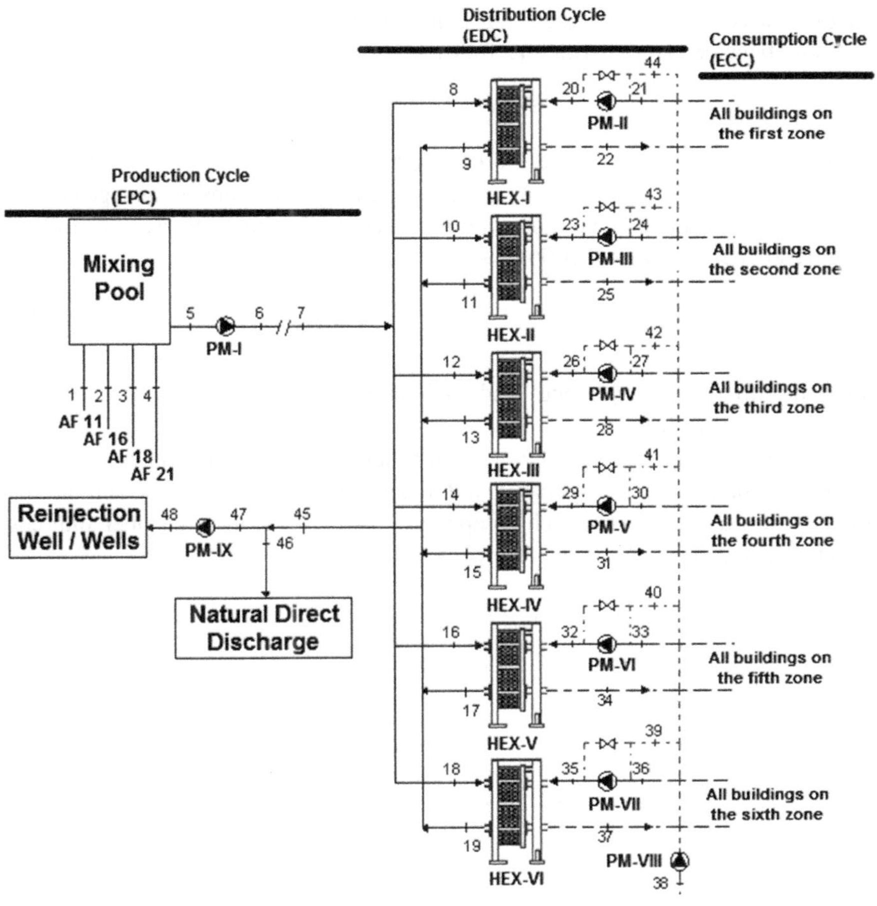

Fig. 28.1 Schematic diagram modified from Keçebaş (2011) and Keçebaş et al. (2011)

remaining fluid is released to the nature direct discharge. For the supply/return water temperatures of the building (energy consumption) cycle, the Afyon GDHS has about temperatures of 70/50 °C. The actual operational data regarding the temperature, pressure, and flow rate of the systems were recorded on February 16, 2012, by the technical staffs based on the state numbers specified in Fig. 28.1.

28.3 Advanced Exergoeconomic Analysis

One of the most interesting features of exergoeconomics is that the exergy destruction cost associated with a component is calculated and compared with the investment cost of the same component, to decide about the design changes that might improve its cost effectiveness. Thus, exergoeconomics is a unique combination of exergy and cost analyses conducted at the component level, to provide the designer or operator of an energy conversion system with information crucial to the design of a cost-effective system. At component level, the conventional exergoeconomic balance can be written as follows:

$$\dot{C}_{\mathrm{P},k} = \dot{C}_{\mathrm{F},k} + \dot{Z}_k \tag{28.1}$$

where $\dot{C}_{\mathrm{P},k}$, $\dot{C}_{\mathrm{F},k}$, and $\dot{Z}_k$ donate the cost rates associated with product, fuel, and capital investment (CI). For a component receiving heat transfer and generating power, the conventional exergoeconomic balance can be expressed as follows (Lazzaretto and Tsatsaronis 2006):

$$\sum_{\mathrm{out}} \dot{C}_{\mathrm{out},k} + \dot{C}_{W,k} = \dot{C}_{Q,k} + \sum_{\mathrm{in}} \dot{C}_{\mathrm{in},k} + \dot{Z}_k \tag{28.2}$$

where the subscripts W, Q, "out," and "in" denote power, heat transfer, outlet, and inlet streams, respectively.

The real cost sources in any component of an energy conversion system are the CI for each component, the operating and maintenance (OM) expenses, and the cost of exergy destruction within each component ($\dot{C}_{\mathrm{D},k}$). Thus, the total operating cost is occurred from these costs, as following

$$\dot{C}_{\mathrm{tot},k} = \dot{Z}_k + \dot{C}_{\mathrm{D},k} \tag{28.3}$$

with

$$\dot{Z}_k = \dot{Z}_k^{\mathrm{CI}} + \dot{Z}_k^{\mathrm{OM}} \tag{28.4}$$

and

$$\dot{C}_{\mathrm{D},k} = c_{\mathrm{F},k} \dot{E}_{\mathrm{D},k} \tag{28.5}$$

The conventional exergoeconomic model of an energy conversion system consists of cost balances and auxiliary costing equations according to the exergy-based monetary costing (Lazzaretto and Tsatsaronis 2006). In determining these auxiliary equations, the F-rule and the P-rule are applied. General information on these rules can be found in Bejan et al. (1996).

Through the above-mentioned conventional exergoeconomic analysis, one cannot assess the mutual interdependencies among the system components neither the real potential for improving the components. To decide about the design changes that might improve cost effectiveness of system component, the investment cost of a component and the exergy destruction cost associated with same component must be determined. This becomes possible in an advanced exergoeconomic analysis (Tsatsaronis 1999), in which the investment cost and exergy destruction cost in each component are split into endogenous/exogenous and avoidable/unavoidable parts. In literature, the methodology for splitting the total operating cost has been discussed in detail in Kelly (2008) and Petrakopoulou (2010). In this study, the terms and equations used to perform the advanced exergoeconomic analysis are given in Table 28.1.

28.4 Results and Discussion

In this study, the concepts of the exergy destruction cost and investment cost for advanced exergoeconomic analysis are investigated, and the Afyon GDHS is evaluated to improve the design and operation of a GDHS for future conditions/projections. To assess total operating costs of the system, the Afyon GDHS as a real case study is considered. The actual operational data obtained from the system were collected at various state points of the systems, as shown in Fig. 28.1. In which, the reference state is considered as 5.8 °C and 101.32 kPa when data were recorded on January 20, 2013. Using these data, the conventional exergetic and exergoeconomic analyses and then the advanced exergetic analysis are carried out for the system.

The general procedure of calculating the exergoeconomic analysis in the thermal conversion systems is described in Tsatsaronis and Park (2002), Cziesla et al. (2006) Tsatsaronis (2008), Wei et al. (2012), Manesh et al. (2013) and Keçebaş and Hepbaslı (2014).An exergoeconomic analyses, which is the specific exergy costing (SPECO) method introduced by Lazzaretto and Tsatsaronis (2006), are used, and the cost data for purchased-equipment cost (PEC) and capital investment cost (CIC) are obtained from the system owners.

Exergy analyses are performed under theoretical and unavoidable process conditions with parameters used in determining unavoidable costs to conduct advanced exergy-based analyses, as illustrated in Table 28.2. These process conditions and parameters have been made based on the authors' and technical staff knowledge as well as experience of the operation. All the calculations are computed with the aid of the EES and GateCycle software packages.

Table 28.1 The terms and equations used to perform the advanced exergoeconomic analysis (Kelly 2008; Petrakopoulou 2010)

Terms	Definition of cost rate	Equations for investment cost rate ($\dot{Z}_k$)	Equations for exergy destruction cost rate ($\dot{C}_{D,k}$)
Endogenous ($\dot{Z}_k^{EN}$, $\dot{C}_{D,k}^{EN}$)	Cost rate of *k*th component associated with the operation of the component itself	$\dot{Z}_k^{EN} = \dot{E}_k^{EN} \times \left(\frac{\dot{Z}}{\dot{E}_P}\right)_k^{REAL}$	$\dot{C}_{D,k}^{EN} = c_{F,k}^{REAL} \times \dot{E}_{D,k}^{EN}$
Exogenous ($\dot{Z}_k^{EX}$, $\dot{C}_{D,k}^{EX}$)	Cost rate within *k*th component caused by the remaining components	$\dot{Z}_k^{EX} = \dot{Z}_k^{REAL} - \dot{Z}_k^{EN}$	$\dot{C}_{D,k}^{EX} = \dot{C}_{D,k}^{REAL} - \dot{C}_{D,k}^{EN}$
Mexogenous ($\dot{Z}_k^{MX}$, $\dot{C}_{D,k}^{MX}$)	Difference between exogenous and sum of split exogenous cost impacts for *k*th component, caused by simultaneous interactions between the component and the remaining components of the system	$\dot{Z}_k^{MX} = \dot{Z}_k^{EX} - \sum_{r=1, r \neq k}^{n} \dot{Z}_k^{EX,r}$	$\dot{C}_{D,k}^{MX} = \dot{C}_{D,k}^{EX} - \sum_{r=1, r \neq k}^{n} \dot{C}_{D,k}^{EX,r}$
Unavoidable ($\dot{Z}_k^{UN}$, $\dot{C}_{D,k}^{UN}$)	Cost rate that cannot be avoided	$\dot{Z}_k^{UN} = \left(\frac{PEC^{UN}}{PEC^{REAL}}\right)_k \times \dot{Z}_k^{REAL}$ for the HEXs % of $\dot{Z}_k^{REAL}$ for the PMs	$\dot{C}_{D,k}^{UN} = c_{F,k}^{REAL} \times \dot{E}_{D,k}^{UN}$
Avoidable ($\dot{Z}_k^{AV}$, $\dot{C}_{D,k}^{AV}$)	Cost rate that can be avoided	$\dot{Z}_k^{AV} = \dot{Z}_k^{REAL} - \dot{Z}_k^{UN}$	$\dot{C}_{D,k}^{AV} = \dot{C}_{D,k}^{REAL} - \dot{C}_{D,k}^{UN}$
UN—EN ($\dot{Z}_k^{UN,EN}$, $\dot{C}_{D,k}^{UN,EN}$)	Unavoidable cost rate within *k*th component associated with the operation of the component itself	$\dot{Z}_k^{UN,EN} = \dot{E}_{P,k}^{EN} \times \left(\frac{\dot{Z}^{UN}}{\dot{E}_P^{REAL}}\right)_k$	$\dot{C}_{D,k}^{UN,EN} = c_{F,k}^{REAL} \times \dot{E}_{D,k}^{UN,EN}$
UN—EX ($\dot{Z}_k^{UN,EX}$, $\dot{C}_{D,k}^{UN,EX}$)	Unavoidable cost rate within *k*th component caused by the remaining components	$\dot{Z}_k^{UN,EX} = \dot{Z}_k^{UN} - \dot{Z}_k^{UN,EN}$	$\dot{C}_{D,k}^{UN,EX} = \dot{C}_{D,k}^{UN} - \dot{C}_{D,k}^{UN,EN}$
AV—EN ($\dot{Z}_k^{AV,EN}$, $\dot{C}_{D,k}^{AV,EN}$)	Avoidable cost rate within *k*th component associated with the operation of the component itself	$\dot{Z}_k^{AV,EN} = \dot{Z}_k^{EN} - \dot{Z}_k^{UN,EN}$	$\dot{C}_{D,k}^{AV,EN} = \dot{C}_{D,k}^{EN} - \dot{C}_{D,k}^{UN,EN}$
AV—EX ($\dot{Z}_k^{AV,EX}$, $\dot{C}_{D,k}^{AV,EX}$)	Avoidable cost rate within *k*th component caused by the remaining components	$\dot{Z}_k^{AV,EX} = \dot{Z}_k^{EX} - \dot{Z}_k^{UN,EX}$	$\dot{C}_{D,k}^{AV,EX} = \dot{C}_{D,k}^{EX} - \dot{C}_{D,k}^{UN,EX}$

Table 28.2 The parameters determined common for the advanced exergetic and exergoeconomic analyses of the Afyon GDHS

Components	Parameter	Unit	Unavoidable conditions	Theoretical conditions	Parameters used in determining unavoidable costs
HEXs	ΔT_{min}	K	0.5	0	20
	ΔP_{min}	kPa	10	0	ΔP_{real}
PMs	η_{is}	%	98	100	–
	η_{mech}	%	100	100	–
	–	–	–	–	60 % of $\dot{Z}_k$

By splitting the investment cost and exergy destruction cost rates into endogenous/exogenous and avoidable/unavoidable parts, the advanced exergoeconomic analysis is capable of providing additional information to the conventional exergoeconomic analysis for improving the design and operation of the system. Better-guided improvement strategies and more accurate evaluations of individual components for both the systems are realized when using the total avoidable cost. The higher this total operating cost rate, the higher the influence of the component on the overall system and thus, the more significant the component is considered. Thus, improvement efforts should be noticed on the avoidable–endogenous and avoidable–exogenous values.

By using the values taken from Keçebaş (2011) and Keçebaş et al. (2011) and the assumptions in Table 28.2, the advanced exergoeconomic analysis is fulfilled for the Afyon GDHS at actual operational data. For all the components of the Afyon GDHS, avoidable investment cost rates associated with the operation of the component itself/caused by the remaining components are shown in Table 28.3. As can be seen in this table, further useful information in improving the performance of a system can be obtained by determining the avoidable–endogenous part. The avoidable–endogenous investment costs indicate that priority for improvement should be given to the HEX-V (6.34 $/h) first, HEX-III second, and HEX-VI third. On the other hand, the difference between the avoidable–endogenous and avoidable–exogenous costs of the PMs is rather small for both the system. The high avoidable investment costs associated with the component itself of both the system show that if we wanted to decrease this cost for a component, changes should relate to the component itself. For instance, it can be accomplished using new construction materials or the manufacturing techniques with less expensive ones. When the total cost associated with a component should be reduced, a more cost-effective operation might be obtained by using the most efficient available component (e.g., for HEXs).

For all the components of the Afyon GDHS, avoidable exergy destruction cost rates associated with the operation of the component itself/caused by the remaining components are shown in Table 28.4. As seen in this table, it is possible to obtain negative values of split costs especially PMs of the both systems. Such results show opposite effects and are related with increased mass flow rates in the simulations

Table 28.3 Avoidable investment cost rates associated with the component itself/caused by the remaining components, for all the components of the Afyon GDHS

Component, k	The Afyon GDHS		
	$\dot{Z}_k$ (\$/h)	$\dot{Z}_k^{AV,EN}$(\$/h)	$\dot{Z}_k^{AV,EX}$(\$/h)
HEX-I	11.990	2.549	−0.218
HEX-II	11.990	2.764	−0.285
HEX-III	11.990	4.001	−0.432
HEX-IV	8.170	3.338	−0.527
HEX-V	8.170	6.342	−2.023
HEX-VI	4.605	3.447	−1.012
PM-I	0.983	0.557	−0.163
PM-II	0.248	0.083	0.016
PM-III	0.248	0.089	0.010
PM-IV	0.248	0.087	0.013
PM-V	0.215	0.075	0.011
PM-VI	0.146	0.035	0.023
PM-VII	0.146	0.033	0.026
PM-VIII	0.030	0.021	0.009
PM-IX	0.983	1.312	−0.915
Overall system	60.164	20.236	−3.077

Table 28.4 Avoidable exergy destruction cost rates associated with the operation of the component itself/ caused by the remaining components, for all the components of the Afyon GDHS

Component, k	The Afyon GDHS		
	$\dot{C}_{D,k}$ (\$/h)	$\dot{C}_{D,k}^{AV,EN}$(\$/h)	$\dot{C}_{D,k}^{AV,EX}$(\$/h)
HEX-I	31.30	6.84	15.89
HEX-II	45.12	14.37	21.25
HEX-III	63.67	29.76	23.14
HEX-IV	33.27	17.34	12.76
HEX-V	29.70	15.07	12.37
HEX-VI	13.84	5.62	6.85
PM-I	24.36	−8.73	15.17
PM-II	4.78	3.52	−1.68
PM-III	3.94	2.92	−1.08
PM-IV	3.94	3.16	−1.32
PM-V	0.63	0.98	−0.85
PM-VI	3.25	3.31	−2.29
PM-VII	2.78	2.76	−2.08
PM-VIII	5.45	−0.78	6.23
PM-IX	35.10	−50.03	64.83
Overall system	1027.25	193.02	565.31

used to calculate the endogenous values, when compared to those of the initial process. Higher mass flow rates result in higher exergy of the product that is correlated with higher costs. The interpretation of these results is that the thermodynamic inefficiency associated with a component with negative exogenous cost of exergy destruction increases when other components operate under theoretical conditions. Thus, to decrease its thermodynamic inefficiencies, inefficiencies of some other components must increase. It should be noticed that the cost rate of avoidable–endogenous exergy destruction within HEX-III of the Afyon GDHS is 29.76 \$/h. Therefore, there is potential for improving these components.

The comparison of the system performance for the conventional and advanced exergoeconomic analyses is shown in Table 28.5. In the conventional exergoeconomic analysis, the larger the absolute value of the total operating cost within a component, the higher its improvement priority must be.

In Table 28.5, the total operating cost rate exposes the components that should have improvement priority, in order to improve the cost effectiveness of the overall system. Here, it can be clearly seen that the economic impact associated with the exergy destruction is the largest contributor to the total operating costs for all components. As can be seen in Table 28.5, the total operating cost rates in the conventional exergetic analysis are all the system with 1087 \$/h and HEX-III with 75.66 \$/h. As a result of improving the overall components at a maximum technical and economic optimization, its total operating cost was found to be 213 \$/h for all the system and 33.76 \$/h for the HEX-III. Here, the HEX-III, HEX-V, and HEX-IV for the Afyon GDHS are ordered as the components with the highest absolute value of total operating cost. However, the highest priority for improvement in the conventional exergoeconomic analysis gave the HEX-III, HEX-II, and HEX-I. This observation clearly demonstrates how to mislead some conclusions from a conventional exergoeconomic analysis can be.

Table 28.5 The performance comparison of the system and its components for the advanced exergoeconomic analysis

Component, k	The Afyon GDHS	
	Conventional	Advanced
	$\dot{C}_{\mathrm{tot},k}$ (\$/h)	$\dot{C}^{\mathrm{AV,EN}}_{\mathrm{tot},k}$ (\$/h)
HEX-I	43.29	9.39
HEX-II	57.11	17.13
HEX-III	75.66	33.76
HEX-IV	41.44	20.68
HEX-V	37.87	21.41
HEX-VI	18.45	9.07
PM-I	25.34	−8.17
PM-II	5.03	3.6
PM-III	4.19	3.01
PM-IV	4.19	3.25
PM-V	0.85	1.06
PM-VI	3.4	3.35
PM-VII	2.93	2.79
PM-VIII	5.48	−0.76
PM-IX	36.08	−48.72
Overall system	1087.41	213.26

28.5 Conclusions

In this study, total operating costs for investment and exergy destruction costs of each component in the Afyon GDHS were evaluated by using the conventional and advanced exergoeconomic analyses. Finally, the following main conclusions may be reached:

- The total operating cost rates for the conventional and advanced exergoeconomic analyses are determined to be 1087 and 213 $/h in the Afyon GDHS, respectively. Thus, it may be concluded that the conventional exergoeconomic analysis can highlight the main components with high cost inefficiencies, but cannot consider the interactions among components or the real potential for improving each component. The approach to the advanced exergoeconomic assessment is a more effective tool in identifying the direction and potential for energy savings in GDHSs.
- The investment cost in the system and its components does not play a very significant role. Therefore, one should focus more on the improvement of total operating costs of other components.
- The Afyon GDHS can be made more cost effectiveness, removing the system components' irreversibilities, technical-economic limitations, and poorly chosen manufacturing methods.

References

Açıkkalp, E., Aras, H., & Hepbasli, A. (2014). Advanced exergoeconomic analysis of an electricity-generating facility that operates with natural gas. *Energy Conversion and Management, 78*, 452–460.

Bejan, A., Tsatsaronis, G., & Moran, M. (1996). *Thermal design and optimization*. New York: Wiley.

Cziesla, F., Tsatsaronis, G., & Gao, Z. (2006). Avoidable thermodynamic inefficiencies and costs in an externally fired combined cycle power plant. *Energy, 31*, 1472–1489.

Hepbasli, A. (2010). A review on energetic, exergetic and exergoeconomic aspects of geothermal district heating systems (GDHSs). *Energy Conversion and Management, 51*, 2041–2061.

Keçebaş, A. (2011). Performance and thermo-economic assessments of geothermal district heating system: A case study in Afyon, Turkey. *Renewable Energy, 36*, 77–83.

Keçebaş, A., & Hepbasli, A. (2014). Conventional and advanced exergoeconomic analyses of geothermal district heating systems. *Energy and Buildings, 69*, 434–441.

Keçebaş, A., Kayfeci, M., & Gedik, E. (2011). Performance investigation of the Afyon geothermal district heating system for building applications: Exergy analysis. *Applied Thermal Engineering, 31*, 1229–1237.

Kelly, S. (2008). *Energy systems improvement based on endogenous and exogenous exergy destruction*. Ph.D. Dissertation, Technische Universität Berlin, Berlin, Germany.

Kelly, S., Tsatsaronis, G., & Morosuk, T. (2009). Advanced exergetic analysis: Approaches for splitting the exergy destruction into endogenous and exogenous parts. *Energy, 34*, 384–391.

Lazzaretto, A., & Tsatsaronis, G. (2006). SPECO: A systematic and general methodology for calculating efficiencies and costs in thermal systems. *Energy, 31*, 1257–1289.

Lund, J. W., & Freeston, D. H. (2001). World-wide direct uses of geothermal energy. *Geothermics, 30*, 29–68.

Manesh, M. H. K., Navida, P., Marigorta, A. M. B., Amidpour, M., & Hamedi, M. H. (2013). New procedure for optimal design and evaluation of cogeneration system based on advanced exergoeconomic and exergoenvironmental analyses. *Energy, 59*, 314–333.

Petrakopoulou, F. (2010). *Comparative evaluation of power plants with CO2 capture: Thermodynamic, economic and environmental performance*. Ph.D. Dissertation, Technische Universität Berlin, Berlin, Germany.

Tsatsaronis, G. (1999). *Design optimization using exergoeconomics, thermodynamic optimization of complex energy systems*. Dordrecht: Kluwer Academic Publishers.

Tsatsaronis, G. (2008). Recent developments in exergy analysis and exergoeconomic. *International Journal of Exergy, 5*, 489–499.

Tsatsaronis, G., & Park, M.-H. (2002). On avoidable and unavoidable exergy destructions and investment costs in thermal systems. *Energy Conversion and Management, 43*, 1259–1270.

Wei, Z., Zhang, B., Wu, S., Chen, Q., & Tsatsaronis, G. (2012). Energy-use analysis and evaluation of distillation systems through avoidable exergy destruction and investment costs. *Energy, 42*, 424–433.

Chapter 29
How the Shadow Economy Affects Enterprises of Finance of Energy

Aristidis Bitzenis, Ioannis Makedos and Panagiotis Kontakos

Abstract The aim of this paper was to present how shadow economy and corruption can affect the enterprises which operate in the sector of finance of energy. The economic damage is extensive in every national economy where increased levels of shadow economy and corruption exist. Accordingly, this study presents possible measures that can decrease shadow economy and corruption. Enterprises in energy finance can provide reliable, competitive, and consistent delivery of customized solutions according to the client's needs. Some of them are finance projects, recapitalizations, single assets, or portfolio credits. Many countries have started to target shadow economy and corruption, since they impede the achievement of their fiscal targets, and harm the overall business environment and the country's attractiveness for foreign investments. In the article, the operation of the energy service companies (ESCOs) is used as a case study.

29.1 Introduction

Companies that are active in the field of finance of energy include the energy service companies (ESCOs), which offer energy services or provide other improvement measures of energy efficiency in the facilities of the user and accept to an extent the financial risk of the procedure. The definition of an ESCO in the international academic and non-academic literature is yet characterized by a lack of consensus. However, many authors largely agree that an ESCO is "a private or a public company that develops, installs, and provides integrated service-based

A. Bitzenis (✉) · I. Makedos · P. Kontakos
University of Macedonia, Thessanoniki, Greece
e-mail: bitzenis@gmail.com

I. Makedos
e-mail: john130570@gmail.com

P. Kontakos
e-mail: pkontakos@uom.gr

A.N. Bilge et al. (eds.), *Energy Systems and Management*,
Springer Proceedings in Energy, DOI 10.1007/978-3-319-16024-5_29

projects with a typical duration of 5–10 years" (Garbuzova-Schlifter and Madlener 2014). The fee for the services provided depends on the extent of achieved energy saving. Financing is provided through sponsoring for the measures undertaken and charges the beneficiaries with an amount equal to the benefits from energy saving, which is achieved as a result of the improvement measures.

The aim of our research was to analyze the interrelation between the levels of shadow economy and the operation of ESCOs. As argued by the authors, considering the range and scope of their activities, ESCOs have a tendency to be mainly active in developed countries and they aim to invest in countries where shadow economy is at low levels. Accordingly, it can be supported that a negative correlation between the two variables exists. The current study is part of the wider European research project "THALES" which targets to measure the various aspects of shadow economy particularly in Greece, including corruption, tax avoidance, social contribution avoidance, undeclared or illegal work, shelf consumption, and illegal acts (black or underground economy). It will cover all economic agents in Greece, such as citizens and corporations (e.g., public and private individuals, companies, and all professional categories). The research is also performed at sector levels, e.g., to identify the extent of tax evasion and corruption practices in the operations of ESCOs in Greece.

The current paper is structured as follows: In the Sect. 29.2, the recent development and services provided by ESCOs are briefly discussed, particularly in countries characterized by high corruption levels; in the Sect. 29.3, selected international indices related to corruption at country and sector levels are presented in order to provide a perspective on the subject under research; in the Sect. 29.4, high-level measures for the reduction of the shadow economy and corruption are proposed by the authors which can be expanded and applied also in the case of ESCOs; in the Sect. 29.5, the conclusions are summarized.

29.2 ESCOs and the Shadow Economy

The first ESCOs were created during the energy crisis in 1970 in the USA and Canada. Later, in the decade of 1980–1990, there was an expansion to Europe and Japan. The European Union with the instruction 2006/32/EC promotes the creation of ESCOs in the countries members "...the use of financial facilitations by others is an innovative practice that has to be encouraged. With these facilitations, the beneficiary avoids investment expenditures, using part of the financial value of the energy saving that arises from the investments of others for the payment of the investment cost and the rates of others..."

A more advanced form of the ESCOs in the UK focuses on the innovative methods of sponsoring while providing the following services:

- Development and planning for the sponsoring of works of energy efficiency;
- Analysis and control of energy data;

- Planning and implementation of energy administration studies;
- Installation and maintenance of the necessary equipment for the works of energy efficiency;
- Facilities administration;
- Assessment, observation, and verification of the energy savings;

These services are included in the cost of the works and repaid through the energy saving that is achieved.

The advantages for the customers of ESCOs can be summarized as following:

- No fund commitment;
- The risk is undertaken by the vehicle of energy;
- Technological update of the facilities;
- Improvement of the competitiveness of the company;
- Upgrading of the financial image /improvement of the financial ratios;
- Reduction of the operational cost of companies;
- Reinforcement of the company's social responsibility as a marketing tool.

ESCOs may further contribute to:

- Achievement of energy goals of a country;
- Energy consumption efficiencies;
- Reduction of gas greenhouse emissions;
- Increase of employment.

On the other side, according to Schneider (1986), shadow economy includes all economic activities that can create added value and should be included in the national income (NI). Moreover, according to Tanzi (2002), shadow economy is the part of GDP that cannot be calculated by the official statistic services.

Corruption is defined as the abuse of the public power for personal benefits (e.g., the bribery of civil servants, misappropriation of public money, etc.). Bribery, in general, is considered a crime and the national legislation imposes severe penalties for violations. Nevertheless, the particular implementation and law enforcement still remains an unsolved problem, particularly in the field of public procurement, where ESCOs are also involved and interact. In the case of Greek government, for example, there is evidence that the political influence and some other factors, such as confidence in the old suppliers, played an important role in the evaluation of offers for the works that have been sponsored or co-sponsored.

Many businesses, such as ESCOs, fail to operate in an environment where corruption and shadow economy are in high levels, and as a result, their investments and services are declined. For example, the main problems faced by ESCOs in Russia are the corruption and the lack of funding (Kozhevnikov 2014). Also, judicial corruption is still a problem despite substantial improvement in efficiency and fairness in the courts in Georgia. Both foreigners and Georgians continue to doubt the judicial system ability to protect private property and contracts also in the sector of ESCOs (Abramia et al. 2009). Political and organizational obstacles primarily exist in developing countries and consist of issues such as insufficient government involvement in energy efficiency and corruption in the companies which operate in this sector (Aminu et al. 2013).

ESCOs' market in Greece is considered to be still in a very primary stage and has not been deployed yet either in the public or in the private sector, although there is a large energy-saving potential mainly in the tertiary sector. Nowadays, only few attempts have been implemented by ESCOs, mainly on renewable energy sources projects especially for PV installations and/or street lighting. Nevertheless, these applications cannot be considered as typical energy performance contract (EPC) examples, but more as guarantee contracts or leasing of equipment.

However, recently under a program deployed by the Center for Renewable Energy Sources, many companies exhibited interest to act as ESCOs and provide EPCs leading to a preliminary interest for the potential implementation of such projects both at public and private sectors.

As main barriers for the development of ESCOs' market in Greece can be identified the lack of knowledge regarding ESCOs and EPCs as mechanisms for the implementation of energy-saving measures in public and private sector, the absence of standard type of EPCs and the necessary monitoring methodology, the existence of corrupted practices, the vague procedure for the applications of EPC in public sector and the lack of confidence for the guarantee of the compliance with the contract's terms and the repayment process among ESCOs and interested parts. Finally, it must be pinpointed the reluctance of financial sector to support ESCOs' actions and measures through EPCs, which hampers the deployment of ESCOs' market (ICLEI 2014).

29.3 International Indices for Shadow Economy and Corruption

Since the mid-1990s, Transparency International has generated a wide range of yearly corruption rankings of countries (Transparency International, 2013a, b). The Index Global Corruption Barometer (GCB) is calculated from the relevant survey of the public opinion of Transparency International. About 1000 people from each country, out of a total of 107 countries, were questioned for GCB 2013. The survey for 2013 indicated that political parties are thought to be the most corrupted institution.

Furthermore, the Bribe Payers Index (BPI), which was first launched in 1999, assesses the "supply side" of corruption—"the likelihood of firms from the world's industrialized countries to bribe abroad." In BPI 2011, the 28 leading economies were ranked according to the perception of thousands of senior business executives from developed and developing countries in the question whether there is bribery abroad (Transparency International, 2013c).

Respectively, according to the evaluation of BPI 2011, companies from Russia and China were considered as the most prone to pay bribes abroad, mainly due to their increased significance in international trade and foreign investments, while companies from Switzerland and Holland were unlikely to do so (ranked in the first places in Table 29.1).

Agriculture and light manufacturing are considered to be the least bribery-prone sectors (Table 29.2). The public works contracts and construction sector ranks last. Other sectors ranked in the bottom quartile of the table include utilities; real estate, property, legal, and business services; mining; and oil and gas (where ESCOs are

Table 29.1 Classification of countries based on BPI in 2011 (adopted from BPI, http://bpi.transparency.org/bpi2011/results/)

Rank	Country/territory	Score	Number of observations	Standard deviation	90 % confidence interval	
					Lower bound	Upper bound
1	The Netherlands	8.8	273	2.0	8.6	9.0
2	Switzerland	8.8	244	2.2	8.5	9.0
3	Belgium	8.7	221	2.0	8.5	9.0
4	Germany	8.6	576	2.2	8.5	8.8
5	Japan	8.6	319	2.4	8.4	8.9
6	Australia	8.5	168	2.2	8.2	8.8
7	Canada	8.5	209	2.3	8.2	8.8
8	Singapore	8.3	256	2.3	8.1	8.6
9	UK	8.3	414	2.5	8.1	8.5
10	US	8.1	651	2.7	7.9	8.3
11	France	8.0	435	2.6	7.8	8.2
12	Spain	8.0	326	2.6	7.7	8.2
13	South Korea	7.9	152	2.8	7.5	8.2
14	Brazil	7.7	163	3.0	7.3	8.1
15	Hong Kong	7.6	208	2.9	7.3	7.9
16	Italy	7.6	397	2.8	7.4	7.8
17	Malaysia	7.6	148	2.9	7.2	8.0
18	South Africa	7.6	191	2.8	7.2	7.9
19	Taiwan	7.5	193	3.0	7.2	7.9
20	India	7.5	168	3.0	7.1	7.9
21	Turkey	7.5	139	2.7	7.2	7.9
22	Saudi Arabia	7.4	138	3.0	7.0	7.8
23	Argentina	7.3	115	3.0	6.8	7.7
24	UAE	7.3	156	2.9	6.9	7.7
25	Indonesia	7.1	153	3.4	6.9	7.7
26	Mexico	7.0	121	3.2	6.6	7.5
27	China	6.5	608	3.5	6.3	6.7
28	Russia	6.1	172	3.6	5.7	6.6
	Average	7.8				

Table 29.2 Classification of sectors based on BPI in 2011 (adopted from BPI, http://bpi.transparency.org/bpi2011/results/)

Rank	Country/territory	Score	Number of observations	Standard deviation	90 % confidence interval	
					Lower bound	Upper bound
1	Agriculture	7.1	270	2.6	6.8	7.4
2	Light manufacturing	7.1	652	2.4	7.0	7.3
3	Civilian aerospace	7.0	89	2.7	6.6	7.5
4	Information tech.	7.0	677	2.5	6.8	7.1
5	Banking and finance	6.9	1409	2.7	6.8	7.0
6	Forestry	6.9	91	2.4	6.5	7.3
7	Consumer services	6.8	860	2.5	6.7	6.9
8	Telecoms	6.7	529	2.6	6.5	6.9
9	Transportation and storage	6.7	717	2.6	6.5	6.9
10	Arms, defense, and military	6.6	102	2.9	6.1	7.1
11	Fisheries	6.6	82	3.0	6.0	7.1
12	Heavy manufacture	6.5	647	2.6	6.4	6.7
13	Pharmaceutical and health care	6.4	391	2.7	6.2	6.6
14	Power generation and transmission	6.4	303	2.8	6.1	6.6
15	Mining	6.3	154	2.7	5.9	6.6
16	Oil and gas	6.2	328	2.8	6.0	6.5
17	Real estate, property, legal, and business services	6.1	674	2.8	5.9	6.3
18	Utilities	6.1	400	2.9	5.9	6.3
19	Public works contracts and construction	5.3	576	2.7	5.1	5.5
	Average	6.6				

active). The main attribute of these sectors is the high value of investment and considerable government involvement and regulation, both of which offer opportunities and motivations for corruption.

Finally, Global Corruption Report (GVR) is one of the most characteristic and important publications of the International Transparency, conveying the experience against corruption. The most recent studies focused on specialized thematic domains such as corruption in the climatic charge and the private sector (among the companies examined were included also ESCOs) (Lawrence and Haas 2008; Komendantova and Patt 2011).

29.4 Proposals Against Shadow Economy

According to Schneider (2013a, b), we pose the question whether shadow economy is a "blessing in disguise." The enlargement of shadow economy leads to a higher added value, while its narrowing may increase social welfare provided it is thoroughly included in the economy. This means that proper financial and tax measures should be undertaken to motivate the transfer of the produced goods and services from the shadow to the official economy.

Schneider (2013a, b) points out that a government that aims at the reduction of activities of shadow economy should first analyze the complex relationship between the official and shadow economy, as well as the consequences of its political decisions. Additionally, he referred to the aim of a significant decrease of corruption in countries such as Greece, proposing direct and powerful measures, such as the prohibition of public contract work for 3–5 years for the businesses that are involved in bribery and/or corruption.

Katsios (2006) points out that the rationalization of administrative costs in combination with the simplification of the legislation frame may bring about a significant decrease in shadow economy. Also, he proposes the geographical relocation of civil servants with the aim to avoid developing their customer relationships with the citizens of the regions.

Some of the measures for the reduction of the shadow economy and corruption, which are suggested by the authors of this working paper and can also be applied in the case of ESCOs, include the following:

- Penalization of the purchaser of undeclared work and in general very severe measures for confirmed violations.
- Stricter penalties for confirmed violations, so as to set an example to avoid repetition of similar cases in the future.
- Decrease of the taxes that involve significant cost of confirmation and low revenues, and reinforcement of taxes with significant revenues, without having an important cost of income.
- Separation of the tax control from the income tax authorities and supervision of both by an independent authority.
- Increased motives for those who indicate tax compliance, as a reward for their tax morality (i.e., exemption).
- Promoting a fair tax system in proportion with the revenues of taxpayers so as to boost the feeling about what is right, as far as taxation is concerned, a fact that increases tax morality.
- Immediate reciprocity of taxes from the state (i.e., investments on security, health, education), to enhance tax morality of taxpayers. In most European countries, this is the focus point of tax morality (i.e., Germany and Austria), as taxpayers can realize that their money is not wasted but invested.
- Banning and discouragement of every form of direct or indirect advertisement that presents the mentality of tax evasion as an ability or increased intelligence.

29.5 Conclusions

In this working paper, the aim was to present the repercussions of shadow economy and business corruption in the field of finance of energy. Many forms of enterprises, such as ESCOs, cannot operate in an environment where corruption and shadow economy are in high levels. Also, large-size companies, in the field of ESCO, that originated from countries characterized by low corruption (such as Switzerland) prefer to select countries with low levels of shadow economy and corruption for investments.

International organisations publish indices that evaluate the levels of shadow economy and corruption, as the BPI, and advice governments on methods to reduce the shadow economy and corruption. The creation of a healthy investment and business environment is currently a top priority for many countries, even for those who present low levels of shadow economy and corruption. Finally, high-level measures were presented that can restrain shadow economy and foster a more efficient business environment.

References

Abramia, G., Jorde, I.K., & Biegert, A. (2009). Renewable energies in central Asia. In regional report on potentials and markets–8 country analyses. Germany: GTZ.

Aminu, U., Khamidi, M. K., Shika, S. A., & Musa, U. (2013). Towards building energy efficiency for developing countries. *Bonfring International Journal of Industrial Engineering and Management Science., 3*(1), 15–28.

Garbuzova-Schlifter, M., & Madlener, R. (2014) Foreign direct investments in carbon footprint reduction projects: The case of the Russian energy market post-2012, E.ON Energy research center series, 6(1).

Katsios, S. (2006). The shadow economy and corruption in Greece. *South-Eastern Europe Journal of Economics, 1*, 61–80.

Komendantova, N., & Patt, A. (2011). *Could corruption pose a barrier to the roll-out of renewable energy in North Africa? In global corruption report: Corruption in the climate change. Transparency international*. UK-USA: Earthscan.

Kozhevnikov, M. (2014). *Approaches to the formation of energy services markets in developing countries, 27, in energy production and management in the 21th century. The quest for sustainable energy* (p. 27). UK: WIT Press.

Lawrence, J., & Haas, M. (2008). Water for energy: Corruption in the hydropower sector. In Global corruption report: Corruption in the water sector. Transparency international, p. 87. UK: Cambridge University Press.

Schneider, F. (1986). Estimating the size of the Danish shadow economy using the currency demand approach: An attempt. *Scandinavian Journal of Economics, 4*, 643.

Schneider, F. (2013a). The shadow economy in Greece and other OECD countries. In A. Bitzenis, V. Vlachos, & I. Papadopoulos (Eds.), *Reflections on the Greek sovereign debt crisis*. UK: Cambridge Scholars Publishing.

Schneider, F. (2013b). Latest developments about the Greek shadow economy and tax evasion. Proceedings of 4th international conference on international business. Thessaloniki, Greece, May 2013.

Tanzi, V. (2002). The shadow economy: Its causes and its consequences. International seminar on the shadow economy index in Brazil, Brazilian institute of ethics in competition. Rio de Janeiro, Brasil, 12 March 2002.

Transparency International. (2013a). Corruption reporting, http://www.transparency.org/whoweare/organisation/secretariat (Accessed January 11, 2014).

Transparency International. (2013b). Who we are, http://www.transparency.org/whoweare/organisation/faqs_on_corruption/2/#measureCorruption (Accessed January 15, 2014).

Transparency International. (2013c). Bribe payers index report 2011. http://www.transparency.org/bpi2011/results, (Accessed January 9, 2014).

ICLEI. (2014), Guidance for local governments and their partners: Toolbox of methodologies on climate and energy, http://toolbox.climate-protection.eu/, (Accessed September 20, 2014).

Chapter 30
Energy Profile of Siirt

Omer Sahin, Mustafa Pala, Asım Balbay, Fevzi Hansu and Hakan Ulker

Abstract Siirt Province has various natural and fossil energy sources such as solar, hydropower, biogas, geothermal, and petrol in terms of energy potential, by contrast with other provinces in Turkey. The data collected in Siirt Province indicate that Siirt has a strong potential for solar energy. The average sunshine duration and total solar radiation in Siirt are about 7.5 h-day and 4.3 kWh/m^2-day, respectively. In Siirt, there are two hydropower plants with a total installed power of 263 MW. In addition, more than 10 hydroelectric power plants with a total installed capacity of 1094 MW of power generation will be established by means of the planned dam to be installed. Considering the establishment of biogas systems, Siirt has an annual biogas production of 20,000 m^3 with around 500,000 small ruminants. Siirt Province is believed to be rich in geothermal care, but there has not been enough research on this topic, yet. And also, petroleum is an important energy source in Siirt Province, lately. As a result, Siirt Province has a rich variety of energy resources, and in case of investment, it would be an energy basin center in the southeast Anatolia region.

O. Sahin
Department of Chemical Engineering, Siirt University, 56100 Siirt, Turkey

M. Pala
Vice Governor of Siirt, 56100 Siirt, Turkey

A. Balbay (✉) · H. Ulker
Department of Mechanical Engineering, Siirt University, 56100 Siirt, Turkey
e-mail: asimbalbay@gmail.com; abalbay@siirt.edu.tr

H. Ulker
e-mail: ulkerhakan@gmail.com

F. Hansu
Department of Electrical Engineering, Siirt University, 56100 Siirt, Turkey

A.N. Bilge et al. (eds.), *Energy Systems and Management*,
Springer Proceedings in Energy, DOI 10.1007/978-3-319-16024-5_30

30.1 Introduction

Energy studies constitute a significant importance for policies of countries in the world. Ease of access to cheap energy resources and reduction of energy dependency on any countries are fundamentals of well-functioning and developed economies. Therefore, energy becomes a very important component of policies in development plans of countries. Also, the depletion of fossil-based resources has forced planners and policy makers to look for alternative energy sources. These energy sources should be the energy originated from resources that are regenerative and clean and do not deplete over time (Toklu 2013). Since, the cleaner environment and high power demands gives rise to an interest in renewable energy sources such as solar, wind etc. (Present Status of PV Industry Environmental Sciences Essay 2014). In recent years, the use of renewable energy sources offers a great potential in Turkey. However, the utilization of energy potential has not reached its desired level. The detailed evaluation will have great benefits for Turkey.

Siirt Province has seen the fastest growth in energy demand in Turkey over the last five years; its economy has avoided the prolonged stagnation that has characterized much of the area. Siirt will continue to act with new structures and arrangements in line with the importance of sustainability. Referring energy profile, this study gives explanation of current situations about renewable energy resources from which Siirt utilize for its domestic energy needs (Balbay and Kizilaslan 2013).

30.2 Solar Energy

Solar energy is a clean, renewable energy that may be used to increase or replace existing power sources. In recent years, the use of solar energy offers a great potential in Turkey. However, the utilization of this energy has not reached its desired level. The evaluation of this potential will have great benefits for Turkey. Turkey has an annual average sunshine duration of approximately 2640 h and average annual solar radiation as 1311 kWh/m^2-year (General Directorate of Electrical Power Resources Survey and Development Administration 2014). According to regional solar energy evaluations in Turkey, the southeastern Anatolia region has the highest solar energy potential. Siirt, which is located between latitude 37–56° North and longitude 41–57° East at an altitude of 895 m above sea level, is one of the highest solar energy potential in this region. The total annual sunshine duration of Siirt is 2738 h (7.5 h a day), and the average annual solar radiation is 1570 kWh/m^2 (4.3 kWh/m^2 a day). The most productive areas and solar potential in terms of solar energy in centrum and towns of Siirt are shown in Fig. 30.1. Although the centrum has the highest solar energy potential, there are remarkable differences between the centrum and towns. The reasons for the differences are highland and lower sunshine duration in some locations. As shown in the figure, Siirt Province should be taken into account on solar energy and the government should encourage the companies in Turkey to benefit from the energy potential.

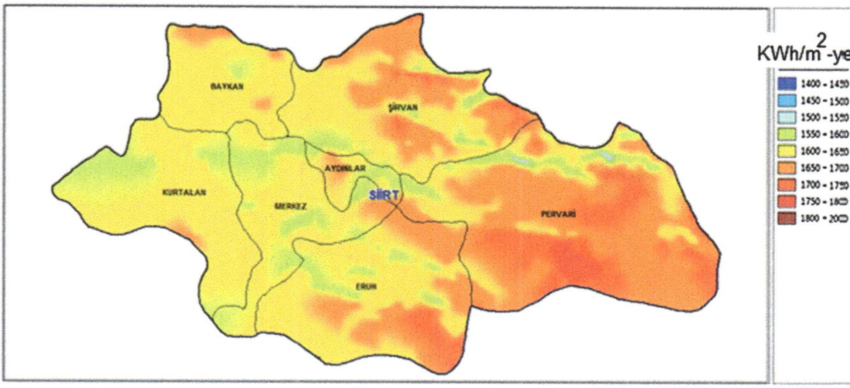

Fig. 30.1 Typical solar radiation for Siirt Province (General Directorate of Electrical Power Resources Survey and Development Administration 2014)

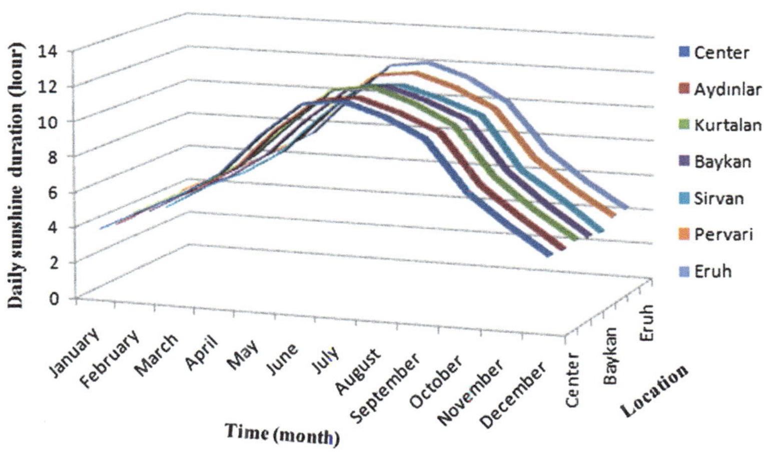

Fig. 30.2 The mean values of sunshine duration (meteorological monthly average data between the years 2000 and 2010 in Siirt Province)

Figure 30.2 shows monthly mean of daily sunshine duration data for Siirt Province. As it can be seen from the figure, the average sunshine duration of Siirt is to change between 11 and 13 h in July and August. Clouds, dust, and other particulate matter in the atmosphere cause variations in radiation absorption and scatter.

Solar radiation data are required to determine the solar potential of a district for solar energy applications such as electricity production, heating, and cooling. Figure 30.3 presents average solar radiation with respect to months in a year. As it can be seen in Fig. 30.3, the highest solar energy potential is obtained between June and August.

Figure 30.4 presents the monthly average temperature in Siirt between 2000 and 2010. As it can be seen in Fig. 30.4, monthly average temperatures are higher in

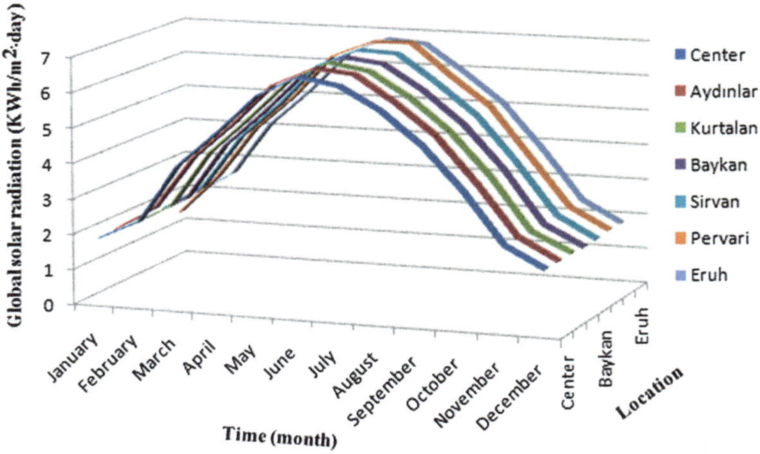

Fig. 30.3 The monthly mean solar radiation (meteorological monthly average data between the years 2000 and 2010 in Siirt Province)

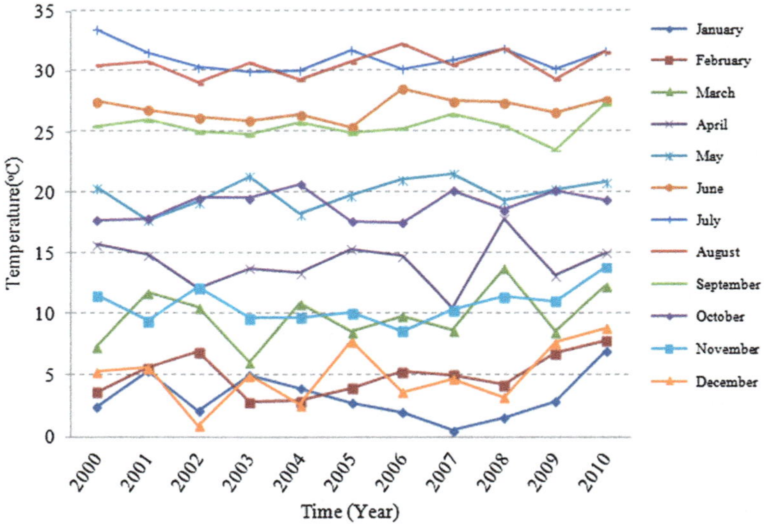

Fig. 30.4 The average monthly temperature depending on years for Siirt Province (meteorological monthly average data between the years 2000 and 2010 in Siirt Province)

summer months and lower in winter ones. The hottest and driest months are July and August; during these months, the daytime high temperatures can hover around 35 °C. The graphs show that the monthly maximum average temperature (33.5 °C) and the monthly minimum average temperature (0.5 °C) for 10 years were recorded on July, 2000 and January, 2007, respectively. January and February are the coldest months of the year, but the conditions are pleasant along these months.

Solar energy has many applications including heating water, generating power to operate remote equipment, and helping to reduce the reliance on overhead, electrical light sources. However, people in Siirt use only the solar energy for domestic hot water, drying fruits and vegetables. There is not any system for generation of electricity using solar energy. Using a small renewable energy system can help to lower electricity costs, achieve energy independence, and reduce pollution. Also, greenhouse can be developed in this province by means of number of sunny days to be more than number of cold days.

30.3 Geothermal Energy

There are three thermal springs called Lif, Billoris, and Hista in Siirt. The hot water of Lif is chlorinated, sulfated, sodium, sulfureted, and calcic and has flow rate at 30 l/s with 41 °C temperature. The hot water of Billoris is muriatic, sulfated, sodium, sulfurous, and calcic and has flow rate at 172–173 l/s with 30–35 °C temperature (The Report of Provincial and Environment 2010). The hot water of Hista is sulfureted sodium and calcium sulfated with flow rate at 60 °C temperature. It is necessary to investigate how much increasing the flow rate and the level of health benefits of the thermal springs. Also, the modern and new facilities should be built in these areas for the tourism.

Temperature values underground are stable winter and summer. Ground source heat pump systems that are used for cooling and heating applications in all around the world can be used in Siirt and around provinces, as well. Therefore, the studies about geothermal energy sources must be supported and encouraged by government.

30.4 A Mobile Power Plant for Electric Energy

There is a mobile power plant in Siirt. It uses fuel oil for energy production. The installed capacity and project generation capacity are 25.6 MW and 190 GWh, respectively.

30.5 Hydroelectric Energy

Siirt Province has the richest conditions in terms of hydropower potential among the provinces of region. Siirt Alkumru Dam began power generation in 2011, while Kirazlık Dam started in 2014 among the existing four hydroelectric plants, as well as the ongoing construction of 4 total planned constructions. And also, in Siirt Province, a total of 13 Hydroelectric Power Plant (HEPP) projects are available in

Table 30.1 The hydroelectric power projects in operation (Regional Directorate of DSI 2013)

Order no.	Project name	District	Energy benefits	
			Installed capacity (MW)	Generated energy (GWh/Year)
1	Alkumru Dam and HEPP	Aydınlar	261.27	817.26
2	Kirazlık HEPP	Aydınlar	37.62	139.55
3	Botan HEPP	Merkez	2	6
	Total		300.89	962.81

survey and design stages. Ahead in the period, when the security problems disappear in the area and according to 6446 numbered new Electricity Market Law with the licensing mechanisms and the provided changes and limitations related to licenses duration in the region, the productions of hydroelectric power plants in the phase studies and projects are expected to begin in the near future.

In Siirt Province, there was only 2 MW installed HEPP capacity in the past. However, in the context of 4628 Electricity Market Law, by the build–operate model, the private sector started Alkumru Dam construction in early 2008 and completed in 2011. Furthermore, by the same company, Kirazlık Dam construction was started in 2010 and completed in 2014 for electricity production. Therefore, the total energy production was approximately 300 MW and this status is given in Table 30.1.

30.5.1 The HEPP's Found in Construction Phase in Siirt Province

The major HEPP construction projects in Siirt Province are composed of main Cetin Dam with 401 MW installed power and Lower Cetin Dam with 116 MW installed power HEPP projects. With 7 HEPP projects at planning stage on Botan, this project has the largest installed capacity of power with 517 MW and the project has the size of 86 % total installed capacity of the construction phase, and this project has being developed by a Norwegian company as Statkraft Energy Inc. The hydroelectric projects under ongoing construction in Siirt Province are given in Table 30.2.

30.5.2 Planned Hydropower Plants in the Province of Siirt

In the Siirt province, there are totally 13 projects of dams and hydropower plants which have 1.094 MW power in planning, construction and operation. The detailed information about these projects is given in Table 30.3.

Table 30.2 The hydroelectric projects under construction (Regional Directorate of DSI 2013)

Order no.	Project name	District	Energy benefits	
			Installed capacity (MW)	Generated energy (GWh/Year)
1	Cetin Dam and HEPP	Pervari	517	1460
2	Sirvan Dam and HEPP	Sirvan	30	104.62
3	Baran 1-2 Reg. and HEPP	Eruh	56.27	153.63
	Total		603.27	1718.25

Table 30.3 Planned HEPP projects (Regional Directorate of DSI 2013)

Order no.	Project name	District	Energy benefits	
			Installed capacity (MW)	Generated energy (GWh/Year)
1	İncir Dam and HEPP	Sirvan	140	323.87
2	Pervari Dam and HEPP	Pervari	400	889.12
3	Eruh Dam and HEPP	Eruh	44.3	150.59
4	Oran Dam and HEPP	Pervari	71.02	211.55
5	Narlı Dam and HEPP	Pervari	36	168
6	Keskin Dam and HEPP	Pervari	164	740.6
7	Baykan Energy Group and HEPP	Baykan	81.5	273.61
8	Tarihler I-II Reg. and HEPP	Sirvan	47.6	169.01
9	Kezer Reg. and HEPP	Merkez	16.16	48.43
10	Hizan Reg. and HEPP	Sirvan	38.5	100.16
11	Karasu Reg. and HEPP	Aydınlar	22.65	85.9
12	Başoren Reg. and HEPP	Sirvan	9.74	33.24
13	Balcılar Reg. and HEPP	Pervari	22.3	72.93
	Total		1094	3267

As a result, when examined all HES project planning in the region, Siirt Province is seen in the front row between the provinces of the region, particularly in terms of the number of projects. It has been concluded that upon completion of all planned HEPP in terms of HEPP projects, Siirt Province will have the largest share of power generation among other provinces in the region, and this will constitute 4 % of the total installed capacity in Turkey.

30.6 Oil Energy

Turkey has invested heavily in the oil industries and with continued population growth and the ever-increasing need for energy in recent years. As a result of these investigations, oil with an API gravity of 42.3° was discovered at Dogu Sadak-1 well in the mountainous region of Siirt, at a depth of 2371–2384 m in 2013. The Dogu Sadak-1 well in Siirt is capable of producing 150 barrels of oil in a day. Also, 250 barrels of oil in a day are produced at Magrip oil field in Kurtalan district of Siirt Province. On the other hand, totally 270 barrels of oil from nine wells in a day are produced at Celikli oil field in Kurtalan district of Siirt Province. However, all areas of the province have not been completely explored or exploited for oil.

30.7 Wind Energy

The wind power is proportional to the cube of the wind speed, which means with lower wind speeds, the power extracted from the wind will decrease drastically. Therefore, the economic feasibility of a wind energy conversion (WECS) system will be open to discussion. Studies have found that average wind speeds in a particular location need to exceed at least 3 m/s for a small wind turbine to be economically viable (Araujo and Freitas 2008). However, the average wind speed is 0.8–1 m/s in Siirt Province (Siirt State Meteorological Station 2014). Therefore, WECS will not be economically feasible in Siirt Province.

30.8 Bio Energy

Bioenergy is a form of renewable energy derived from biomass (organic materials) to generate electricity and heat and to produce liquid fuels for transport. The potential bioenergy resources (livestock and algae) in Siirt Province are large and diverse. Since livestock is an important source of income in Siirt, most of the people living in countryside work with livestock. Thus, people living in squatter's house and countryside in Siirt use dried dung, wood, and coal as fuel for domestic heating. Wood is provided from coppice forest and dried dung from families working with cattle breeding or from rangelands. Although livestock is an important source of income in Siirt Province, there is no biogas production system.

Biogas is a fuel obtained by converting any organic substance into carbon dioxide and marsh gas (methane) as a result of agricultural wastes or animal manure to be allowed to decay in an oxygen-free environment (anaerobic). Therefore, the animal manure has an important potential for production of biogas.

Approximately, there are 500,000 ovine within the Siirt provincial boundaries. Ovine leave manure about 0.7 tones/year and biogas can be obtained about 58 m^3/ton, which results approximately 20 million m^3 biogas per year. So, electricity can be produced by means of this biogas to be used as fuel in power plant.

30.8.1 Algae Oil Energy

In Siirt, there are favorable conditions for production of algae and obtaining biodiesel from algae oil by means of tanning time to be higher and number of stream to be much more comparing with other cities. Algae are living creature like plant with dozens of species, which evolve quickly with photosynthesis in sun-exposed and aquatic environments. Although algae contain high amount of oil, it is impossible to be consumed as food. It is possible to obtain an alternative fuel to fossil fuels with conversion of oil obtained from algae to diesel as fuel and be used in vehicles.

30.9 Results

In Turkey, southeastern Anatolia region is famous for energy sources such as solar energy, petroleum energy, and hydroelectric energy. Siirt Province that is located in south east of this region is one of the rich provinces of Turkey in the way of especially solar energy, hydroelectric energy, and other energy sources. However, Siirt cannot make use of available energy sources sufficiently and is dependent on external energy sources until the years of 2010. Results of this study can be listed as follows:

- Although solar energy potential in Siirt is high, there is no solar power plant. A solar power plant must be built in this province as soon as possible.
- Greenhouse can be developed in this province by means of number of sunny days to be more than number of cold days.
- Siirt is among the important provinces in Turkey in the way of hydroelectric energy potential. If HEPP projects are completed, they provide a solution to energy demand of especially Siirt and Turkey. However, irregular rain and periodical drought cause installed power plant to be unable to be used with full efficiency. Absolutely, projects must be designed about increasing the efficiency of power plants.
- Searching geothermal energy sources must be supported and encouraged by government. Temperature values underground are stable winter and summer. Ground source heat pump systems that are used for cooling and heating applications in all around the world can be used in Siirt and around provinces, as well.
- In 2013, petroleum with a gravity of 42.3° API was found in between 2371 and 2384 m underground as a result of searching in Doğu Sadak-1 oil exploration well in Eruh, which is a town in Siirt. That is why, Siirt to be an important center for petroleum energy is possible by means of oil exploration.
- In the area of spas, researches must be made about increasing the current flow rates and temperatures of spas and how they are healthy.

- A research must be made about whether there is potential of animal manure to be used in production of biogas or not.
- Since Siirt is a plenty of sun-exposed province, algae production and biodiesel production from algae oil must be researched.

References

Araujo, M. S. M., & Freitas M. E. V. (2008). Acceptance of renewable energy innovation in Brazil —case study of wind energy. *Renewable and Sustainable Energy Reviews, 12*, 584–591.

Balbay, A., & Kizilaslan, E. (2013). Energy resources and potential analysis in Siirt Province (in Turkish), 2. In *Anatolian Energy Symposium, Diyarbakir, Turkey.*

General Directorate of Electrical Power Resources Survey and Development Administration. (2014). Retrieved from http://www.eie.gov.tr

Present Status of PV Industry Environmental Sciences Essay. (2014). Retrieved from http://www.ukessays.com/essays/environmental-sciences

Regional Directorate of DSI. (2013). Siirt

Siirt State Meteorological Station. (2014). *Records for weather data's in Siirt, Turkey*. Retrieved from http://www.mgm.gov.tr

The Report of Provincial and Environment. (2010). Siirt. Governor of Siirt, Provincial Directorate of Environment and Forestry.

Toklu, E. (2013). Overview of potential and utilization of renewable energy sources in Turkey. *Renewable Energy, 50*, 456–463.

CPSIA information can be obtained at www.ICGtesting.com
Printed in the USA
LVOW01*2127120515

438213LV00001B/20/P

9 783319 160238